I0047941

Hans Landolt

Handbook of the Polariscope and Its Pracitcal Applications

Hans Landolt

Handbook of the Polariscope and Its Pracitcal Applications

ISBN/EAN: 9783337321635

Printed in Europe, USA, Canada, Australia, Japan

Cover: Foto ©berggeist007 / pixelio.de

More available books at **www.hansebooks.com**

HANDBOOK

OF THE

POLARISCOPE

AND ITS PRACTICAL APPLICATIONS

ADAPTED FROM THE GERMAN EDITION OF

H. LANDOLT

PROFESSOR OF CHEMISTRY AT THE POLYTECHNICUM, AACHEN

BY D. C. ROBB B.A.

AND

V. H. VELEY B.A. F.C.S.

WITH AN APPENDIX BY I. STEINER F.C.S.

London

MACMILLAN & CO.

1882

AUTHOR'S PREFACE.

—·⟶·—

THE importance so long assigned to the property, possessed by many organic substances, of rotating the plane of polarization, alike in its theoretical and practical aspects, makes it rather surprising that hitherto the only help available for the study of the subject was to be found in the various memoirs, in which the necessary information lay scattered, in scientific journals. The present work was undertaken with a view to relieve the inconvenience arising from the want of a comprehensive treatise. It is based upon a paper of mine published some time ago in *Liebig's Annalen*, Bd. 189, in which I discussed the mode of determining specific rotation, and, by way of introduction, gave a brief general account of optical activity. Since then requests have frequently reached me, urging the desirability of extending that paper by including an account of all the recent instruments and the practical applications, so as to make it a complete monograph of the subject. I was the more readily induced to undertake the task by the fact that, of late years, the increased attention given to the phenomena of rotation has brought to light such a mass of facts as makes it possible to present the material in a more or less complete form.

In its theoretical aspect the optical activity of organic substances possesses high interest. As it is a consequence of peculiarity of arrangement of the atoms in the molecule, it must assuredly afford some assistance in determining the constitutional formula to be assigned to the substances. The investigation of this connection between optical power and chemical constitution is, indeed, of the utmost importance, and gives promise of a rich harvest for future workers.

For this purpose it is, above all, necessary that determinations of specific rotation should be made with the most rigid accuracy. Special attention has therefore been paid in the present work to the needs of the scientific investigator, by affording a detailed account of the different polarimetric instruments, and the methods of observation of specific rotation, as well as of other data connected therewith. The methods described are those which ensure the highest accuracy, and care has been taken to indicate in each case the limits of accuracy attainable. Where only a rougher estimate is required, it will readily be seen that steps may be omitted so as to simplify the process.

The importance of the subject, from a practical point of view, has been long acknowledged in its application to the determination of sugar, and recently of other substances, more especially the cinchona bases. The methods of observation in these special cases have been fully treated, and the sugar-chemist in particular will find interest and novelty in the account of the different saccharimeters, and the corrections to be applied to the result.

The introductory chapter on the optical principles of the subject may, perhaps, be not unwelcome to many a chemist. It has been made as elementary and succinct as possible. The relation between rotatory power and crystalline form, as belonging rather to crystallographic physics, has only been briefly touched upon.

For any further account of the work the reader need only be referred to the table of contents, which has been made as complete as possible.

HANS LANDOLT.

AACHEN, *January*, 1879.

TRANSLATOR'S PREFACE.

A FEW words will suffice to introduce the present edition to the English reader. Some months ago, Mr. Frank Faulkner, the energetic and intelligent brewer of St. Helens and Beeston, to whom the English public are already indebted for the appearance of Pasteur's "Studies on Fermentation," placed in my hands, for revision and editing, a manuscript translation[1] of Dr. Landolt's work; and for what now appears as a corrected version, although I cannot claim all the credit, yet I alone am responsible. What has been kept in view throughout was to make the English edition, as far as possible, an exact reproduction, in all respects, of the original work. A few notes have been ventured where it was thought they would be useful. The longer note at the end of the introductory chapter is placed there for the sake of practical people who, wishing to understand more fully the physical explanation of the fundamental phenomena, do not have the time or opportunity to read up the subject in special treatises on physics. The advanced student will, it is hoped, indulgently consider it in that light.

A feature of the English edition, which will doubtless render it specially valuable to the technical chemist, is the appendix contributed by Mr. Faulkner's assistant, Mr. Ignatius Steiner, of Vienna University.

Lastly, I would take this opportunity to claim for Mr. Faulkner publicly the merit due to a busy, practical man, who is, notwithstand-

---

[1] The translation here alluded to was executed at Mr. Faulkner's request by Mr. H. M. Chichester.

ing, intelligent and far-sighted enough to see that *in the end* knowledge wins the day over empiricism in all departments of human activity, and who has already shown, and now once more by the appearance of the present translation shows, that he is also disinterested and spirited enough to seek to disseminate among his brethren in trade a similar persuasion as to the value of knowledge, whilst at the same time he affords them the best help that he knows of for attaining it.

D. C. ROBB.

BLAIRGOWRIE, N.B., *March*, 1881.

POSTSCRIPT.—The lamented decease of Mr. Robb delayed the publication of this work, and the final revision was entrusted to the writer, who cannot but allude to the appreciation of Mr. Robb's cotemporaries at Oxford, for one whose scientific work was so exact and unobtrusive, and whose character was so honest and sincere.

V. H. VELEY.

UNIVERSITY COLLEGE, OXFORD, *December*, 1881.

TABLE OF CONTENTS.

		PAGE
AUTHOR'S PREFACE	.. ..	v
TRANSLATOR'S PREFACE	.. ..	vii

I. INTRODUCTION.

§ 1.	Difference between ordinary and polarized light	1
§ 2.	Polarization of light by reflection. Planes of polarization and of vibration	1
§ 3.	Polarization by double reflection in calc-spar.—Nicol's prism	2
§ 4.	Polarizer and analyzer	4
§ 5.	Simple polariscope	6
§ 6.	Rotation of the plane of polarization by active substances. Circular polarization	7

II. GENERAL ASPECTS OF OPTICAL ACTIVITY.

A. Classification of Active Substances.

§ 7.	Rotation by crystals. Table of active crystals	10
§ 8.	Rotation by organic liquids or solutions. Table of all such substances known	11
§ 9.	Substances active, both as crystals and in solution	16

B. Nature of Rotatory Power.

§ 10.	Difference between rotation by crystals and that by liquids. Rotation by vapours. Molecular rotation	16
§ 11.	Magnetic rotation	18
§ 12.	Fresnet's theory of circular polarization. Structure of active crystals ..	19
§ 13.	Constitution of active liquids. Asymmetrical structure of molecules. Optically different modifications	20

C. Dependence of Optical Activity upon Chemical Constitution.

PAGE

§ 14. Theory of Hoppe-Seyler and Mulder. Active radicles. Theory of Le Bel and Van't Hoff. Asymmetrical carbon. Tables to illustrate influence of asymmetrical carbon 24

§ 15. Artificial production of active substances. Explanation of inactivity of most artificial substances 31

§ 16. Optical properties of derivatives of active bodies. Explanation by Van't Hoff's hypothesis. Rotation influence by presence of acids, alkalies, or neutral salts 35

III. PHYSICAL LAWS OF CIRCULAR POLARIZATION.

§ 17. Amount of rotation dependent on thickness of medium. Standard thickness for solids and liquids 42

§ 18. Amount of rotation dependent on wave-length. Rotatory dispersion. Dispersion formulæ. Relation of a_0 to a_j 43

§ 19. Abnormal rotatory dispersion in tartaric acid 47

IV. SPECIFIC ROTATORY POWER.

A. Definition of Specific Rotation.

§ 20. Specific rotation of active liquids. Influence of density and temperature .. 49
§ 21. Specific rotation of active solids in solution. Method of observation and formulæ for calculating [a] 50
§ 22. Influence of temperature on specific rotation 51

B. Dependence of Specific Rotation on Nature and Amount of Solvent.

§ 23. Earlier observations on the variability of the specific rotation of substances in solution 53
§ 24. Variation of specific rotation with amount of inactive solvent. Calculation of law of variation for solutions in a single liquid 55
§ 25. Substitution of weight per cent. of active substance p, or concentration c in the formulæ, instead of weight per cent. of inactive liquid q 59
§ 26. Simultaneous influence of two inactive bodies upon a single active one .. 59
§ 27. Phenomenon of bi-rotation in milk-sugar and certain glucoses 61
§ 28. Theories as to the cause of variation of specific rotation 62

C. Determination of the True Specific Rotation of Active Substances from the Rotatory Power of their Solutions.

§ 29. Earlier experiments of Biot. Recent researches by Landolt on turpentine-oils, nicotine, and ethyl tartrate 64
§ 30. Determination of the specific rotation of left-handed oil of turpentine and of its mixtures with alcohol, benzene, and acetic acid respectively .. 66
§ 31. Do. of right-handed oil of turpentine and of its mixtures with alcohol .. 70
§ 32. Do. of nicotine and of its mixtures with alcohol and water 72

PAGE

§ 33. Determination of the specific rotation of tartrate of ethyl and of its mixtures with alcohol, wood-spirit, and water 77
§ 34. Inferences from the preceding observations 81
§ 35. Method of determining the specific rotation of active solids 83
§ 36. Details of the author's determination of the true specific rotation of ordinary camphor 84
§ 37. Details of determination by Tollens and Schmitz of true specific rotation of cane-sugar 88
§ 38. Do. of glucose hydrate and anhydride 90
§ 39. Worthlessness of many existing statements of specific rotation values .. 91
§ 40. Conditions necessary that specific rotation values may serve as characteristic marks of substances 92
§ 41. Molecular rotation—two uses of the expression 93

V. PROCESS OF DETERMINING SPECIFIC ROTATION.

§ 42. General statement of data necessary 95

A. Determination of the Angle of Rotation.

§ 43. Apparatus for the qualitative examination of rotatory power 96
§ 44. Classification of instruments used for accurate determinations 98

(a.) MITSCHERLICH'S POLARISCOPE.

§ 45. Description of the instrument 98
§ 46. Description of lamp for sodium-flame. Mode of observation with homogeneous light 99
§ 47. Mode of observation with white day or lamp-light 102
§ 48. Larger form of instrument for taking observations at constant temperature 104

(b.) WILD'S POLARISCOPE.

§ 49. Description of the instrument 107
§ 50. Arrangement for constant temperature 110
§ 51. Process of taking observations 110
§ 52. Mode of ascertaining the direction of the rotation 112
§ 53. Do. in dealing with solutions of high rotatory power 113
§ 54. Degree of accuracy attainable. Examples of actual observations .. 115

(c.) HALF-SHADE INSTRUMENTS OF JELLETT, CORNU, AND LAURENT.

§ 55. Jellett's instrument 117
§ 56. Cornu's instrument 117
§ 57. Laurent's instrument 118
§ 58. Optical principle of Laurent's instrument 120
§ 59. Process of taking observations. Examples of actual observations .. 121

PAGE

(d.) Comparison of Mitscherlich's, Wild's, and Laurent's Instruments.

§ 60. Experiments to determine the degree of concordance among the results
       obtained by different observers and with different instruments   ..   122

(e.) Determination of the Angle of Rotation for Different Rays.
                          Broch's Method.

§ 61. By using homogeneous light of different kinds   124
§ 62. By using sun-light (Broch)   124
§ 63. V. Lang's method by using artificial sources of light   125

B. Measurement of the Length of Tubes and their Adjustment.

§ 64. Description of tubes, and mode of closing their ends   128
§ 65. Different water-bath arrangements for constant temperature   129
§ 66. Process of measuring tube-lengths   130

C. Estimation of percentage Composition of Solutions.

§ 67. Preparation of solutions by weighing the active and inactive constituents   133
§ 68. Alteration of percentage in filtration   134
§ 69. Reduction of weighings to weight *in vacuo*   135

D. Determination of the Specific Gravity of Liquids.

§ 70. Mode of using the pycnometer   138
§ 71. Details of calculation of specific gravity   142

E. Estimation of the Concentration of Solutions.

§ 72. Preparation of solutions in graduated flasks   144
§ 73. Standardizing the flasks.   Table of density of water from 0° to 50° Cent.   145
§ 74. Mohr's graduation   148

F. Influence of the several Observation-Errors on Specific Rotation Values.

§ 75. Amounts of error from different sources, and their respective influence on
       the result   149

VI. PRACTICAL APPLICATIONS OF ROTATORY POWER.

A. Determination of Cane-sugar. Optical Saccharimetry.

§ 76. Principles on which the method is based   154

(a.) The Soleil-Ventzke-Scheibler Saccharimeter.

§ 77. Description of the instrument   155
§ 78. Explanation of the action of the several parts   156
§ 79. External form of the instrument   159
§ 80. Mode of taking observations   160

PAGE

§ 81. Graduation of Ventzke's scale. Calculation of concentration, or of per cent. composition of saccharine solutions 162
§ 82. Correction of saccharimeter readings for slight disproportionality between rotation and concentration. Schmitz's table of corrections for each degree on the scale.. 163
§ 83. Conversion of degrees (Ventzke) into degrees of angular measurement .. 169
§ 84. Correction of errors due to imperfect construction. Scheibler's method of double observation 169
§ 85. Influence of temperature on determinations by the saccharimeter. Mategczek's table of corrections 172

(b.) SOLEIL-DUBOSCQ SACCHARIMETER.

§ 86. Explanation of the scale and mode of calculating results 174
§ 87. Tables of correction for imperfect proportionality between deviation and concentration, and for changes of temperature.. 175

(c.) WILD'S POLARISCOPE WITH SACCHARIMETRIC SCALE.

§ 88. Mode of graduating the scale 176
§ 89. Table of correction for imperfect proportionality between rotation and concentration 177

(d.) SACCHARIMETER WITH ANGULAR GRADUATION ON MITSCHERLICH'S, WILD'S, OR LAURENT'S PRINCIPLE.

§ 90. Calculation of sugar percentage, assuming the specific rotation as constant 178
§ 91. Schmitz's table for correction of error due to variability of specific rotation 179

(e.) PREPARATION OF SOLUTIONS FOR THE SACCHARIMETER.

§ 92. Standardizing the 50 and 100 cubic centimetre flasks. Weighing the sugar-samples 183
§ 93. Process of clearing solutions. Error due to presence of precipitates formed 184
§ 94. Decoloration of solutions with animal charcoal. Caution in the use of .. 186

(f.) DETERMINATION OF CANE-SUGAR IN PRESENCE OF OTHER ACTIVE SUBSTANCES.

§ 95. Removal of optically-active impurities 187
§ 96. Estimation of cane-sugar in presence of invert-sugar. Inversion method of Clerget 187

B. Determination of Glucose.

§ 97. Calculation of the percentage of glucose in dilute solutions from the angle of rotation 190
§ 98. Table of corrected values for solutions of higher concentration 191
§ 99. Use of saccharimeters with Ventzke's or Soleil's scale, or with special scale 193
§ 100. Determination of grape-sugar in diabetic urine 194
§ 101. Detection of potato-sugar in " chaptalized " wine 195

C. Determination of Milk-Sugar.

PAGE

§ 102. Rotatory power of milk-sugar. Determination of its amount in milk .. 197

D. Determination of Cinchona Alkaloids.

§ 103. Rotation constants of quinine, cinchonidine, quinidine (conchinine), and
        cinchonine, and their salts, according to Hesse   198
§ 104. Oudemans' investigations on the rotatory power of quinine, cinchoni-
        dine, and quinidine, and their salts   203
§ 105. Determination of the purity of samples by means of their rotation-
        angle 205
§ 106. Optical analysis of mixtures of cinchona alkaloids 209
§ 107. Estimation of quinine mixed with cinchonidine   212

VII. ROTATION CONSTANTS OF ACTIVE SUBSTANCES.

§ 108. Preliminary remarks   214
§ 109. Sugars having the formula $C_{12}H_{22}O_{11}$ 216
§ 110. Sugars having the formula $C_6H_{12}O_6$ 221
§ 111. Mannite group   223
§ 112. Carbohydrates $(C_6H_{10}O_5)_n$ 224
§ 113. Glucosides   225
§ 114. Derivatives of the sugars (amyl-alcohol, valerianic acid, para-lactic acid) 226
§ 115. Vegetable acids 230
§ 116. Terpenes $(C_{10}H_{16})$ 233
§ 117. Ethereal oils   235
§ 118. Resins   235
§ 119. Camphors   235
§ 120. Alkaloids 236
§ 121. Unclassified vegetable substances   244
§ 122. Bile constituents   245
§ 123. Gelatinous substances 247
§ 124. Albumins   247

APPENDIX   249

REESE LIBRARY
OF THE
UNIVERSITY
OF
CALIFORNIA.

ILLUSTRATIONS.

NO. OF ILLUSTRATION.		PAGE
1.	TRANSVERSE SECTION OF RAY OF LIGHT	1
2.	LINEAR POLARIZED RAY	1
3, 4.	POLARIZATION BY REFLECTION	2
5.	RHOMBOHEDRON OF ICELAND SPAR	3
6.	REFRACTION OF LIGHT BY ICELAND SPAR	3
7—10.	CONSTRUCTION OF NICOL'S PRISM	4
11, 12.	POLARIZER AND ANALYZER	5
13.	INSTRUMENT WITH POLARIZER AND ANALYZER	6
14.	PLANE OF POLARIZATION OF FIXED POLARIZER	7
10A.	CONSTRUCTION OF NICOL'S PRISM	9
15.	GRAPHIC REPRESENTATION OF SPECIFIC ROTATIONS	68
16.	GRAPHIC REPRESENTATION	71
17.	GRAPHIC REPRESENTATION	76
18.	GRAPHIC REPRESENTATION	80
19.	GRAPHIC REPRESENTATION	86
20.	APPARATUS FOR QUALITATIVE EXAMINATION OF ROTATORY POWER	96
21.	MITSCHERLICH'S INSTRUMENT	99
22.	LAMP FOR INSTRUMENT	99
23.	VERTICAL BAND OF LIGHT	100
24.	CHANGE OF DIRECTION OF PLANE OF POLARIZATION	101
25.	LAMP FOR INSTRUMENT	102
26.	MITSCHERLICH'S LARGER INSTRUMEN	104
27, 28.	WILD'S POLARISCOPE	107
29.	ARRANGEMENT OF APPARATUS FOR OBSERVATION	109
30.	PARALLEL FRINGES IN FIELD OF VISION	110
31—34.	RELATIVE DIRECTIONS OF ROTATION	113
35, 36.	RELATIVE DIRECTIONS OF ROTATION	114

NO. OF ILLUSTRATION.		PAGE
37, 38.	LAURENT'S HALF-SHADE INSTRUMENT	118
39—42.	QUARTZ PLATE OF LAURENT'S INSTRUMENT	120
43.	ARRANGEMENT OF APPARATUS FOR OBSERVATION	126
44.	TUBE WITH WATER BATH	129
45.	TUBE WITH LEAD SPIRAL	130
46.	INSTRUMENT FOR MEASUREMENT OF TUBE LENGTHS	131
47, 48.	MEASUREMENT OF TUBE LENGTHS BY CATHETOMETER	132
49.	FLASK FOR PREPARATION OF SOLUTIONS	134
50.	PYCNOMETER AND INDIA RUBBER BALL	139
51, 52.	SPRENGEL'S PYCNOMETER	139
53.	MODIFIED FORM OF PYCNOMETER	141
54.	GRADUATED FLASK	144
55.	SOLEIL-VENTZKE-SCHEIBLER SACCHARIMETER	155
56.	SOLEIL'S SACCHARIMETER WITH SCHEIBLER'S IMPROVEMENTS	159

I.

INTRODUCTION.

§ 1. In an ordinary ray of light the vibrations of the particles of ether take place successively in all possible directions perpendicular to its axis. Fig. 1 shows a transverse section of a ray, projected on a vertical plane.

By certain means it is possible to restrict the vibrations to some one particular direction (Fig. 2). The ray is then called a *linear polarized ray.* Its behaviour is no longer identically the same all around its axis, as in an ordinary ray ; on the contrary, it displays two distinct *sides,* one in the plane of its vibrations, the other in a plane at right angles thereto.

Fig. 1.

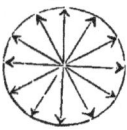

Fig. 2.

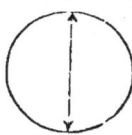

§ 2. This conversion of common into polarized light may be effected, first, by *reflection,* for which purpose a glass mirror, inclined to the perpendicular at a certain angle (35° 25′), as *L M* (Figs. 3 and 4), will be found best. Rays falling in the direction *a b,* so as to make an angle of 55° with the normal *x y,* are reflected in the direction *b c,* and at the same time are polarized. This becomes manifest when the reflected rays meet the mirror *P Q,* which has a rotatory movement about *b c* as an axis. When the second mirror is parallel to the first, as in Fig. 3, the ray *b c* is wholly reflected in the direction *c d;* but as the mirror turns on its axis, the intensity of the light reflected from it diminishes, until at 90° from the starting-point there is no longer any reflection, and the mirror appears dark. Continuing the rotation, we find that at 180°,

B

i.e., when the mirrors are inclined, so that the planes *a b c* and *b c d* are again coincident, as in Fig. 4, there is another maximum of reflection, and, lastly, at 270° another minimum. Thus the ray

Fig. 3.   Fig. 4.

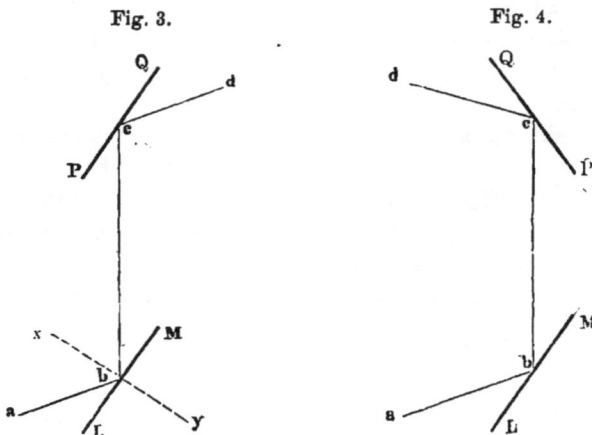

behaves differently in two different directions at right angles to each other, one direction being in the plane of incidence or reflection, *a b c*, the other in a plane at right angles therewith. The ray is *polarized*, and the former of the two planes is its *plane of polarization*.

We can therefore recognize a ray as polarized, and determine the position of its plane of polarization, by permitting it to fall upon a glass mirror at an angle of 55°. If the mirror be turned about the ray as an axis, light and darkness will alternate at intervals of 90°, and if the mirror be set so that the light emitted by it is at its brightest, the plane passing through it and the incident polarized ray is the plane of polarization of the latter. Again, if the mirror is at its darkest, the plane at right angles to the plane of incidence of the polarized ray is coincident with the plane of polarization.

By the undulatory theory of light it can be proved that the plane of polarization is either the plane in which the vibrations of ether take place, or a plane at right angles thereto. Which of these is really the case is still an open question among physicists; but for simplicity's sake we shall adopt the former view, and in these pages assume that the planes of vibration and polarization are coincident.

§ 3. But a pencil of light can also be polarized by *repeated single refractions* or by *double refraction* in certain crystals, as of calc-spar—the latter being the most suitable means.

In a natural rhombohedron of Iceland spar (Fig. 5), the princi-pal axis lies in the line joining the points *d* and *f*, where three obtuse angles meet.

Fig. 5.

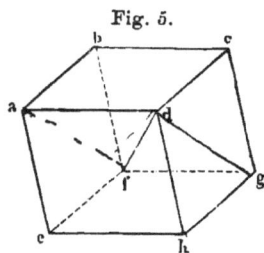

Suppose a plane to pass through the shorter diagonals of two opposite faces of the rhomb, *i.e.*, either *d g*, *a f*, or *d b*, *h f*, or *d e*, *c f*, it will invariably contain the principal axis *d f*. Any such plane, and all planes parallel thereto, are called *principal sections* of the prism. If a pencil of light, *m n* (Fig. 6), falls on one

Fig. 6.

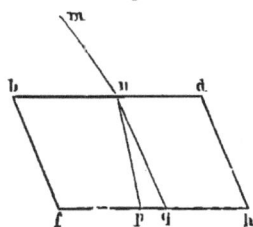

face, as *a b c d* (of which *d b h f* represents the principal section), it will, on entering the crystal, divide into two refracted rays unequally bent. Both are polarized, and application of the mirror will show that the plane of polarization (or plane of vibration) of the less refracted or *extra-ordinary* ray *n q* is perpendicular to the principal section *d b h f*, while that of the more highly refracted or *ordinary* ray *n p* coincides with the plane of the said section.

For polariscopic purposes it is best to give exit to one only of the two polarized rays ; that, namely, in the direction of the incident light, and to eliminate the other ray. This can be done in various ways, but most completely by converting the calc-spar into a Nicol's prism.[1]

For this purpose a piece of Iceland spar is split up into an elon-gated rhombohedron, as in Fig. 7, in which the plane passing through the points *a b c d* represents a principal section. The natural ends of the prism *a f b e* and *d g c h*, the former of which is inclined to *a d* and the latter to *c b* at an angle of 71°, are ground so as to reduce these angles to 68° (*see* Fig. 8). The prism is then divided in the direction *b' d'*, which is perpendicular to *a b'* and *c d'*, and the halves[2] —after polishing the faces of the section—are united as before with

[1] For other forms of calc-spar polarizing prisms (Sénarmont's, Foucault's, and "achromatized" prism) see Wüllner's *Lehrbuch der Physik*, 3 Aufl. 2, 528—530.

[2] Smaller Nicols may be prepared by grinding two separate crystals.

Canada balsam. Finally, the sides are blackened and the Nicol
(Fig. 9) is fixed with cork into a brass case. (The principal section
of this prism passes through the shorter diagonals of the two
rhombic ends. If a pencil of light, *l m* (Fig. 10), parallel to the
edges of the longer side, falls on the face *a b'*, it divides into two
rays, which are polarized at right angles to one another. The less
refracted (or extraordinary) ray, *m p q*, traverses the film of balsam
at *p*, and emerges in the direction *q s*, parallel to *l m*. The more
refracted (or ordinary) ray, *m o*, meets the balsam at *o*, which, from

Fig. 7.   Fig. 8.   Fig. 9.   Fig. 10.

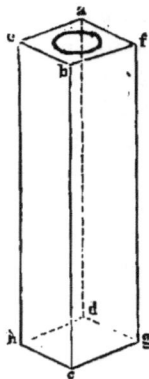 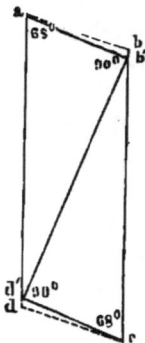 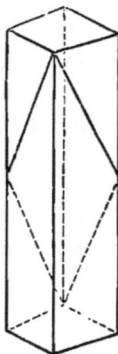 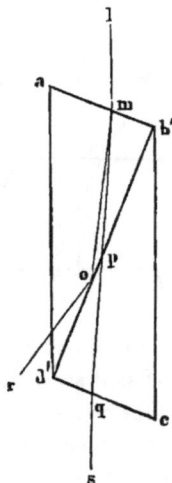

its being a medium of so much feebler refractive power, causes *total
reflection* of the ray in the direction *o r*, whereby it becomes absorbed
by the case of the prism. The other ray emerges in the direction of
the incident pencil, but possesses only half of its luminous power.
The plane of polarization (and vibration) of this ray is at right angles
to the principal section, and therefore passes through the longer
diagonals of the end faces of the prism.

§ 4. We have next to consider the behaviour of a polarized ray
from a fixed Nicol, in passing through a second Nicol having a movement of rotation about its longitudinal axis. The first prism we will
call the *polarizer*, the second the *analyzer*.

Figs. 11 and 12 show the two prisms in their cases, the principal sections being indicated by $b\,b\,m$, $b'\,b'\,n$, and the planes of polarization at right angles thereto, by $a\,a\,m$, $a'\,a'\,n$.

If we turn the analyzer A so that its plane of polarization $a'\,a'\,n$ is parallel with the plane of polarization $a\,a\,m$ of the polarizer P (Fig. 11), whereby, too, the principal sections $b\,b\,m$ and $b'\,b'\,n$ of the two prisms are brought into the same direction, the ray m, which enters P as ordinary light and emerges polarized at n, is not decomposed on passing through A. It is merely slightly refracted in

Fig. 11.

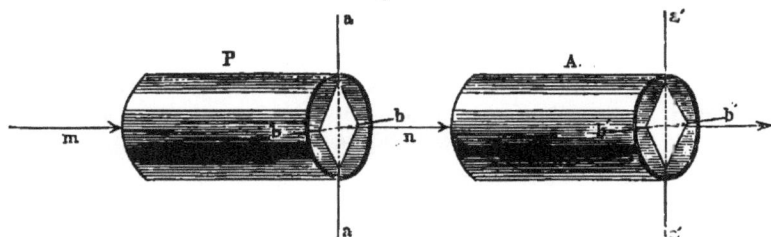

Fig. 12.

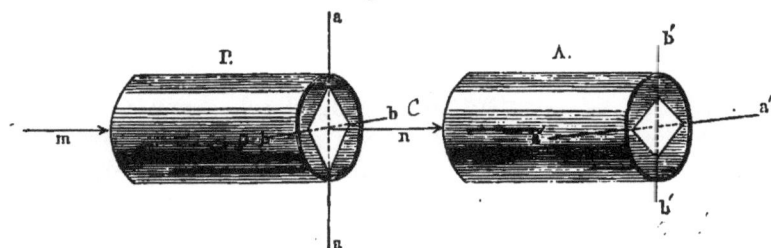

the direction of an extraordinary ray $m\,p\,q$ (Fig. 10), and emerges so at the opposite end of the analyzer. The same happens if the latter be turned through an angle of 180°, so as to bring the planes again parallel. But, if the analyzer is adjusted so that its plane of polarization is at right angles to that of the polarizer—that is, when the prisms are arranged as in Fig. 12—then the ray entering A takes the direction of an ordinary ray, like $m\,o\,r$ (Fig. 10), and is eliminated by the film of balsam. No light leaves the analyzer, and, accordingly, the field of vision appears dark. The same thing happens at a distance of 180°.

In all cases where the planes of polarization (or the principal

sections) of the two Nicols are neither parallel nor at right angles
to each other, the polarized light entering the analyzer is separated
into an ordinary and an extraordinary ray, varying in intensity with
the angle at which the planes of polarization of the two prisms are
inclined to each other. Suppose them at first to be crossed, that is,
set for darkness. Then, if we turn the analyzer through a small
angle, the luminous intensity of the laterally-deflected ordinary
ray will greatly exceed that of the transmitted extraordinary ray.
Nevertheless, the latter suffices to slightly illuminate the field of
vision. If the angle be increased to 45°, the ordinary and extraordinary rays will be of equal intensity, so that the light leaving
the analyzer will have exactly half the luminous power of the
total entering light. By increasing the angle further, the luminous
power of the transmitted ray will gradually exceed that of the
eliminated ray, until at an angle of 90° the latter ceases altogether
to exist, and the field of vision exhibits the *maximum* of brightness.
Continuing the rotation in the same direction, we find another *minimum* of light at 180°, and another *maximum* at 270°.

§ 5. To observe these phenomena, the instrument shown in

Fig. 13.

Fig. 13 may be used. The horizontal bar *d*, secured on a stand,
carries at one end the fixed polarizing Nicol *a*, and at the other
the analyzing Nicol *b*, which, by
means of the lever *c*, can be
turned with its frame about its
axis. In connection with the
lever, a single or a double index
moves round the divided disc,
which is fixed to the bar. Between the Nicols can be inserted
the tube *f*, the ends of which can
be closed with glass plates.

We first direct the polarizer *a*
towards some luminous source,
and the phenomena are simplified
by using monochromatic light,
e. g., a Bunsen flame playing on a
bead of common salt. The tube *f* is left empty or filled with water.

Now, if we look through the analyzing Nicol while revolving it, we shall be able readily to find a position in which the field of vision appears darkest. Suppose the index now to be at the zero-point on the disc. Then, as explained above, on continuing the rotation we shall find, in a complete revolution, another maximum of darkness at 180°, and the two maxima of light at 90° and 270° respectively. For purposes of scientific observation, the points of greatest darkness are to be preferred as marks of reference, since at these points the least movement of the Nicol produces a perceptible change in the appearance of the field of vision.

§ 6. Now, if the tube be filled with a solution of cane-sugar instead of water and put in its place, the analyzer having been previously set to darkness (0° or 180°), it will be found that the field

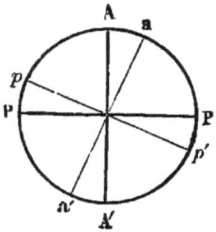

Fig. 14.

of vision now appears bright, and to obtain the maximum of darkness we must turn the analyzer to the right through a certain angle. If the plane of polarization of the fixed polarizer of the instrument has the direction $P P'$ (Fig. 14), so long as the tube is empty the rays cannot pass through the analyzer, since its plane of polarization $A A'$ is at right angles to $P P'$. But, if after the introduction of the sugar solution the field of vision exhibits the maximum of darkness when the plane of polarization of the analyzer is revolved into the position $a a'$, we are bound to conclude that the rays originally vibrating in the plane $P P'$, in their passage through the solution, have experienced a certain deflection of their plane of vibration, and that their vibrations are now perpendicular to $a a$— that is, they take place in the plane $p p'$.

The angle a, through which the analyzer has to be turned to bring a recurrence of darkness in the field of vision, and which can be read off on the graduated rim of the disc, is called the *angle of rotation*, and is the measure of the deflection experienced by the plane of polarization.

A number of other substances behave in the same way as cane-sugar—that is to say, the analyzer requires to be turned to the *right* hand from zero to reach the point where the light vanishes. Again, if the tube be filled with nicotine or a solution of amygdalin, the phenomenon of reappearance of light occurs as before; but, in this case, to set the instrument back to darkness the analyzer has to be turned to the *left*. These substances, therefore, cause a deflection of the plane of polarization to the left.

This rotation of the plane of vibration, or of polarization, is called *circular polarization*. Substances which exhibit this power are said to be circular-polarizing or optically *active*, and are distinguished as *right-rotating* (dextro-gyrate) or *left-rotating* (lœvo-gyrate), whilst those substances which have not this power are said to be *inactive*.

Circular polarization was first noticed in 1811 by Arago, in rock-crystal. In 1815 Biot discovered the optical activity of organic bodies, and in a series of important investigations, extending over more than forty years, he deduced the laws and explained the nature of the phenomena. His observations form the basis of our present knowledge of the subject.[1]

[1] Biot: *Mém. de l'Acad.* 2, 41; 3, 177; 13, 39; 15, 93; 16, 299. *Ann. Chim. Phys.* [2] 9, 372; 10, 63; 52, 58; 69, 22; 74, 401; [3] 10, 5, 175, 307, 385; 11, 82; 28, 215, 351; 29, 35, 341, 430; 36, 357, 405; 59, 206.

NOTE BY TRANSLATOR.

For the sake of some readers, it may be as well to add here a rather more explicit account of the action of a piece of apparatus so fundamental in polariscopic work as the Nicol prism. Taking Fig. 10 in the text, let us add to the author's construction by drawing through m a both-ways perpendicular (*normal*) to $a\,b'$, as also through p and o similar normals to $d'\,b'$. Now, in their passage through the first half of the prism, the rays are both bent *towards the normal* $m\,n'_1$, (*i.e.*, outward from the balsam), to extents due to their different *refractive indices*, the ordinary ray $m\,o$ (refr. ind. 1·66) more than the extraordinary ray $m\,p$ (refr. ind. 1·52). The refractive index for Canada balsam for mean light being 1·54, the extraordinary ray on meeting the line $b'\,d'$ (which, to represent a layer of sufficient thickness, must be broadened into a rhomboid) here encounters a medium of refractive power almost identical with that of the calc-spar which it left, so that this ray passes on with but a minute deviation *inwards*, due to the balsam being slightly more refractive than the spar for this ray. On the other hand, the balsam being very considerably less refractive than the calc-spar for the ordinary ray, causes that ray to diverge *outwards from the normal*, $o\,n'_3$, and so much so, that it cannot hold a course through the balsam at all, but, inasmuch as the angle of incidence $m\,o\,n_3$ in the more refractive medium exceeds the so-called *critical angle*, the ray suffers *total reflection* from the surface, so that the angle $r\,o\,n_3$ equals angle $m\,o\,n_3$. The *critical angle*, or angle at which a ray, issuing from a more refractive into a less refractive medium, emerges just parallel to the bounding surfaces, depends on the relative index of refraction. Simple geometrical considerations show that if the angle to the normal in the more refractive medium has a sine whose value is greater than the ratio of the absolute indices, the ray cannot emerge into the less refractive medium. Now, in the case before us the ratio in question for balsam and spar is

$$\frac{1\cdot 54}{1\cdot 66} = 0\cdot 928 = \sin 68°.$$

Hence the limiting value of $m\,o\,n_3$, so that $m\,o$ might just emerge in direction $o\,d'$, is 68°. If now $m\,o$ were parallel to $a\,d'$, the angle $m\,o\,n_3$ would be just 68°, being opposite to

$b'a d'$, which has been ground down to 68°—the figure $m\,a\,d'\,n_3$ in that case forming a parallelogram. But in passing through the first half of the prism, the ray is refracted so that the angle $m\,o\,n_3$ is always greater than $b'a d'$. Thus we see that by grinding the end-faces of the prism so that $b'a d' = 63°$, we secure that the angle $m\,o\,n_3$ shall always exceed the critical angle, and the importance of this procedure in the construction of the prism becomes apparent. As to what happens when a second Nicol is used to receive the extraordinary ray emerging from the first, the following considerations may perhaps be found useful:—In all uniaxial crystals there are two directions at right angles to each other, the one of greatest resistance to the propagation of luminous vibrations, the other of least resistance. These planes are in the direction of the principal axis (see Fig. 5) and at right angles thereto. Calc-spar transmits only such light, the vibrations of which take place in either of these two directions; and all incident light propagated by vibrations in a plane at any other angle to the principal section is resolved into two such component rays. But the velocities of transmission in the two directions are unequal; that is, since amount of refraction depends on velocity of transmission, the refractive index of the spar is, as we have already said, different in the two directions. Now if the second Nicol be arranged behind the first, so that corresponding planes in the two prisms be coincident (as in Fig. 11), the extraordinary ray, coming into a plane of the same resistance as that which it left, is propagated with the same velocity as it had in the first prism, and is, therefore, similarly refracted, i.e., it takes a course similar to $l\,m\,p\,q$ (Fig. 10) in the first Nicol, emerging unaltered.

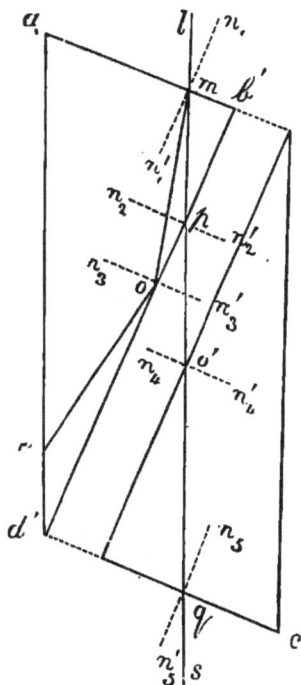

FIG. 10A.

If, however, the second prism be arranged so that corresponding planes shall cross, then the extraordinary ray, coming into a new plane in which it travels with greater velocity than before, is refracted accordingly, taking a course similar to $l\,m\,o\,r$ (Fig. 10) in the first Nicol, i.e., is totally reflected, so that no light emerges. Lastly, if the planes of the Nicols be crossed at any other angle, the light cannot pass in the plane it encounters in the second Nicol, but is resolved into two components in the two directions at right angles to each other, in which alone (as we have said) calc-spar transmits light-vibrations—that component which takes the course of the principal section being eliminated by the Canada balsam, whilst that which takes a course at right angles thereto alone emerges. Now that reduction of intensity in luminous power which may be effected on a polarized ray emerging from one Nicol by opposing to its course an *impassable* plane of a second Nicol, is also effected by opposing to its course a rotatory substance. The ray is made to vibrate in a different plane; in other words, the plane of polarization is rotated, and the resulting phenomenon is the same as if the first Nicol had been rotated to the same extent. The physical explanation of this rotatory power is discussed in succeeding chapters.—[D. C. R.]

GENERAL ASPECTS OF OPTICAL ACTIVITY.

A. Classification of Active Substances.

§ 7. Circular polarizing substances may be divided into two classes :—

(1) Bodies which only in a crystalline state possess the property of rotating the plane of polarization, and which lose this property entirely when brought (either by solution or fusion) into an amorphous condition. Up to the present, only a few such active crystalline substances are known. In the table annexed these substances are recorded, along with the angle a, through which the ray D, or mean yellow light j (jaune), is rotated on passing through a plate of each substance 1 millm. in thickness.

Substances.	Formula.	a per 1 millm.	Observer.
Cinnabar	HgS	$a_D = 32\cdot5°$	Descloizeaux
Rock crystal	SiO_2	$a_D = 21\cdot67$	Biot, Broch, &c.
Sodium chlorate	$NaClO_3$	$a_j = 3\cdot67$	Marbach
,, bromate	$NaBrO_3$	$a_j = 2\cdot80$	,,
,, periodate ..	$NaJO_4 + 3 H_2O$	$a_D = 23\cdot3$	Ulrich, Groth
Potassium dithionate ..	$K_2S_2O_6$	$a_D = 8\cdot39$	Pape
Strontium ,, ..	$SrS_2O_6 + 4 H_2O$	$a_j = 1\cdot64$	,,
Calcium ,, ..	$CaS_2O_6 + 4 H_2O$	$a_j = 2\cdot09$	,,
Plumbic ,, ..	$PbS_2O_6 + 4 H_2O$	$a_D = 5\cdot53$	,,
Sodium sulphantimoniate	$Na_3SbS_4 + 9 H_2O$	$a_j = 2\cdot7$	Marbach
Uranium sodium acetate..	$(UrO)_2.Na . 3 C_2H_3O_2$	$a_j = 1\cdot8$	,,
Matico-camphor ..	$C_{10}H_{16}O$	$a_D = 2\cdot4$	Hintze
Benzil	$C_{14}H_{10}O_2$	$a_D = 24\cdot92$	Descloizeaux
Ethylene-diamine sulphate	$(N_2H_4 . C_2H_4)H_2SO_4$	—	v. Lang
Guanidine carbonate ..	$(CH_5N_3)_2 . H_2CO_3$	$a_D = 14\cdot35$	Bodewig
Diacetyl-phenol-phtaleïn	$(C_{20}H_{12}O_4)(C_2H_3O)_2$	$a_D = 19\cdot7$	,,

The crystals of these substances are, without exception, either single-refracting (regular) or uniaxial double-refracting (hexagonal or quadratic). In the latter, the optical power is only displayed in the direction of the principal axis, and we have therefore to use plates cut perpendicularly thereto. Moreover, every one of the above specified substances occurs both in right-rotating and left-rotating crystals, and the amounts of deviation are exactly the same for plates of equal thickness.

Several of these substances—quartz, sodium periodate, dithionates, guanidine carbonate, and matico-camphor, give external evidence of the possession of this property by the existence of hemihedral or tetartohedral faces, which are right-handed or left-handed according as the crystal rotates to the right or left.

§ 8. (2) Bodies which display rotatory power when dissolved, and, consequently, in the amorphous state. The substances of this class are, without exception, carbon-compounds, and either occur naturally in vegetable or animal organisms, or as derivatives from these by simple metamorphoses. No inorganic substance is known which in solution exhibits rotatory power, and it would seem that this property is a peculiar attribute of the carbon-atom.

The substances in this class are either right-rotating (+) only or left-rotating (−) only, with the exception of a few which manifest the power in both directions. The subjoined table contains as complete a list as possible of all active substances known up to the present time, with their most important derivatives, and also, in the last column, a list of compounds which, although nearly related to these substances, are inactive :—

Substances.	Lævo-rotatory.	Dextro-rotatory.	Inactive.
Sugars $C_{12}H_{22}O_{11}$		Cane-sugar, Milk-sugar, Mycose, Melitose, Melezitose, Maltose	Synanthrose
Sugars $C_6H_{12}O_6$	Lævulose Invert-sugar Inverted Synanthrose	Dextrose (Honey-sugar, Grape-sugar, Starch-sugar, Salicin-, Amygdalin-, Phlorhizin-sugar, Gum-sugar).	
	Sorbin	Galactose, Eucalyn	Inosite ?

Substances.	Lævo-rotatory.	Dextro-rotatory.	Inactive.
Mannite Group	Mannite Lævo-mannitan Mannitone Matezite	Nitro-mannite Dextro-mannitan Nitro-mannitan Quercite Pinite Iso-dulcite Bornesite	Mannitose Quercitose Sorbite Dulcite, Nitro-dulcite Erythrite Dambonite, Dambose
Carbohydrates $C_6 H_{10} O_5$	Inulin, Inuloid Gum Arabic Beetroot Gum	Starch, Xyloidin Dextrin, Glycogen Gum Arabic Dextran (Fermentation Gum)	Cellulose Nitro-cellulose Pectin
Glucosides	Amygdalin, Amygdalic Acid, Mandelic Acid Salicin, Populin Phlorhizin, Digitalin, Cyclamin, Coniferin	Quinovin Apiin	Glycyrrhizin Phloretin Tannic Acids?
Derivatives of the above Groups.	Acetyl derivatives of Inulin Gummic Acid Fermentation Amyl-alcohol Para-lactic Acid Salts and Ether Anhydrides	Acetyl derivatives of Dextrose, Milk-sugar, Mannite, Mannitan, Dulcite and Starch Glucosan, Saccharic Acid Amyl-alcohol from Dextro-amyl Chloride Derivatives of Lævo-amyl-alcohol (Di-amyl, Ethyl-amyl, Amyl chloride, iodide and cyanide, Amylamine, Amyl-valerate, Valeric aldehyde, Valerianic Acid, Capronic Acid) Para-lactic Acid	Levulinic Acid Mucic Acid Fermentation Butyl-alcohol Octyl-alcohol from Ricinus Oil Methyl-amyl Amyl hydride Amylene from active Amyl-alcohol

Substances.	Lævo-rotatory.	Dextro-rotatory.	Inactive.
Vegetable Acids and Allied Substances	Lævo-tartaric Acid ,, Salts Lævo-tartramide Natural Malic Acid Acid Malate of Ammonia in Water Malamide from Lævo-malic Acid Asparagin in Aqueous and Alkaline Solutions Aspartic Acid in Alkaline Solutions Glutaric * Acid Quinic Acid Lactonic Acid Atractylic Acid	Dextro-tartaric Acid ,, Salts Dextro-tartramide Meta-tartaric Acid Di-tartaric Acid Malic Acid from Dextro-tartaric Acid or Asparagine Malate of Ammonia in Nitric Acid Neutral Malates(Zinc- and Antimon-Ammonium Malates)in Water Asparagin in Acid Solutions Aspartic Acid in Acid Solutions Glutamic Acid Quinovic Acid in Alkaline Solutions Dextronic (Gluconic) Acid	Para-tartaric Acid Synthetic Tartaric Acid Pyro-tartaric Acid Nitro-tartaric Acid Synthetic Malic Acid Maleic Acid Fumaric Acid Succinic Acid Citric Acid Citro-malic Acid Aspartic Acid from Fumaric or Maleic Acid
Terpenes $C_{10} H_{16}$	Lævo-oil of Turpentine or Terebenthene (French, from *Pinus maritima* ; Venetian, from *P. Larix* ; Templin-oil, from *P. Picea* or *Pumilio*). Terebenthone hydrochlorate Terecamphene Liquid Terpinhydrate Iso-terebenthene Terpene from Parsley Oil	Dextro-oil of Turpentine or Australene (English or American, from *Pinus balsamica, Australis* and *Taeda;* German, from *P. sylvestris, nigra,* and *Abies*). Australene hydrochlorate Austracamphene Tetra-terebenthene Terpene from Oils of Lemon, Orange, and Poplar Cicutene	Camphene Camphilene Terebene Terebilene Polyterebenes Terpinol Solid Terpinhydrate
Ethereal Oils	Oils of Copaiba, Cubebs, Lavender, Parsley, Rue, Roses, Tansy, Thyme, Juniper, and Crisped mint	Anethol, Oils of Cascarilla, Chamomile, Coriander, Fennel, Nutmeg, Myrtle, Sassafras	Oils of Anise, Cassia, Cloves, Cinnamon, Gaultheria, Bitter Almonds, Mustard, Thymol

The following are both Dextro- and Lævo-rotatory: Oils of Peppermint, Cumin, Rosemary, Salvia, Savine, Elemi, Cascarilla

* Probably this should be *Glutanic acid* (*vide* p. 232).—[D. C. R.]

Substances.	Lævo-rotatory.	Dextro-rotatory.	Inactive.
Resins	Sylvic Acid Pimaric Acid Guaiacum Acid	Podocarpic Acid Dextro-pimaric Acid Euphorbone	
Camphors and Allied Substances	Matricaria-camphor Menthol, Patchouli-, Blumea-(Ng ai)and Rubia-camphor Borneene Camphoric Acid, from Matricaria-camphor Camphoric Anhydride and CamphoronicAcid,from Dextro - camphoric Acid Carvol, from Crisped mint Oil Citronellol	Ordinary Camphor Borneol, Amber-camphor, Rosemary-(Ledum) camphor Ethyl- and Amyl-camphor Camphor Bromide Camphoric Acid,from ordinary Camphor Camphic Acid Cymol, from Oil of Cumin and Cuminol Carvol, from Cumin-oil Absinthol Myristicol	Lavandula-camphor Camphrene Sulpho-camphoric Acid All other Cymols Safrol Geraniol
Alkaloids	Quinine, Cinchonidine Homocinchonidine Paytine, Cusconine Aricine Morphine, Codeine Narcotine in Alcohol Pseudomorphine Thebaine, Papaverine Laudanine Strychnine, Brucine Nicotine Atropine Aconitine Geissospermine	Quinidine (Conchinine), Quinicine, Cinchonine, Cinchonidine, Quinamine Quinidamine,* Quinamicine, Dihomo-cinchonicine* Apodiquinicine* Narcotine in Hydrochloric Acid Laudanosine Conine Cicutine Pelosine (Buxine) Pilocarpine	Meconine Narceine Hydrocotarnine Cryptopine Aribine, Betaine Paraconine Berberine Veratrine Emetine Piperine Sanguinarine
Indifferent Substances	Santonin Santonic Acid Picrotoxin Jalapin	Hydrosantonic Acid Hæmatoxylin Echicerin, Echitin Echitein, Echiretin	Ostruthin Leucotin Oxyleucotin Hydrocotoin

Substances.	Lævo-rotatory.	Dextro-rotatory.	
Bile Constituents	Cholesterin (Phytosterin)	Glycocholic Acid Taurocholic Acid Cholalic Acid Choloidic Acid Hyoglycocholic Acid Hyocholoidic Acid Lithofellic Acid	* These are the names by which the three substances are best known to English chemists; but the author gives Hesse's names, viz., *conchinamine*, *dihomo-cinchonine*, and *dicinchonine* respectively. For the last, *apodiquinicine* was suggested by Wright, on the ground of its greater resemblance to *quinine*, of which it may be regarded as a first anhydride, thus: $2 (C_{20}H_{24}N_2O_2) - H_2O = C_{40}H_{46}N_4O_3$.—[D. C. R.]
Gelatinous Substances.	Gelatin, Chondrin		
Albumins	Serum-albumin Egg-albumin Paralbumin Sodium-albuminate Casein, Syntonin Peptones		

According to the foregoing list, the number of natural active substances known amounts to about 140, of which 65 are left-rotating, 60 right-rotating, and 15 both right and left-rotating.

Of active derivatives, counting all hitherto examined salts of the alkaloids and vegetable acids, there are at least as many known, thus bringing the total number of optically active carbon-compounds up to close on 300, and no doubt many other substances hitherto unexamined possess the power of rotating the plane of polarization.

Substances which display the rotatory power when in a state of solution, and are crystallizable, are not found to exhibit optical activity in the crystalline state, as when a polarized ray is passed through plates cut from them. This is the case with cane-sugar, tartaric acid, asparagin, camphors, etc. (see Biot,[1] Descloizeaux[2]). Now the phenomenon of circular polarization is only observable in single-refracting or in uniaxial double-refracting crystals, and in the latter only in the direction of the optic axis. But the substances just referred to are all biaxial, and thus in no direction single-refracting ; consequently, *circular double-refraction*[3] could not in any case be observed in them, as it would be over-powered by the more marked phenomenon of ordinary double-refraction. Whether they are really inactive in the crystalline state is undecided.

[1] Biot : *Mém. de l'Acad.* 13, 39.   [2] Descloizeaux : *Pogg. Ann.* 141, 300.

[3] The expression refers to Fresnel's theory of circular polarization, in which the two rays are supposed to vibrate in opposite circular paths. *See* § 12.—[D. C. R.]

But when these substances are brought into the amorphous solid form their optical activity is retained—a fact first observed by Biot with cast plates of sugar and tartaric acid.[1]

§ 9. As a third and distinct class are regarded those substances which are known to exhibit rotatory power both in the crystalline state and in solution. At present only two such substances are known, viz., strychnine sulphate crystallizing with water in quadrate octahedra,[2] and regular amylamine-alum.[3]

B. Nature of Rotatory Power.

§ 10. The fact that substances in the first of the above classes manifest rotatory power only in the crystalline state and lose it directly they are brought into solution, is proof that the rotation is dependent on crystalline structure—that is, upon a particular arrangement in the groups of molecules (forming the crystal). Dissolution or fusion breaks up this arrangement, and the optical power is consequently lost. In this case then the phenomenon is purely physical.

The second class of substances, on the contrary, exhibit rotatory power in the liquid state. Now there is every reason to believe of matter in this form that the smallest quantities, capable of independent motion as units consist, not of individual molecules, but of groups, and it may therefore be conjectured that the solution of a solid in a liquid does not entail a complete separation of the molecules from each other, but that they still exist in composite groups.[4] Whenever, therefore, we find liquids exhibiting rotatory power, we might assume that, as in the case of crystals, the cause lies in the mode in which the molecules group themselves. Thus again the phenomenon would be purely physical.

But for this supposition to be correct the rotatory properties of active substances should vanish when these groupings are really broken up—that is, when the substances are brought into the norma gaseous state. This important point was first investigated by Biot,[5] in 1817. He filled a tin tube, fitted at both

[1] Biot: *Mém. de l' Acad.* 13, 126. *Ann. Chim. Phys.* [3] 10, 175; 28, 351.

[2] Descloizeaux : *Pogg. Ann.* 102, 474.

[3] Le Bel: *Ber. d. deutsch. chem. Gesell.* 5, 391.

[4] See, on this point, Naumann : *Ueber Moleculverbindungen nach festen Verhältnissen* Heidelberg, 1872, pp. 37—49.

[5] Biot : *Mém de l' Acad.* 2, 114.

ends with glass plates, and 30 metres in length, with vapour of oil of turpentine, which he found had still the property of producing a certain amount of deviation in a ray of polarized light. Unluckily, before the observations were completed, the vapour accidentally caught fire, and the apparatus was destroyed. The experiment was next tried by D. Gernez,[1] in 1864, who, with the aid of instruments of a superior kind, determined the rotatory powers of various active substances at rising temperatures, and eventually in the gaseous state. The substances thus examined were orange-peel oil (+), bitter orange oil (+), turpentine oil (−), and camphor (+). In each the specific rotation [a], that is, the angle of rotation calculated for equal densities, = 1, and equal lengths of layer = 1 decim. diminished as the temperature increased; and when the same substances were tested in the gaseous state they gave a specific rotation merely reduced in proportion to the temperature to which they had been exposed. The table appended shows the results obtained with oil of turpentine and camphor :—

State of Aggregation.	Temp. (Cent.)	Density compared with water, d.	Observed Angle of Rotation, a.	Length of Tube in decim., l.	Specific Rotation $[a] = \dfrac{a}{d \cdot l}$
Oil of Turpentine (left-rotating).					
Liquid {	11°	0·8712	15·97°	0·5018	36·53
	98°	0·7996	14·47°	0·50215	36·04
	154°	0·7505	13·50°	0·50237	35·81
Vaporized	168°	0·003987	5·76°	40·61	35·49

Observed Density of Vapour at 168° Cent. = 4·981
Calculated   ,,   ,,   ,,   = 4·700

Camphor (right-rotating).					
Melted	204°	0·812	31·46?	0·5509	70·33
Vaporized	220°	0·003843	10·98°	40·63	70·31

Observed Density of Vapour at 220° Cent. = 5·369
Calculated   ,,   ,,   ,,   = 5·252

It will be seen that the observed densities of the vapours used

---

[1] Gernez : *Ann. Scient. de l'école norm. sup.* 1, 1.

in the experiment agreed very nearly with their calculated densities. The polarized ray must therefore have been influenced almost entirely by individual molecules, not by groups. The rotatory power [a] was, however, manifested to its full extent, and the conclusion is that here optical activity must be a property resident in the molecule itself, and dependent on its atomic structure. The phenomenon is thus seen to be really chemical.

The optical activity of crystals and that of liquids are, therefore, wholly distinct phenomena, and to the latter Biot has given the name of *moleculur rotation*, indicating that it is a property resident in the individual molecule.

§ 11. *Magnetic Rotation.*—A rotatory movement of the plane of vibration of a ray of polarized light can be produced in all transparent isotropic bodies, solid or liquid (as glass, water, &c.), by placing them between the poles of a magnet, or within the helix of an induction-coil. This so-called magnetic rotation differs altogether from rotation as seen in naturally active substances. It lasts so long only as the electric influence is continued; it varies in degree with the intensity of the latter; and it takes a right- or left-handed direction, irrespective of the medium, according to the position of the poles of the magnet or the direction of the electric current. There is also this further characteristic difference between the two. Let a polarized ray be transmitted through a naturally active substance, which, for the sake of example, we will say is right-rotating. Then the deviation of the plane of polarization will always be such that, to follow the movement of the ray, the instrument must be turned towards the right of the observer—that is to say, the direction of rotation, with reference to that of propagation of the ray, is invariable. If after passing through the refractive medium the ray is returned into it by reflection, and the analyzer brought round to the same side as the entering ray, it will be found that rotation is annulled. The rotation dependent on magnetism is of a quite different character. The ray transmitted, let us say, from south to north pole, in the direction of the observer, will appear deflected towards the right hand, and, transmitted from the opposite end of the tube, towards the left. If the ray transmitted from south to north pole be reflected back, it will appear farther deflected to the left, so that an analyzing Nicol placed to receive it must be rotated to the left through an angle equal to double the previous angle of rotation. If again brought back to the north end by a second reflection, this third transmission of the ray through the refractive medium will carry the analyzer placed at the north end through an angle equal to three times the original angle of rotation, and so on. The same thing happens when circular polarization is induced by an electric current, the rotation always taking the direction of the induction-current from the observer's stand-point.

Magnetic rotation, not being a property of the chemical molecule, need not be further discussed here.

§ 12. The optical theory of circular polarization in quartz is due to Fresnel.[1] According to him there occurs in quartz, in a direction parallel to the main axis of the crystal, a peculiar kind of double refraction, whereby a linear polarized ray on entering is decomposed into two rays, each of which pursues a helical course, the one turning to the right, the other to the left. On emerging, the two circular-polarized rays unite into a single linear-polarized ray again, but if the velocities with which they have traversed the refractive medium have been unequal, the plane of vibration of the emergent ray will have a different direction to what it had originally. It will follow the hands of a watch—that is to say, it will have rotated to the right—if the circular-polarized ray turning in that direction has had the superior velocity, and *vice versâ*. The existence of these divided rays in rock-crystal was experimentally established by Fresnel, and subsequently by Stefan,[2] and also by Dove,[3] who found that in coloured quartz (amethyst) they were unequally absorbed. The theory of circular polarization has since been treated mathematically by Clebsch,[4] Eisenlohr,[5] Briot,[6] v. Lang[7] and others.

Regarding the structure requisite in a crystalline medium to produce rotation of the plane of polarization, a theory has been proposed of an unequal condensation in certain directions of the

[1] Fresnel: *Ann. Chim. Phys.* [1] **28**, 147; Wüllner's *Lehrbuch der Phys.* 3 Aufl., **2**, 589.

[2] Stefan: *Pogg. Ann.* **124**, 623.

[3] Dove: *Pogg. Ann.* **110**, 284.

[4] Clebsch: *Crelle's Journ. f. Math.* **57**, 319.

[5] Eisenlohr: *Pogg. Ann.* **109**, 241.

[6] Briot: *Comptes Rend.* **50**, 141.

[7] v. Lang: *Pogg. Ann.* **119**. 74. *Erg. Bd.* **8**, 608.

ether surrounding the molecules, considerable enough in reference
to the wave-length of the transmitted ray, and conditioning, of
course, a particular molecular structure of the substance.   The
connection between the direction of rotation and the appearance
of right or left-handed hemihedric planes in active crystals has
led to the supposition that their ultimate parts are superposed so as
to form right-handed or left-handed helices.   This view, suggested
by Pasteur,[1] Rammelsberg,[2] and others, appears highly probable
from experiments first instituted by Reusch,[3] and more recently
further extended by Sohncke.[4]   If a number (12 to 36) of thin
laminæ of optically biaxial mica be superposed in the form of a
spiral, so that the principal section of each may form a certain angle
(45°, 60°, 90°, or 120°) with that of the preceding one, an optical
combination is produced, which causes rotation in a ray of polarized
light precisely like an active crystal, the direction of the rotation
being to the right or left hand according as the plates are arranged
in a right or left-handed spiral.   The optical properties of such
mica-combinations were minutely investigated by Sohncke, who has
arrived at the conclusion that, provided we use sufficiently thin
laminæ, we shall obtain combinations exhibiting rotation-phenomena
more nearly obeying the laws found to hold good for quartz and
other active crystals.   Hence, Sohncke considers as, to say the
least, probable, that rotatory crystals possess a structure analogous
to that of these mica-combinations.

§ 13. As to *the constitution of active liquids*, we are driven to
seek for the peculiarity of structure on which their power of rotation
depends in the arrangement of atoms in the molecule. Now, Pasteur[5]
supposes that molecules—like all other material objects—may be
divided, in respect of shape and the repetition of their symmetrical
parts, into two great classes, viz.:—1. Those whose images are
superposable by the bodies themselves (as straight flights of steps,
dice, &c.). 2. ·Those whose images are not superposable (as winding
stairs, screws, irregular tetrahedrons, &c.), and which may possess

[1] Pasteur : *Recherches sur la dissymétrie moléculaire des produits organiques naturels.*
Leçons de Chimie professées en 1860.   Paris, 1861.

[2] Rammelsberg : *Ber. d. deutsch. chem. Gesell.* 2, 31.

[3] Reusch : *Pogg. Ann.* 138, 628.

[4] Sohncke : *Pogg. Ann. Erg.* Bd. 8, 16.

[5] Pasteur : *Recherches*, &c. p. 27.

either of *two* structurally opposed (or enantiomorphous) shapes. Molecules of the former class possess symmetry of structure; those of the latter class have their atoms disposed asymmetrically, and accordingly exhibit optical activity. In 1848, Pasteur[1] made the important discovery that inactive para-tartaric acid is separable into right-rotating and left-rotating tartaric acids; and the sodium-ammonium salts of these two acids are distinguishable from each other by the presence of dextro-hemihedric and lævo-hemihedric planes respectively. Moreover, these salts retain their opposite characters in solution, by exhibiting opposite rotatory powers. Hence we may suppose that the property of asymmetrical structure of opposite kinds, such as we have seen in crystals, may occur in molecules also, and the precise nature of the arrangement of the atoms, or rather atom-groups, may reasonably be assumed to be here also of a helical kind. Whether the phenomenon of circular double-refraction, as exhibited by crystals, occurs also in active liquids is still an undecided point, several experiments made by Dove[2] on sugar solutions and on oil of turpentine having led to no conclusive result.

Hence, according to Pasteur's[3] views, the different optical modifications of tartaric acid may be explained on the supposition that in dextro-tartaric acid the atoms which go to form the molecule are grouped in right-handed helices, whilst in lævo-tartaric acid they are grouped in helices, equal in size, but left-handed in direction: and hence, too, the inactivity of racemic (para-tartaric) acid on the ground of its being formed by the union of equal molecules of the two former modifications. But besides these, other forms are well known, optically inactive, but not separable into the two optically active acids. To explain the existence of these, some other assumption is necessary, either that the helical structure is in their case abolished (untwisted), as Pasteur suggests, or that their inactivity arises from compensation within the molecule which is composed of two atom-groups possessing opposite rotatory powers. As to chemical structure, however—that is, the distribution of affinities between the atoms—they do not differ from the other isomeric acids.

Analogous optical modifications have been observed in a few other substances, which have been brought together in the table on page 22.

[1] Pasteur: *Ann. Chim. Phys.* [3] 24, 442; 28, 56; 38, 437.
[2] Dove: *Pogg. Ann.* 110, 290.   [3] Pasteur: *Recherches*, &c. p. 38.

		INACTIVE.	
ACTIVE.			
Dextro-rotatory.	Lœvo-rotatory.	By Combination of equivalent Molecules of right and left-handed Modifications.	By Difference in the Atomic Structure of the Molecule.
Dextro-tartaric Acid	Lœvo-tartaric Acid	Para-tartaric Acid	Meso-tartaric Acid
Malic Acid from Dextro - tartaric Acid[1]	Natural Malic Acid	Malic Acid from Para-tartaric Acid[1]	Malic Acid from Succinic Acid
Laurel or ordinary Camphor	Matricaria Camphor	Racemoid [or Para] Camphor[2]	—
Camphoric Acid from Laurel Camphor	Camphoric Acid from Matricaria Camphor	Para - camphoric Acid from Para Camphor or Lavender Camphor	Meso - camphoric Acid[3]

Similar conditions are found to exist in other substances, with the difference that the two oppositely active isomers exhibit unequal rotatory powers.

Thus, the following occur in dextro-rotatory (+), lœvo-rotatory (−), and inactive (0) forms:—glucose (as dextrose +, lœvulose −, and glucose obtained by heating cane-sugar with water to a temperature of 160° Cent., 0); terpenes (australene, the English oil of turpentine +, terebenthene, the French oil−, terebene 0); amyl-alcohol (that formed from the lœvo-alcohol by conversion into the chloride and reconversion into alcohol +, fermentation-amyl-alcohol −, and the modification obtainable from either by distillation with caustic potash 0). See Le Bel[4]; Balbiano.[5]

In many substances one of the two active modifications is unknown. For example: ethylidene-lactic acid (from muscle juice +, by fermentation or synthesis 0); cymol (from oil of Roman cumin +,

[1] Bremer: *Ber. d. deutsch. chem. Gesell.* 1875, 1594.
[2] Chautard: *Comptes Rend.*, 38, 166; 56, 698. Erdmann : *Journ. für prakt. Chem.* 90, 251.
[3] Chautard: *Jahresb. für Chem.* 1863, 394 ; Jungfleisch: *Jahresb. für Chem.* 1873, 631. Wreden: *Liebig's Ann.* 167, 302.
[4] Le Bel: *Bull. soc. chim.* [2] 25, 545.   [5] Balbiano: *Jahresb. für Chem.* 1876, 348.

synthetic cymols 0); mandelic acid (from amygdalin —, from benzoic aldehyde 0); aspartic acid (from active asparagin in acid solutions +, from fumaric or maleic acids 0).

Lastly, in a few substances the inactive form is still unrecognized, as, for example, in borneol (as dryobalanops camphor +, as blumea (Ngai) camphor, and camphor from fermentation of madder-sugar —); carvol (from cumin-oil and dill-oil +, from mint-oil —). See Flückiger.[1] A whole series of ethereal oils exist in both dextro-rotatory and lævo-rotatory modifications.

Optically different modifications of particular substances are found, in some cases, to exhibit differences in their other properties. Thus, the salts of active para-lactic acid are distinguished from those of inactive fermentation-lactic acid by different amounts of water of crystallization and somewhat different degrees of solubility (Wislicenus). Para-tartaric acid is more difficult of solution than the active tartaric acids. In their behaviour with inactive substances, Pasteur finds no difference between dextro- and lævo-tartaric acid; thus their potassium, sodium, and ammonium salts, tartar emetics and tartramides exhibit no difference beyond opposite rotatory powers and the occurrence of incongruous hemihedry in the crystals. But it is otherwise when the two acids are allowed to react with active substances, as asparagin, quinine, strychnine, sugar, &c. Where combination takes place the compounds formed differ from each other in crystalline form, specific gravity, water of crystallization, and in the readiness to decompose under the action of heat. Dextro-tartaric acid forms with asparagin a highly crystallizable substance, lævo-tartaric acid does not: lævo-acid malate of ammonia combines with dextro-acid tartrate of ammonia to form a crystallizable double salt, but not with the lævo-tartrate: lævo-tartrate of cinchonine is more difficultly soluble in water than the dextro-tartrate: dextro-tartrate of ammonia is decomposed by ferment-action, whilst lævo-tartrate undergoes no fermentation, and lævo-tartaric acid can, in consequence, be obtained in this way from para-tartaric acid, and so on.[2] To illustrate these peculiarities, Pasteur suggests the case of two screws—one right-handed, the other left-handed—driven into separate pieces of wood. When the fibres of the wood are rectilinear (inactive substance), two systems of the same kind will be produced; but this will no longer be the

[1] Flückiger: *Ber. d. deutsch. chem. Gesell.* 1876, 468.
[2] Pasteur: *Comptes Rend.* 46, 615.

case when the fibres are themselves arranged helically, and especially when the helices take opposite directions in the two pieces.

Incongruous hemihedric faces are found in most crystallizable active substances. Pasteur[1] has observed them not only in the tartaric acids, tartrates, tartramides, and amic acids, but also in the crystals of acid malates of lime and ammonia, valerianate, and chloride of morphine, &c. In other cases, however, they are absent, as in active amyl-sulphate of barium.[2] Incongruous hemihedry, moreover, is found to occur in some crystals exhibiting no rotatory power, as formiate of strontium and magnesium sulphate.[3] The two characteristics are, therefore, not inseparable.

C. Dependence of Optical Activity upon Chemical Constitution.

§ 14. On this question Hoppe-Seyler[4] and also Mulder,[5] proceeding on the ground that the rotatory properties of natural organic substances appear to be to some extent inherited by their derivatives, have expressed an opinion that optical activity is not dependent on the whole atomic structure of the molecule, but only on a particular part of it. The original compound they assume to include one or more *active radicles*, which in the derivatives may either appear unchanged or transformed into new but still active groups, or are eliminated altogether.

A theory has lately been proposed by Le Bel,[6] and nearly at the same time by van't Hoff,[7] which is much more plausible, and inasmuch as it brings into direct connection the rotatory powers and the constitutional formulæ of substances, is of special significance to chemistry. Le Bel first suggested, that when a carbon-atom occurs in combination with four different radicles, a molecule of asymmetrical shape is constituted, which, as such, should exhibit rotatory properties. Van't Hoff, proceeding on a hypothesis of his own respecting

[1] Pasteur: *Ann. Chim. Phys.* [3] 38, 437 ; 42, 418.   *Comptes Rend.* 35, 176.

[2] Pasteur: *Comptes Rend.* 42, 1259.

[3] Pasteur: *Ann. Chim. Phys.* [3] 31, 67.

[4] Hoppe-Seyler: *Journ. für prakt. Chem.* 89, 274.

[5] Mulder: *Zeitsch. für Chem.* 1868, 58.

[6] Le Bel: *Bull. Soc. Chim.* [2] 22, 337 (1874).

[7] J. van't Hoff: *Bull. Soc. Chim.* [2] 23, 295 (1875). *La chimie dans l'espace.* Rotterdam, 1875. German ed. by F. Hermann, *Die Lagerung der Atome im Raum.* Braunschweig, 1877.

the ultimate arrangement of atoms in space, was led to the same idea, in the working out of which a quite new stand-point has been reached in the subject of optical activity, as to the number of possible active and inactive isomers. This may briefly be explained as follows:—Let us suppose a substance formed on the type CR_4 to be represented by a tetrahedron, in which the carbon-atom occupies the centre, and one of the radicles (simple or compound) combined with it, each of the four summits. It follows that when the radicles are all different, and exhibit different affinities for the central carbon-atom, their proximity to the latter will also be different. Such a combination $CR_1R_2R_3R_4$ corresponds to an irregular tetrahedron destitute of planes of symmetry, and may always exist in two enantiomorphous forms. These two tetrahedrons will each exhibit, with reference to an axis parallel to a given side, a helical arrangement of the four summits, following a right-handed direction in one and a left-handed direction in the other. Such a carbon-atom, in combination with four different radicles, which van't Hoff denotes by the term *asymmetrical*, admits the possibility of two modifications of optical activity, the rotatory powers being equal in degree but manifested in opposite directions.

In substances possessing two such asymmetrical carbon-atoms, and having their molecules composed of two similarly formed atom-groups, we may have, according as the groups themselves possess like or opposite activities, not only right- and left-handed modifications of the compound, but also an inactive form resulting from intra-molecular compensation. Of this we have an instance in the case of the tartaric acids (COOH . CHOH) (CHOH . COOH). If the number of asymmetrical carbon-atoms be further augmented, we shall obtain from the combined effect of the several atom-groups, partly positive and partly negative in their action, a still larger series of differently active modifications, consisting of pairs having equal but opposite activities; the existence of several inactive forms becomes at the same time possible. (Bodies of the mannite group, glucoses, &c.)

If we consider the chemical formulæ of active compounds, the chemical constitution of which is known, we shall find that these always contain one or more asymmetrical atoms of carbon, and that no active organic substance can be adduced in which such atoms are wanting.

On the other hand, we find that there exists a large number of substances, containing asymmetrical carbon-atoms, in which the power of optical rotation has not been observed, and it becomes a question how this absence of power is to be accounted for. The following suggestions are made by van't Hoff :—

1. When such substances have a symmetrical chemical constitution, the occurrence of internal compensation (as already mentioned) may explain their inactivity (as in erythrite, dibromo-succinic acid, &c.).

2. The inactive substances in question may in reality be compounds or mixtures of two isomers, with rotatory powers of equal intensity but opposite directions (para-tartaric acid), and in many cases these isomers, owing to the similarity of their other properties, are difficult to distinguish from one another, and have not yet been separated. In forming artificially substances containing asymmetrical carbon, it is probable that an equal number of molecules of the right-rotating and left-rotating modifications is always produced.

3. Many substances have not, up to the present time, been subjected to optical examination, or have been examined in a very superficial manner, and it is possible that owing to feebleness of action, inherent in or caused by the difficultly soluble nature of the substances, their rotatory powers may have been overlooked. (Maunite, for instance, which at one time was regarded as inactive, has lately been shown by Bouchardat[1] and by Vignon[2] to possess rotatory power.)

To test as fully as possible the value of van't Hoff's hypothesis of the connection between asymmetrical carbon-atoms and the occurrence of optical activity, the following tabular arrangement has been prepared. The substances are grouped as follows :—

a. A list of active substances containing asymmetrical carbon-atoms, the latter being indicated by the prefix *.

b. A list of substances closely related to the above, but which are inactive and contain no asymmetrical carbon-atoms.

c. A list of substances containing asymmetrical carbon-atoms, but which, so far as is known at present, exhibit no optical activity.

The formulæ have been expressed so as to show the similarity or otherwise of the four radicles in combination with the centrical carbon-atom.

---

[1] Bouchardat: *Comptes Rend.* **80**, 120; **84**, 34.
[2] Vignon: *Comptes Rend.* **78**, 148.

C_3 Group.

Active.

a. Ethylidene-lactic Acid $\qquad (CH_3)—*CH(OH)—(CO.OH).$

Inactive.

b.
Ethylene-lactic Acid	$(CH_2.OH) — CH_2 — (CO.OH).$
Propionic Acid	$(CH_3) — CH_2 — (CO.OH).$
Glycerine	$(CH_2.OH) — CH(OH) — (CH_2.OH).$
Tartronic Acid	$(CO.OH) — CH(OH) — (CO.OH).$
Malonic Acid	$(CO.OH) — CH_2 — (CO.OH).$

c.
Propylene-glycol	$(CH_3) — *CH(OH) — (CH_2.OH).$
Glyceric Acid	$(CH_2.OH) — *CH(OH) — (CO.OH).$
α-Bromo-propionic Acid	$(CH_3) — *CHBr — (CO.OH).$
β-Dibromo-propionic Acid	$(CH_2Br) — *CHBr — (CO.OH).$
Propylene-dichloride	$(CH_3) — *CHCl — (CH_2Cl).$
Propylene-chlorhydrin	$(CH_3) — *CH(OH) — (CH_2Cl).$

C_4 Group.

Active.

a.
Tartaric Acid	$(CO.OH) — *CH(OH) — [*CH(OH) — (CO.OH)].$
Tartramide	$(CO.NH_2) — *CH(OH) — [*CH(OH) — (CO.NH_2)].$
Malic Acid	$(CO.OH — CH_2) — *CH(OH) — (CO.OH).$
Malamide	$(CO.NH_2 — CH_2) — *CH(OH) — (CO.NH_2).$
Asparagin	$(CO.NH_2 — CH_2) — *CH(NH_2) — (CO.OH).$
Aspartic Acid	$(CO.OH — CH_2) — *CH(NH_2) — (CO.OH)$

Inactive.

b.
Succinic Acid	$(CO.OH) — CH_2 — (CH_2 — CO.OH).$
Normal Butyric Acid	$(CH_3 — CH_2) — CH_2 — (CO.OH).$
Iso-butyric Acid	$(CH_3) — CH(CH_3) — (CO.OH).$
Butyramide	$(CH_3 — CH_2) — CH_2 — (CO.NH_2).$
Fumaric Acid	$(CO.OH) — CH = (CH — CO.OH).$

c.
Erythrite	$(CH_2.OH) — *CH(OH)—[*CH(OH) — (CH_2.OH)].$
Monobromo-succinic Acid	$(CO.OH) — *CHBr — (CH_2 — CO.OH).$
Dibromo-succinic Acid	$(CO.OH) — *CHBr — [*CHBr — (CO.OH)].$
Dimethyl-succinic Acid	$(CO.OH) — *CH(CH_3) — [*CH(CH_3)—(CO.OH)].$
Pyro-tartaric Acid	$(CO.OH) — *CH(CH_3) — (CH_2 — CO.OH).$
Secondary Butyl-alcohol	$(CH_3) — *CH(OH) — (CH_2 — CH_3).$
β-Butylene Glycol	$(CH_3) — *CH(OH) — (CH_2 — CH_2.OH).$
α-Hydroxy-butyric Acid	$(CH_3 — CH_2) — *CH(OH) — (CO.OH).$
β-Ditto ditto	$(CH_3) — *CH(OH) — (CH_2 — CO.OH).$
Aldol	$(CH_3) — *CH(OH) — (CH_2 — CHO).$
Chlorobutyl-aldehyde	$(CH_3 — CH_2) — *CHCl — (CHO).$
Normal Butylene Dibromide	$(CH_3-CH_2) — *CHBr — (CH_2Br).$

C_5 Group.

Active.

a. Active Amyl-alcohol $(CH_3 - CH_2) - {}^*CH(CH_3) - (CH_2 . OH)$.
   ,,   Valerianic Acid $(CH_3 - CH_2) - {}^*CH(CH_3) - (CO.OH)$.
Oxy-glutaric Acid $(CO.OH - CH_2 - CH_2) - {}^*CH(OH) - (CO.OH)$.
Glutamic Acid $(CO.OH - CH_2 - CH_2) - {}^*CH(NH_2) - (CO.OH)$.

Inactive.

b. 1. Primary Isoamyl-alcohol $(CH_3 - CH.CH_3) - CH_2 - (CH_2.OH)$.
Tertiary Isoamyl-alcohol $(CH_3 - C.OH.CH_3) - CH_2 - (CH_3)$.
Iso-valerianic Acid $(CH_3 - CH.CH_3) - CH_2 - (CO.OH)$.
Oxypyro-tartaric Acid $(CO.OH - CH_2) - CH.(OH) - (CH_2 - CO.OH)$.
Citric Acid $(CO.OH - CH_2) - C(OH)(CO.OH) - (CH_2 - CO.OH)$.

c. 1. Sec. Normal Amyl-
      alcohol $(CH_3 - CH_2 - CH_2) - {}^*CH(OH) - (CH_3)$.
Sec. Isoamyl-alcohol $(CH_3 - CH.CH_3) - {}^*CH(OH) - (CH_3)$.
Isoamylene-glycol $(CH_3 - CH.CH_3) - {}^*CH(OH) - (CH_2 OH)$.
a-Hydroxyiso-valerianic
     Acid $(CH_3 - CH.CH_3) - {}^*CH(OH) - (CO.OH)$.
Ethometh-oxalic Acid $(CH_3 - CH_2) - {}^*C(OH)(CH_3) - (CO.OH)$.

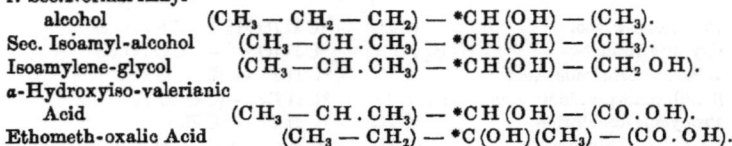

C_6 Group.

Active.

a. Active Capronic Acid $(CH_3 - CH_2 - CH_2) - {}^*CH(CH_3) - (CO.OH)$.
Mannite $CH_2.OH - ({}^*CH.OH)_4 - CH_2.OH$.
Dextrose $CH_2.OH - ({}^*CH.OH)_4 - CHO$.
Saccharic Acid $CH_2.OH - ({}^*CH.OH)_4 - CO.OH$.
Dextrin[1] $CH_2OH - {}^*CH.OH - {}^*CH.OH - {}^*CH - {}^*CH - CHO$.

                                           O

Appended we may place :—

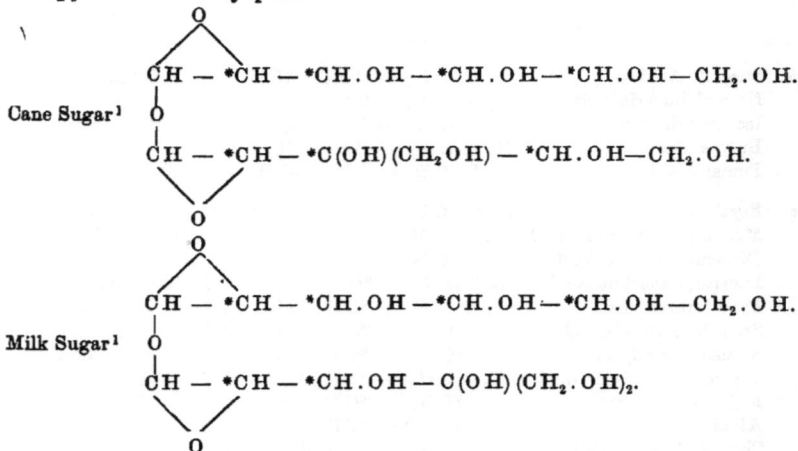

                 O
                ∕ ╲
$CH - {}^*CH - {}^*CH.OH - {}^*CH.OH - {}^*CH.OH - CH_2.OH$.

Cane Sugar[1] $\overset{|}{O}$
          $CH - {}^*CH - {}^*C(OH)(CH_2OH) - {}^*CH.OH - CH_2.OH$.

                   ╲ ∕
                   O

                O
               ∕ ╲
$CH - {}^*CH - {}^*CH.OH - {}^*CH.OH - {}^*CH.OH - CH_2.OH$.

Milk Sugar[1] $\overset{|}{O}$
          $CH - {}^*CH - {}^*CH.OH - C(OH)(CH_2.OH)_2$.

                   ╲ ∕
                   O

[1] Fittig: *Zeitsch. f. Rübenzucker-Ind.* 1871, 288.

REESE LIBRA
UNIVER SI
OF
CALIFORNIA

Inactive.

b.   Normal Caproic Acid $(CH_3 - CH_2 - CH_2 - CH_2) - CH_2 - (CO.OH)$.

     Iso-caproic Acid $(CH_3) - CH(CH_3) - (CH_2 - CH_2 - CO.OH)$.

     Diethyl-acetic Acid . $(C_2H_5) - CH(C_2H_5) - (CO.OH)$.

c.   Methylisopropyl-acetic Acid $(CH_3 - CH.CH_3) - {}^*CH(CH_3) - (CO.OH)$.

     Leucic Acid $(CH_3 - CH.CH_3 - CH_2) - {}^*CH(OH) - (CO.OH)$.

The substances in C_1 and C_2 Groups are all of them inactive, though many possess asymmetrical carbon-atoms, as the following :—

Sodium Nitroethane	${}^*C.H.Na.(NO_2).(CH_3)$.
Aldehyde Ammonia[1]	${}^*C.H.(OH).(NH_2).(CH_3)$.
Chloral Sulphydrate	${}^*C.H.(OH).(CCl_3).(SH)$.
,, Alcoholate[1]	${}^*C.H.(OH).(CCl_3).(O.C_2H_5)$.
,, Hydrocyanide	${}^*C.H.(OH).(CCl_3).(CN)$.
Bromoglycollic Acid	${}^*C.H.(OH).Br.(CO.OH)$.
Hydrogen Silver-fulminate	${}^*C.H.Ag.(NO_2).(CN)$.
Ethylidene Iodo-bromide	$CH_3 - {}^*C.H.J.Br$.
Ethylidene Methethylate	$CH_3 - {}^*C.H.(O.CH_3).(O.C_2H_5)$.
Ethylidene Chloracetate	$CH_3 - {}^*C.H.Cl.(O.C_2H_3O)$.
Ethylidene Chloro-sulphonic Acid	$CH_3 - {}^*C.H.Cl.(SO_2H)$.
Ethylidene Oxychloride	$CH_3 - {}^*C.H.Cl.(O - {}^*CHCl - CH_3)$.

In aromatic substances precisely similar conditions may be observed as in the fatty series. All ordinary benzene derivatives, in which the three double bonds of the C_6 nucleus remain intact, and which, in consequence, contain no asymmetrical carbon, are inactive. On the other hand, when one or more of these bonds are broken, the occurrence of asymmetrical carbon-atoms becomes possible. Where these are present the rotatory power will appear in some of the substances, as we find it does in turpentine-oil, camphor, borneol, campholic acid, camphoric acid, &c., while in others it may be wanting, as in benzene dichloride, benzene tetrachloride, dihydrophthalic acid, &c. Thus :—   *a.*   Active (with asymmetrical C).

Camphor.   Camphoric Acid.

      C_3H_7

      ${}^*CH$

$H_2C \quad\quad CO$

$HC \quad\quad CH_2$

      C

      CH_3

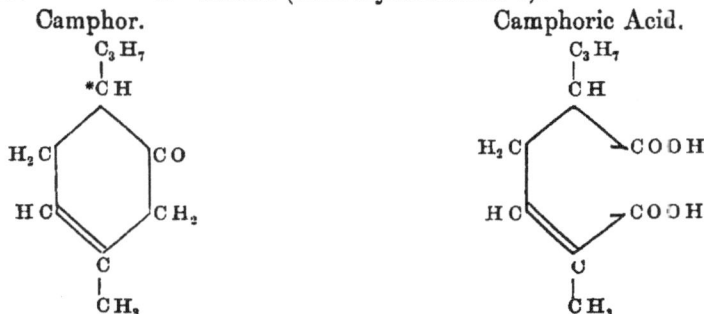

      C_3H_7

      CH

$H_2C \quad\quad COOH$

$HC \quad\quad COOH$

      C

      CH_3

---

[1] These two substances, in the form of highly concentrated solutions in tubes one metre long, I have found to produce no perceptible deviation of the plane of polarization. As regards the others no experiments have been recorded, but there can be no question that they are, like all artificial substances, beyond doubt inactive.

b. Inactive (without asymmetrical C).

Bitter Almond Oil.   Resorcin.

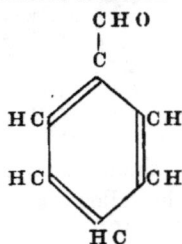

$$CHO$$
$$|$$
$$C$$

HC⟨   ⟩CH   C.OH

HC⟨   ⟩CH   HC⟨   ⟩CH

HC   HC⟨   ⟩C.OH

                                                CH

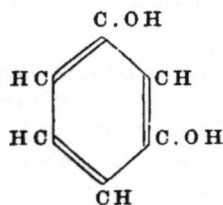

c. Inactive (with asymmetrical C).

Benzene dichloride.   Dihydrophthalic Acid.

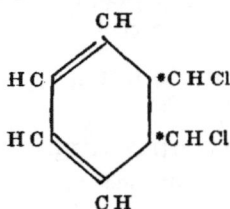

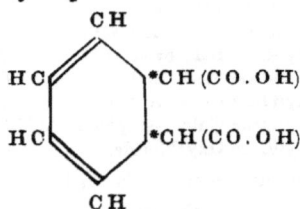

CH   CH

HC⟨   ⟩•CHCl   HC⟨   ⟩•CH(CO.OH)

HC⟨   ⟩•CHCl   HC⟨   ⟩•CH(CO.OH)

CH   CH

The active terpenes, $C_{10}H_{16}$ cannot be formulated as No. I. of the annexed formulæ, since it contains no asymmetrical C-atom, but only as either II. or III.

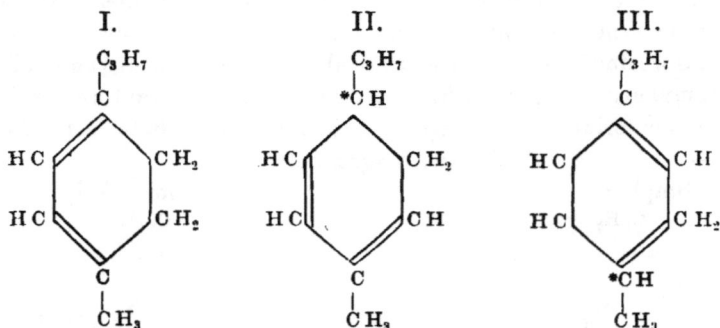

I.   II.   III.

C_3H_7 C_3H_7 C_3H_7
$|$ $|$ $|$
C $•CH$ C

HC⟨   ⟩CH₂   HC⟨   ⟩CH₂   HC⟨   ⟩CH

HC⟨   ⟩CH₂   HC⟨   ⟩CH   HC⟨   ⟩CH₂

C C $•CH$
$|$ $|$ $|$
CH_3 CH_3 CH_3

Lastly, the asymmetrical carbon-atoms may be found in a lateral series, and then, as before, we get compounds, some active, some inactive, as the following :—

. Active.

Mandelic Acid $(C_6H_5) -- •CH(OH) — (CO.OH)$.

Undetermined.

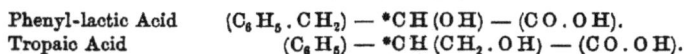

Phenyl-lactic Acid $(C_6H_5.CH_2) - {}^*CH(OH) - (CO.OH)$.

Tropaic Acid $(C_6H_5) - {}^*CH(CH_2.OH) - (CO.OH)$.

Inactive.

Benzoin $(C_6H_5.CO) - {}^*CH(OH) - (C_6H_5)$.

Cinnamic Acid dibromide $(C_6H_5) - {}^*CHBr - [{}^*CHBr - (C_6H_5)]^1$

Diphenyl-succinic Acid $(C_6H_5) - {}^*CH(CO.OH) - [{}^*CH(CO.OH) - C_6H_5]$.

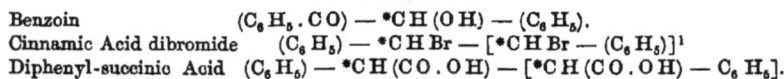

By comparisons of this kind, which might easily be extended, we find that up to the present no substance can be indicated with certainty[2] as disproving van't Hoff's theory or the following statements based thereupon :—

1. That optically active substances invariably contain one or more asymmetrical carbon-atoms.

2. That substances containing no asymmetrical carbon-atoms exhibit no optical activity.

On the other hand, as van't Hoff has pointed out, and as may be seen from the above examples, the converse is not necessarily true, that " bodies containing asymmetrical carbon-atoms are always optically active." There are, in fact, numerous substances which contain asymmetrical carbon-atoms and yet exhibit no optical activity, and further research is needed to decide whether or not this inactivity can be referred to the causes before specified. But, even should further inquiry prove that asymmetrical carbon-atoms are not the sole condition but merely one of the conditions of optical activity, the foregoing statements, unless disproved by fresh discoveries, remain of great importance to chemistry, as they not only afford some sort of control over active substances, but may also yield definite indications as to the proper structural-formulæ.

§ 15. *Artificial Production of Active Substances.*—The carbon-compounds in which optical activity has hitherto been observed, are all of them found in vegetable or animal organisms or as derivatives from these by simple decomposition. Many of these substances can be prepared artificially, but even when all the chemical attributes,

---

[1] Presumably this should stand $(C_6H_5) - {}^*CHBr - [{}^*CHBr - CO.OH]$, the formula in the original being that for stilbene dibromide.—[D.C.R.]

[2] The activity observed by Berthelot (*Comptes Rend.* **63**, 818; **85**, 1191) in styrol and metastyrol, $C_6H_5.CH = CH_2$, obtained from liquid storax, which substances contain no asymmetrical C-atoms, is referred by van't Hoff (*Ber. d. deutsch. chem. Gesell.* 1876, 5) to the presence of another substance, probably corresponding to the formula $C_{10}H_{16}O$. In like manner, the supposed optical activity of iodide of trimethylethyl-stibin is attributed to chemical impurity. Le Bel (*Bull. Soc. Chim.* **27**, 444).

and, consequently, the chemical constitution of such compounds, agree with those of the natural substances, a difference is nevertheless found to exist in their optical properties. Direct synthesis of substances from inactive components has hitherto resulted in the production of *inactive* modifications only.

As already indicated, this inactivity of artificial substances may be apparent (*i.e.* latent) only and dependent on the following causes, which at the same time indicate the means to its possible removal :—

1. In synthesis, it is probable that an equal number of dextro-rotatory and lævo-rotatory molecules may always be formed, which mix or combine together and so produce optical neutrality. As an instance of the kind, Jungfleisch[1] has shown that by converting ethylene—through the intermediate products ethylene bromide, ethylene cyanide, succinic acid, and dibromo-succinic acid—into tartaric acid, the inactive form is obtained, which by crystallization of its sodium-ammonium salt may be separated into dextro-tartaric and lævo-tartaric acids. This, at present, is the only instance that can be alleged of an artificial active substance ; and even here it must be observed that the direct result of the synthesis, the para-tartaric acid, exhibits no optical power.[2]

As the physical and chemical properties of such opposite-rotating modifications may differ but little, they cannot in general be separated without great difficulty, and the more so, when they form not merely mechanical mixtures but true chemical combinations, as is the case with para-tartaric acid.

Indeed the only substance as yet, whose inactivity depends on neutralization, which has been separated into right- and left-rotating modifications is para-tartaric acid. The separation can be effected by one of the following methods :—

a. By crystallization of the sodium-ammonium salt, and separation by selection of the crystals with dextro-hemihedric from those with lævo-hemihedric planes (Pasteur[3]). This method has been somewhat simplified by Gernez,[4] who found that by bringing into contact with a supersaturated solution of para-tartrate of sodium and

[1] Jungfleisch : *Bull. Soc. Chim.* [2] 19, 194. *Comptes Rend.* 76, 286.

[2] Pasteur (*Comptes Rend.* 81, 128) rejects the instance unconditionally ; he maintains that up to the present time no active substance has been derived from inactive substances, and that, therefore, optical activity affords a definite distinguishing characteristic between natural and artificial substances.

[3] Pasteur : *Ann. Chim. Phys.* [3] 24, 442 ; 28, 56 ; 38, 437.

[4] Gernez : *Liebig's Ann.* 143, 376.

ammonium a crystalline fragment of the dextro-tartrate, the latter alone crystallizes out, the lævo-tartrate remaining in solution, and *vice versâ*.

b. By causing the para-tartaric acid to combine with some other active substance, whereby two modifications with unequal solubilities are produced. Thus if cinchonicine (dextro-rotatory) be dissolved in para-tartaric acid, lævo-tartrate of cinchonicine separates out first on evaporation. On the other hand, from a solution of quinicine (dextro-rotatory) in para-tartaric acid, crystals of the dextro-tartrate of quinicine are first separated (Pasteur).[1]

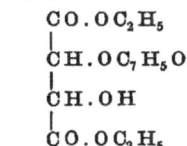

c. By mixing a solution of para-tartrate of ammonia with ferment (yeast-extract), the dextro-tartrate is eliminated by fermentation, while the lævo-tartrate remains unchanged (Pasteur).[2]

Mixtures of right- and left-rotating molecules are probably produced in the synthesis of all substances containing asymmetrical carbon, the molecules of which are not made up of two similarly constituted parts. This is probably the case with the malic acid obtained from monobromo-succinic acid, $CO.OH-CH_2-CH.OH-CO.OH$, with ethylidene-lactic acid from propionic acid, $CH_3-CH.OH-CO.OH$, with mandelic acid prepared by treating benzoic aldehyde with prussic and muriatic acids, $C_6H_5-CH.OH-CO.OH$, &c. These products are all inactive, whereas the malic acid of plants, the ethylidene-lactic acid of muscle juice, and mandelic acid obtained by decomposition of active amygdalin with muriatic acid, despite the similarity of chemical constitution, all exhibit rotatory power. No attempts to dissociate any of these inactive preparations have yet been made.

2. The molecule produced by synthesis may be inactive by virtue of internal compensation—that is, it may be made up of equal halves with opposite rotatory powers. In such cases it may reasonably be expected that optical inactivity will disappear when the symmetry of the molecule is disturbed, as for instance, in the case of inactive indivisible tartaric acid, by converting it into benzo-tartrate of ethyl or simply into an acid tartrate, thus:—

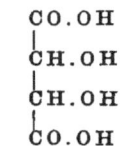

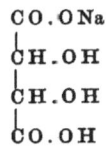

$CO.OH$	$CO.OC_2H_5$	$CO.ONa$
$CH.OH$	$CH.OC_7H_5O$	$CH.OH$
$CH.OH$	$CH.OH$	$CH.OH$
$CO.OH$	$CO.OC_2H_5$	$CO.OH$
Tartaric Acid.	Benzo-tartrate of Ethyl.	Acid Tartrate of Soda.

[1] Pasteur: *Comptes Rend.* 37, 162.   [2] Pasteur: *Comptes Rend.* 46, 615.

In these derivatives the halves being dissimilarly constituted, there ought to be rotatory power. Experiments in this direction have not yet been made, and it is quite possible that even such derivatives may prove inactive. This might easily be the case if the substitution took place in the right-rotating and left-rotating groups respectively of the two halves of the molecule, the result being an optically neutral mixture.

Substances containing several asymmetrical C-atoms are, according to van't Hoff's view, capable of forming by partial internal compensation certain isomers, with very feeble rotatory powers. Such a substance is mannite, $CH_2 . OH-(*CH . OH)_4-CH_2 . OH$, which exhibits an extremely slight lævo-rotation. Converted into mannite dichlorhydrin, mannite hexacetate, mannite hexanitrate or into mannitan we obtain strongly dextro-rotatory substances.[1] By conversion, for example, into hexanitrate (nitro-mannite), $CH_2 . O . NO_2$ $-(*CH . ONO_2)_4-CH_2 . O . NO_2$, the symmetry of the chemical formula is not destroyed, but some alteration apparently takes place in the four asymmetrical groups, whereby they, or at any rate three of them, assume the right-rotating position.

In this way Müntz and Aubin[2] converted an inactive substance, viz., the glucose obtained by exposing cane-sugar with a little water to a temperature of 160° Cent., into an active substance, by transforming it first into mannite (with sodium-amalgam and water), and then into nitro-mannite.

So, likewise, inactive dulcite exhibits rotatory powers when converted into the diacetate, or into dulcitan diacetate (Bouchardat).[3]

3. Some of the resulting products do possess rotatory power, but only of a very feeble kind; thus, for example, in the case of mannite, a layer 3 to 4 metres in depth is requisite for its determination. In such cases aid may be furnished by an observation of Biot, who found that the rotation of tartaric acid was considerably increased by the addition of boric acid. Similar results have been observed in other substances. Thus, Vignon[4]

[1] Loir: *Jahresb. für Chem.* 1861, 729. Tichanowitsch: *Jahresb.* 1864, 582. Grange: *Jahresb.* 1869, 752. Schützenberger: *Liebig's Ann.* 160, 94 (1871). Krecke: *Arch. Néerland*, vii. (1872). Vignon: *Jahresb.* 1874, 884. Bouchardat: *Jahresb.* 1875, 790, and *Comptes Rend.* 84, 34 (1877). Müntz and Aubin: *Jahresb.* 1876, 149; *Ann. Chim. Phys.* [5] 10, 553 (1877).

[2] Müntz and Aubin: *Ann. Chim. Phys.* [5] 10, 553.

[3] Bouchardat: *Ann. Chim. Phys.* [4] 27, 68, 145.

[4] Vignon: *Ann. Chim. Phys.* [5] 2, 433.

has shown that nearly perfectly inactive mannite solutions become strongly dextro-rotatory on the addition of borax. Again, according to Müntz and Aubin,[1] inactive glucose becomes strongly dextro-rotatory on the addition of sodium sulphate, and still more so of borax; addition of sodium carbonate produces, on the contrary, left-rotation. Moreover, the activity of organic acids and also of alkaloids is in most cases increased by their conversion into salts, and sometimes very considerably. The reverse also happens, as, for example, in the case of chloride of laudanine, which Hesse[2] reports as inactive, although the free base is lævo-rotatory. Papaverine exhibits the same property.

4. In the preceding cases we have spoken only of the synthesis of bodies the chemical structure of which is identical with that of the natural substances, but where structural differences exist there will obviously be dissimilarity of optical power. Thus inactive paraconine from butyric-aldehyde is not an imide base like the active natural conine (Schiff[3]), and accordingly differs from it in its other structural properties.

§ 16. *Optical Properties of Derivatives of Active Substances.*— It has long been observed that when optically-active substances undergo chemical changes, some of their derivatives exhibit rotatory powers whilst others do not. Where the molecular constitution remains unaltered, as in the conversion of acids into salts, ethers, amides, &c., or in the combination of alkaloids with acids, the activity is usually retained.[4] On the other hand, where there is actual chemical decomposition, active and inactive derivatives are obtained with an apparent absence of all rule. If, however, the hypothesis of asymmetrical carbon-atoms be called into aid, the apparent irregularity vanishes, and the conditions, as van't Hoff[5] has shown, resolve themselves into the following :—

1. The rotatory powers of active substances are obliterated in such of their derivatives as are wanting in asymmetrical carbon-atoms.

[1] Müntz and Aubin : *Ann. Chim. Phys.* [5] 10, 564.
[2] Hesse : *Liebig's Ann.* 176, 198, 201.
[3] Schiff : *Liebig's Ann.* 157, 352 ; 166, 94.
[4] Exceptions exist in laudanine and papaverine, which, as mentioned in § 15, when free are optically-active, but in the form of chlorides are inactive (Hesse).
[5] J. van't Hoff : see the two pamphlets mentioned in note 7, p. 24 of the present work ; also *Ber. d. deutsch. chem. Gesell.* 1877, 1620.

As examples may be cited :—

1. From Active Tartaric and Malic Acids is derived Inactive Succinic Acid.
          (by treating with hydriodic
          acid or by fermentation)

2. ,,   ,,   Cane-sugar and Tartaric Acid   ,,   ,,   Oxalic Acid.
          (by treating with nitric
          acid)

3. ,,   ,,   Tartaric Acid (with phos-   ,,   ,,   Chloro-maleic Acid.
          phoric chloride)

4. ,,   ,,   Lævo-malic Acid (by heating)   ,,   ,,   Maleic Acid and Fu-
                                                             maric Acid.

5. ,,   ,,   Lævo-malate of Ammonia   ,,   ,,   Fumarimide.
          (by heating)

6. ,,   ,,   Asparagin (by fermentation)   ,,   ,,   Succinic, Maleic, and
                                                             Fumaric Acids.

7. ,,   ,,   Glucose (by fermentation)   ,,   ,,   Ethyl-alcohol.

8. ,,   ,,   Carbo-hydrates (with sul-   ,,   ,,   Furfurol.
          phuric acid)

9. ,,   ,,   Amyl-alcohol   ,,   ,,   Amyl hydride, Amy-
                                              lene,and Methyl-amyl.

10. ,,   ,,   Amygdalin (by fermentation)   ,,   ,,   Bitter Almond Oil.

In none of the above derivatives do any asymmetrical C-atoms occur. See formulæ on a previous page (p. 27 and following pages).

2. Derivatives of active substances, when such derivatives contain asymmetrical carbon-atoms, usually exhibit rotatory power.

As examples may be taken :—

1. From Active Amyl-alcohol (by oxi-   is derived Active Valerianic Acid, &c.
          dation)

2. ,,   ,,   Camphor (with nitric acid)   ,,   ,,   Camphoric Acid.

3. ,,   ,,   Amygdalin (by boiling with   ,,   ,,   Amygdalic Acid.
          baryta-water)

4. ,,   ,,   Amygdalin (with hydro-   ,,   ,,   Mandelic Acid.
          chloric acid)

5. ,,   ,,   Asparagin (by boiling with   ,,   ,,   Aspartic Acid.
          acids or alkalies)

6. ,,   ,,   Aspartic Acid (with nitrous   ,,   ,,   Dextro-malic Acid.
          acid)

7. ,,   ,,   Dextro-tartaric Acid (with   ,,   ,,   Dextro-malic Acid.
          hydriodic acid)

8. ,,   ,,   Cane-sugar, Mannite, Glu-   ,,   ,,   Saccharic Acid.
          cose, and Lævulose (with
          nitric acid)

9. ,,   ,,   Saccharic Acid (with nitric   ,,   ,,   Dextro-tartaric Acid.
          acid)

10. ,,   ,,   Milk-sugar (with nitric   ,,   ,.   Dextro-tartaric Acid.
          acid)

On the other hand, cases occur in which such derivatives, the presence of asymmetrical carbon-atoms notwithstanding, exhibit no rotatory power, as the subjoined (compare with the formulæ previously given, page 27 and following pages) :—

1. From Active Malic Acid (with   is derived Inactive Monobromo-succinic Acid.
                 hydrobromic acid)
2.   ,,   ,,   Dextro-tartaric Acid   ,,   ,,   Pyro-tartaric Acid.
                 (by heating)
3.   ,,   ,,   Lactose (with nitric   ,,   ,,   Mucic Acid.
                 acid)

A conversion of active substances into inactive isomers frequently results from exposure to a higher temperature. Dextrotartaric acid, heated in presence of water to a temperature of 160° Cent., is converted chiefly into inactive, undecomposable tartaric acid ; whereas, at a temperature of 175° Cent., a mixture of the latter with a predominance of para-tartaric acid is obtained. Conversely, at a temperature of 160° Cent. para-tartaric acid is converted into the other inactive tartaric acid (Jungfleisch).[1] Again, dextrocamphoric acid mixed with a little water, and raised to a temperature of 170° to 180° Cent., passes into inactive camphoric acid (Jungfleisch[2]) ; and active amyl-alcohol when heated in sealed tubes becomes inactive (Le Bel[3]). The conversion into the inactive state is often assisted by the presence of other substances. Thus fermentation-amyl-alcohol loses its rotatory power when distilled under ordinary pressure in presence of caustic potash, caustic soda, or chloride of calcium (Balbiano[4]). By merely heating in a sealed tube for a quarter of an hour at 250° Cent. with a few drops of concentrated sulphuric acid, active valerianic acid is rendered inactive (Erlenmeyer and Hell[5]). Repeated agitation with one-twentieth of its volume of concentrated sulphuric acid in the cold, and subsequent distillation, converts active turpentine-oil into inactive terebene (Riban[6]). The case of tartaric acid indicates the probability that such conversions of active into inactive substances are due to the simultaneous formation of right-rotating and left-rotating modifications, resulting in optical neutrality.

In turpentine-oil a gradual conversion of the right-rotating into the left-rotating form has been noticed. If the oil is placed

[1] Jungfleisch : *Ber. d. deutsch. chem. Gesell.* 1872, 985 ; 1873, 33.
[2] Jungfleisch : *Jahresb. für Chem.* 1873, 631.
[3] Le Bel : *Jahresb. für Chem.* 1876, 347.
[4] Balbiano : *Jahresb. für Chem.* 1876, 348.
[5] Erlenmeyer and Hell : *Liebig's Ann.* 160, 302.
[6] Riban : *Jahresb. für Chem.* 1873, 370.

in a retort provided with an upright condenser, and, whilst the retort is kept filled with carbonic acid at the ordinary pressure of the atmosphere, the contents are raised to boiling, whereby a temperature of 160° to 162° Cent. is attained, no change of rotatory power can be detected at the end of sixty hours. But such change occurs when the temperature passes beyond 250° Cent. in a sealed tube. A sample of English right-rotating turpentine-oil, which gave an original deviation of $a_j = + 18 \cdot 6°$ in a layer of 100 millimetres, showed the following angles of rotation at higher temperatures:—

After	4 hours at 250° Cent.	$a_j. = + 15 \cdot 3°$
After an additional 4	,, ,, 250° to 260° Cent.	$a_j. = + 11 \cdot 8°$
,, ,, 60	,, ,, 250° to 260° Cent.	$a_j. = - 8 \cdot 6°$
,, ,, 42	,, ,, about 300° Cent.	$a_j. = - 5 \cdot 6°$

At a certain stage an inactive mixture of right-rotating and left-rotating molecules must therefore have been present (Berthelot[1]). Polymerization occurs simultaneously with the above changes. This can readily be produced without any change of temperature, by treating with antimony trichloride, whereby, for example, left-rotating terebenthene is converted into right-rotating tetra-terebenthene (Riban[2]).

In certain cases, two isomeric derivatives possessing opposite rotatory powers but of unequal intensity are simultaneously formed from an active substance, the immediate product being therefore also active. By heating mannite to 150° Cent., or treating it with muriatic acid, we obtain a right-rotating mixture, which on evaporation separates out the crystallizable lævo-mannitan, leaving amorphous dextro-mannitan in the mother liquor (Bouchardat[3]). Again, cane-sugar is transformed by the action of ferments or dilute acids into left-rotating invert-sugar, which can be split into right-rotating glucose (dextrose) and left-rotating glucose (lævulose). On the other hand, cane-sugar exposed with one-twentieth part water, in a sealed tube, for the space of two or three minutes, to a temperature of 160° Cent., gives an inactive glucose, which appears to be indivisible (Mitscherlich,[4] Müntz and Aubin[5]).

[1] Berthelot: *Ann. Chim. Phys.* [3], **39**. 10.
[2] Riban: *Bull. Soc. Chim.* [2], **22**, 253. *Jahresb. für Chem.* 1874, 451.
[3] Bouchardat: *Jahresb. für Chem.* 1875, 792.
[4] Mitscherlich: *Lehrbuch der Chem.* 4 Aufl. I, 337.
[5] Müntz and Aubin: *Ann. Chim. Phys.* [5], **10**, 564.

The direction of rotation in derivatives, in relation to that in the parent substance, follows no fixed rule. Most derivatives exhibit their rotatory power in the *same* direction as the parent substance, particularly when no real alteration of molecular constitution has taken place. This is seen in the conversion of active acids and alkaloids into their salts. Still there are exceptions here, as we find that malic acid, which is left-rotating when free, is right-rotating in its neutral salts, especially in the double salt of antimony and ammonium, from which, after precipitation of the antimony with sulphuretted hydrogen, a left-rotating solution of acid malate of ammonia is obtained (Pasteur[1]). Again, right-rotating para-lactic acid has left-rotating zinc and calcium salts (Wislicenus[2]). But even in more profound reactions the original direction is generally maintained. Thus, dextro-tartaric acid gives dextro-malic acid; from dextro-camphor we get dextro-camphoric acid; from lævo-camphor, lævo-camphoric acid; dextro- and lævo-borneols give respectively dextro- and lævo-camphors; and left-rotating amygdalin gives left-rotating amygdalic and mandelic acids (Bouchardat[3]). Nevertheless, some derivatives do exhibit rotatory power in a direction *contrary* to that of the parent substance, as in the following instances:— Dextro-camphoric acid gives left-rotating camphoric anhydride (Montgolfier[4]); dextro-para-lactic acid gives left-rotating ether-anhydrides (Wislicenus[2]); the active compounds derived from lævo-amylalcohol (amyl chloride, amyl iodide, amyl cyanide, diamyl, ethyl-amyl, amylamine, amyl valerate, valerianic aldehyde, valerianic acid, and capronic acid) are all right-rotating; left-rotating santonic acid, with nascent hydrogen passes into right-rotating hydrosantonic acid (Cannizarro[5]); left-rotating terebenthene hydrochlorate, heated with stearate of soda, gives left-rotating terecamphene. By saturating the latter with hydrochloric acid we get right-rotating camphene hydrochlorate, from which, by heating with water, left-rotating camphene, but with rotatory power much feebler than at first, can be recovered (Riban[6]). Again, mannite, with weak lævo-rotatory power, gives both right- and left-rotating derivatives. Of these nitro-mannite, mannite hexacetate, amorphous mannitan, nitro-

[1] Pasteur : *Ann. Chim. Phys.* [3] 31, 67.
[2] Wislicenus : *Liebig's Ann.* 167, 322.
[3] Bouchardat : *Comptes Rend.* 19, 1174.
[4] Montgolfier : *Jahresb. für Chem.* 1872, 569.
[5] Cannizarro : *Jahresb. für Chem.* 1876, 619.
[6] Riban : *Bull. Soc. Chim.* [2], 24, 10.

mannitan, mannitan tetracetate, mannitan monochlorhydrin are right-rotating ; whilst mannite dichlorhydrin, crystallizable mannitan, and mannitone are left-rotating (Vignon,[1] Bouchardat[2]). By oxidation of dextro-camphor with nitric acid we obtain, simultaneously with right-rotating camphoric and camphic acids, left-rotating camphoronic acid as well (Montgolfier[3]).

Lastly, the remarkable fact must be mentioned that in certain cases active substances exhibit changes in the direction of their rotatory powers, when dissolved in various liquids or when certain substances are added to their solutions. Asparagin and aspartic acid are left-rotating in alkaline (*i.e.* sodic, or ammoniacal) solutions ; but in acid (*i.e.* hydrochloric or nitric acid) solutions, on the contrary, they are right-rotating (Pasteur[4]). The acid ammonium salt of lævo-rotatory malic acid exhibits left-handed rotation in aqueous and ammoniacal solutions; in nitric acid it forms a right-rotating solution (Pasteur[4]). The calcium salt of dextro-tartaric acid is dextro-rotatory in aqueous and lævo-rotatory in hydrochloric acid solution ; whilst, on the other hand, lævo-tartrate of lime is dextro-rotatory in hydrochloric acid solution (Pasteur[5]). An aqueous solution of mannite, which, as such, manifests a very feeble left-handed rotation, becomes strongly lævo-rotatory, on addition to the solution of certain alkalies (as caustic potash, caustic soda, magnesia, lime, baryta), and dextro-rotatory, in presence of salts of the alkalies (as borax, chloride of sodium, sodium sulphate, potassium hydrarseniate). Ammonia renders it feebly dextro-rotatory, but acids have no effect upon it (Vignon,[6] Bouchardat,[7] Müntz and Aubin[8]).

Similar properties are exhibited to a remarkable extent by ordinary tartaric acid. In aqueous solution it is dextro-rotatory, whilst in the solid state it may assume a lævo-rotatory power (§ 19). The rotatory power of its aqueous solutions is very considerably increased by the addition of even small quantities of boric acid, or borax ; on the other hand, it is diminished by the addition of sulphuric, hydrochloric, or citric acid, and also of alcohol or wood-spirit (Biot[9]).

[1] Vignon : *Jahresb. für Chem.* 1874, 885.
[2] Bouchardat : *Jahresb. für Chem.* 1875, 790—792.
[3] Montgolfier : *Jahresb. für Chem.* 1872, 569.
[4] Pasteur : *Ann. Chim. Phys.* [3], 31, 67.   *Jahresb. für Chem.* 1851, 176.
[5] Pasteur : *Ann. Chim. Phys.* [3], 28, 56.   *Jahresb. für Chem.* 1849, 128.
[6] Vignon : *Ann. Chim. Phys.* [5], 2, 433.   *Jahresb. für Chem.* 1874, 884.
[7] Bouchardat : *Comptes Rend.* 80, 120.   *Jahresb. für Chem.* 1875, 145.
[8] Müntz and Aubin : *Ann. Chim. Phys.* [5], 10, 553.   *Jahresb. für Chem.* 1876, 149.
[9] Biot : *Mém. de l'Acad.* 16, 229.

If, moreover, tartaric acid is dissolved in acetic ether or acetone, the resulting solutions, which either are inactive or may exhibit a feeble lævo-rotation, become immediately dextro-rotatory, on the addition of a little water. Malic acid appears to possess similar peculiarities. Phenomena of this sort differ from those exhibited by derivatives, in that the changes in the direction of rotation are temporary only, and disappear with the removal of the substances which induce them.

The whole series of phenomena of this kind require, however, fuller investigation before the subject can be properly explained.

III.

PHYSICAL LAWS OF CIRCULAR POLARIZATION.

§ 17. *Amount of Rotation dependent on the Thickness of the Medium.*—From experiments with quartz, Biot,[1] in 1817, deduced the following laws :—

1. The angle, through which the plane of polarization of a ray of given wave-length rotates, is directly proportional to the thickness of the quartz plate.

2. Dextro-rotatory and lævo-rotatory plates of quartz of equal thickness, cause the plane of polarization to deviate through equal angles. If the ray be transmitted through more plates than one, the final deviation is equal to the sum of the individual deviations if the plates all rotate in the same direction, and to their collective differences if they rotate in different directions.

To allow comparison to be made of the rotatory powers of different active crystals, Biot proposed to adopt as a standard the angle of rotation afforded by plates 1 millimetre (·039 in.) in thickness.

In active liquids (as, for example, oil of turpentine, or solutions of active solid substances in inactive liquids) the following laws have been observed :—

1. The angle of rotation is directly proportional to the thickness of layer traversed by the ray.

2. When a ray is transmitted through several media, the observed rotation will be the algebraical sum of the individual rotations.

As the rotatory power is generally much weaker in liquids

[1] Biot: *Mém. de l'Acad.* 2, 41.

than in crystals, Biot proposed that the angle of rotation for thicknesses of 1 decimetre (3·9 in.) should be taken as the standard of comparison for the former.

§ 18. *Amount of Rotation dependent on the Wave-length of Transmitted Ray. Rotatory Dispersion.*—If a series of polarized rays of different colours be transmitted successively through an active medium, it is found that the amount of rotation which the plane of polarization experiences, varies with the wave-length of the ray, being least for red, and greatest for violet. From his experiments with quartz, Biot arrived at the conclusion, that the angle of rotation a varies inversely almost exactly as the square of the wave-length of ray λ, but this has not been confirmed by later observers. The formula $a = A + \dfrac{B}{\lambda^2}$, in which A and B are two constants, has also been found unsatisfactory; whereas the equation since proposed by Boltzmann,[1]

$$a = \frac{B}{\lambda^2} + \frac{C}{\lambda^4},$$

which, like the preceding, contains two constants only (B and C), and which is based upon the assumption that rotation in a ray of infinite wave-length $= 0$, has been shown to agree very closely with the results of observation.

For a complete determination of the rotatory power of a given substance, it is necessary to take the angles of rotation for a number of different rays of known wave-lengths. This can only be fully accomplished with the aid of Fraunhofer's lines; that is, by the use of solar light, the deviations being determined by the method of Broch (§§ 61, 62, 63). V. von Lang[2] has lately employed artificial light, the lithium, sodium, and thallium rays, for the same purpose.

In the case of quartz, the angles of rotation for different lines have been accurately determined by Broch,[3] Stefan,[4] Soret and Sarasin,[5] and von Lang.[6] The table on page 44 contains the observations of Stefan and von Lang, along with the corresponding wave-lengths λ, expressed in millimetres.

[1] Boltzmann : *Pogg. Ann. Jubelband*, p. 128.
[2] von Lang : *Pogg. Ann.* **156**, 422.
[3] Broch : *Dove's Repert. d. Phys.*, **7**, 115.
[4] Stefan : *Sitzungsber. der Wiener Acad.* **50**. *Pogg. Ann.* **122**, 631.
[5] Soret and Sarasin : *Pogg. Ann.* **157**, 447.
[6] von Lang : *ut supra.*

Ray.	λ.	a for 1 millim.
B	0·0006871	15·55°
C	6560	17·22°
D	5888	21·67°
E	5269	27·46°
F	4860	32·69°
G	4309	42·37°
H	3967	50·98°
.	.	.
Li.	0·0006703	16·43°
So.	5888	21·64°
Th.	5346	25·59°

From the foregoing measurements Boltzman has deduced the dispersion-formula

$$a = \frac{7·07018}{10^6 . \lambda^2} + \frac{0·14983}{10^{12} . \lambda^4}$$

which does not differ from the results of actual observation by more than the hundredth part of a degree.

For ray D different observers have obtained the following values :—Biot, 20·98°; Broch, 21·67° ± 0·11; Stefan, 21·67°; Wild,[1] 21·67° ; von Lang, 21·64° at a temperature of 13·3° Cent.; Soret and Sarasin 21·80°, at about 35° Cent. According to Scheibler[2], quartz from different places, and even from the same place, is not quite constant in its rotatory power.

As von Lang has shown, the rotatory power of quartz increases with rise of temperature, the relative alteration for different rays remaining the same. At 0° we have

	Li.	So.	Th.
a_0 =	16·402°	21·597°	26·533°

and for any given temperature t,

$$a_t = a_0 (1 + 0·000149 \, t).$$

According to the later researches of Sohncke,[3]

$$a_t = a_0 (1 + 0·0000999 \, t + 0·000000318 \, t^2)$$

expresses the value more accurately.

Extending the investigation to other substances, we find that not only do the absolute values of rotation in any one substance vary for different rays, but also in different substances the values vary by a different series of proportions—that is, different bodies have different powers of rotatory dispersion.

---

[1] Wild : *Ueber ein neues Polaristrobometer.* Berne, 1865, p. 55.
[2] Scheibler : *Zeitsch. des Vereins für Rübenzuckerindustrie.* 1869, p. 388.
[3] Sohncke : *Wied. Ann.* 3, 516.

For example, taking the following observations:—[1]

			B	C	D	E	F	G
Quartz	.	α 1 millim.	15·50°	17·22°	21·67°	27·46°	32·69°	42·37°
Sugar	.	[α] 1 decim.	+ 47·56°	52·70°	66·41°	84·56°	101·18°	131·96°
Cholalic Acid	.	[α] ,,	+ 28·2°	30·1°	33·9°	44·7°	52·7°	67·7°
Cholesterin	.	[α] ,,	− 20·63°	25·54°	31·59°	39·91°	48·65°	62·37°
Oil of Turpentine	α	,,	− 21·5°	23·4°	29·3°	36·8°	43·6°	55·9°
Oil of Lemon	.	α ,,	+ 34·0°	37·9°	48·5°	63·3°	77·5°	106·0°

and calculating the ratios of rotation experienced by rays $C, D, E, F, G,$ as compared with that by ray B, we get the following results :—

			B	C	D	E	F	G
Quartz	.	.	1	1·11	1·39	1·77	2·10	2·72
Sugar	.	.	1	1·11	1·40	1·78	2·13	2·77
Cholalic Acid	.	.	1	1·07	1·20	1·59	1·87	2.40
Cholesterin	.	.	1	1·24	1·53	1·93	2·36	3·02
Oil of Turpentine	.	.	1	1·09	1·36	1·71	2·03	2·60
Oil of Lemon	.	.	1	1·11	1·43	1·86	2·23	3·12

It will be seen that the ratios in the case of sugar and quartz agree very closely. These two substances have thus equal powers of rotatory dispersion, while the others have either less or more than quartz. This fact has been turned to account in the construction of the Soleil (and Ventzke-Scheibler) saccharimeter, the principle of which supposes the rotatory dispersion of the active substance to be equal to that of quartz. But, so far as we know, this is only the case with cane-sugar, so that other substances cannot be properly examined with instruments of this description.

In the majority of cases, the determination of the rotation for several rays would be too troublesome, and it is considered sufficient to determine it for a single ray. In Wild's polariscope, and in the so-called "half-shade" instruments of Jellett, Cornu, and Laurent, the light is supplied by a sodium flame, thus giving the angle of rotation for ray D of the solar spectrum. If a polariscope consisting simply of two Nicol prisms be used, it is requisite to employ monochromatic light (sodium flame) in order to get reliable results. With Soleil's saccharimeter, as well as those of Ventzke, Scheibler, and Hoppe-Seyler, which are all made on the same optical principle, white (gas-lamp) light is used, and the rotation given is for the so-called *transition tint*—that is to say, the colour complementary

---

[1] The values for oils of turpentine and lemon (given by Wiedemann, *Pogg. Ann.* 82, 222) are the angles of rotation α, directly determined with a layer 1 decimetre in depth ; whilst those for cane-sugar (Stefan, *Sitzungsber. der Wiener Acad.*, 52, 486, II^ᵗᵃ Abth.), anhydrous cholalic acid (Hoppe-Seyler, *Journ. für prakt. Chem.* 89, 257), and cholesterin (Lindenmeyer, *Journ. für prakt. Chem.* 90, 323), express the specific rotation [α]. For cholesterin the solvent used was alcohol ; for sugar and cholalic acid, water.

to mean yellow light—the wave-length of which may be taken at about 0·00055 millimetre. The angle of rotation and the specific rotation thus obtained are indicated by a_j and $[a]_j$ (*jaune moyen*) respectively, as proposed by Biot.

Now the wave-length of this mean yellow light is less than that of the ray D, which lies on the border between orange and yellow, so that the value of a_j is always less than that of a_D[1]. For example, with quartz, according to Broch, $a_D = 21·67°$, and $a_j = 24·5°$, so that, to express the one in terms of the other, we have—

$$a_j = \frac{24·5}{21·67} a_D = 1·1306 a_D, \text{ or approximately} = {}^9/_8 a_D;$$

$$a_D = \frac{21·67}{24·5} a_j = 0·8845 a_j, \text{ or approximately} = {}^8/_9 a_j.$$

The proportion, however, between a_j and a_D varies in different substances, according to their different rotatory dispersions. J. de Montgolfier[2] has determined it in the following :—

Quartz (according to Broch)	$a_D : a_j$ =	1 :	1·131
Aqueous solutions of Sugar	,, ,,	1 :	1·129
Alcoholic solutions of Camphor	,, ,,	1 :	1·198
Oil of Turpentine	,, ,,	1 :	1·243

According to L. Weiss,[3] in aqueous solutions of sugar containing 5 to 19 grammes in 100 cubic centimetres, the proportion is $a_D : a_j = 1 : 1·034$.

In any other substance, the rotation for one ray can be estimated from that for the other only approximately, by assuming that the rotatory dispersion of the substance agrees with that of some one of the preceding.

As the *transition tint* corresponds to no sharply-defined ray, its use is attended with inconvenience, and latterly has been mostly abandoned.

Many older observations are in existence, made by Biot with red light, obtained by transmission through glass coloured by suboxide of copper, with a refrangibility about equal to that of Fraunhofer's line C. Assuming that, on passing through a quartz plate 1 millimetre thick, it experienced rotation through an angle of 18·414°, Wild[4] calculates its wave-length to be 0·000635 millimetre. The ratio between this red ray and the *transition tint* Biot[5] gives as 23 : 30.

---

[1] By many observers the values of a_j and a_D have been taken as equal, a confusion which has been remarked upon by J. Montgolfier, *Bull. Soc. Chim.* 22, 487, and Riban, *idem*, 22, 492.   [2] Montgolfier: *Bull. Soc. Chim.* 22, 489.

[3] Weiss: *Sitzungsber. der Wiener Acad.* 69, 157, III^te Abth.

[4] Wild: *Polaristrobometer*, p. 35.   [5] Biot: *Mém. de l'Acad.* 3, 177.

§ 19. *Abnormal Rotatory Dispersion.*—Although in ordinary cases the angle of rotation increases *pari passu* with the refrangibility of the ray, there are exceptions, as Biot,[1] and later Arndtsen,[2] have noticed in aqueous solutions of dextro-tartaric acid. This acid exhibits, in a remarkable degree, the property that its specific rotation $[a]$ (§ 24) increases with the diluteness of the solution from observation of which it is calculated. The increase is directly proportional to the dilution, and may be expressed by the formula $[a] = A + Bq$, where q represents the percentage by weight of water in the tartaric acid solution. Arndtsen has determined by Broch's method the deviations for the Fraunhofer lines C, D, E, b, F, e, in a series of solutions of different degrees of concentration, at a temperature of 24° Cent., from which he deduces the subjoined values for the constants A and B in the preceding formula :—

$$
\begin{aligned}
[a]_C &= + 2\cdot748 + 0\cdot09446\,q \\
[a]_D &= + 1\cdot950 + 0\cdot13030\,q \\
[a]_E &= + 0\cdot153 + 0\cdot17514\,q \\
[a]_b &= - 0\cdot832 + 0\cdot19147\,q \\
[a]_F &= - 3\cdot598 + 0\cdot23977\,q \\
[a]_e &= - 9\cdot657 + 0\cdot31437\,q
\end{aligned}
$$

Computing from these formulæ the specific rotation for solutions containing from 10 to 90 per cent. of tartaric acid, we get the annexed results :—

In 100 parts Solution		$[a]_C$	$[a]_D$	$[a]_E$	$[a]_b$	$[a]_F$	$[a]_e$
Water q.	Tartaric Acid.	Red.	Yellow.	Green.	Green.	Blue-green.	Blue.
90	10	11·25°	13·68°	15·92°	16·40°	17·98°	(18·64°)
80	20	10·30°	12·37°	14·16°	14·49°	(15·59°)	15·49°
70	30	9·36°	11·07°	12·41°	12·57°	(13·19°)	12·35°
60	40	8·42°	9·77°	10·66°	10·66°	(10·79°)	9·21°
50	50	7·47°	8·47°	(8·91°)	8·74°	8·39°	6·06°
40	60	6·53°	(7·16°)	(7·16°)	6·83°	5·99°	2·92°
30	70	5·58°	(5·86°)	5·41°	4·91°	3·60°	− 0·23°
20	80	(4·64°)	4·56°	3·66°	3·00°	1·20°	− 3·37°
10	90	3·69°	3·25°	1·90°	1·08°	− 1·20°	− 6·51°

[1] Biot : *Mém. de l'Acad.*, 15, 93.

[2] Arndtsen : *Ann. Chim. Phys.* [3], 54, 403. *Pogg. Ann.* 105, 312.

From this table it appears that every solution exhibits a maximum of rotatory power for some one particular colour (as will be seen from the bracketed figures). In the most dilute solutions, containing 10 per cent. only of tartaric acid, the maximum is found, in its normal position, under the most refrangible ray *e* ; but as the concentration of the solution is increased it shifts towards the red end of the spectrum, and in the case of solutions containing 40 to 50 per cent. is found under the green rays. In 80 and 90 per cent. solutions the rotatory dispersion is entirely changed; dextro-rotation is there at its maximum for the red rays, and decreases as the refrangibility of the rays increases, until under the blue ray it passes into lævo-rotation. Such is the case also with anhydrous tartaric acid. Now, according to the preceding formula of Arndtsen, this should be dextro-rotatory for the rays *C, D, E,* lævo-rotatory for *b, F, e,* and inactive for light between *E* and *b,* and Biot [1] actually observed these phenomena in cast plates of tartaric acid ; moreover, Arndtsen [2] observed that the left-handed rotation for highly refrangible rays occurs when concentrated alcoholic solutions of this acid are used.

The anomalies of rotatory dispersion in tartaric acid disappear when the solutions are exposed to higher temperatures (Krecke [3]), or when mixed with a small quantity of boracic acid (Biot) ; moreover, they do not occur in the tartrates.

Similar conditions may be produced artificially by mixing dextro- and lævo-rotatory solutions together in certain proportions. Biot [4] in this way obtained an achromatic compensation of the rotation for certain rays.

Lastly, as regards the influence of temperature on rotatory dispersion, Gernez [5] discovered that the application of heat, even to the extent of complete evaporation, produces no change in the dispersive powers of oils of turpentine, orange, bigaradia, and camphor.

[1] Biot : *Ann. Chim. Phys.* [3], **28,** 351.
[2] Arndtsen : *Ann. Chim. Phys.* [3], **54,** 415.
[3] Krecke : *Arch. Néerland,* Bd. 7 (1872).
[4] Biot : *Ann. Chim. Phys.* [3], **36,** 405.
[5] Gernez : *Ann. de l'école norm.* 1, 1.

IV.

SPECIFIC ROTATORY POWER.

A. Definition of Specific Rotation.

§ 20. In the discussions that follow certain abbreviations have been adopted viz. :—

a, the observed angle of rotation for a given ray.

l, the length of liquid column used, in decimetres.

d, the density of the rotatory liquid.

p, the weight of active substance in 100 parts by weight of solution (per cent. composition).

q, the weight of inactive liquid in 100 parts by weight of solution.

$c = p\,d$, the number of grammes of active substance per 100 cubic centimetres of solution (concentration).

In comparisons of the rotatory powers of different substances it is not enough, as Biot[1] first showed, merely to take account of the angles of rotation observed in layers of a uniform length of 1 decimetre. It must be remembered that when different active substances, ordinarily liquids (as oil of turpentine, amylic alcohol, nicotine, &c.), are placed in turn in the tube of a polariscope, very different masses of molecules will be brought to bear upon the transmitted rays in accordance with the different densities of the liquids. Therefore, before any just comparison can be made, the observed angles of rotation must be calculated to some common standard of density. If the density 1 be taken, the angles of rotation a of the several liquids must be divided by their specific gravity d. The angle of rotation given by a length of 1 decimetre of any active liquid, corrected to the standard density 1, is denominated by Biot the *specific rotation* $[a]$ of that substance, and may be found from the observed data a, l, and d, by the formula

I.
$$[a] = \frac{a}{l \cdot d}.$$

[1] Biot: *Mém. de l'Acad.* 13, 116 (1835). *Ann. Chim. Phys.* [3], 10, 5.

By taking as unit-density that of water at 4° Cent., we make density synonymous with weight in grammes of 1 cubic centimetre of the substance, and the specific rotation of any active substance may then be defined as the deviation produced by 1 gramme of the substance when occupying a space of 1 cubic centimetre, and forming a column of 1 decimetre in length for the ray to traverse. The diameter of the column is thus immaterial.

Moreover, since not only the density, but apart from that the rotatory power of an active liquid, is affected by changes of temperature, the specific rotation will vary with the temperature, and it therefore becomes necessary to record the readings of the thermometer when a and d are observed. At any given temperature, the specific rotation of an active liquid in a state of purity is always constant.

§ 21. For active *solid* substances brought into the liquid state by solution in optically inactive and chemically indifferent solvents, the specific rotation is determined as follows :—Let P grammes of the active substance be dissolved in E grammes of inactive liquid, and d be the density of the resulting solution. Then the latter will contain in unit volume (cubic centimetre) $\dfrac{P}{P + E} d$ grammes of active substance.

If a solution of the above composition, in a tube l decimetres long, gives an angle of rotation a, then the deviation for a solution containing 1 gramme of active substance in 1 cubic centimetre of solution—*i.e.*, the specific rotation $[a]$—is deducible from the proportion

$$\frac{P}{P + E} d : \frac{a}{l} = 1 : [a],$$

whence,

$$[a] = \frac{a (P + E)}{l . P . d}.$$

If, with Biot, we indicate $\dfrac{P}{P + E}$, that is, the amount of active substance in the unit weight of solution, by ϵ, then

$$[a] = \frac{a}{l . \epsilon . d}.$$

Lastly, if the proportion of active substance be stated for 100 parts by weight of solution, so as to make the numbers more convenient for reference, and this proportion be indicated by p, the formula becomes

II.
$$[a] = \frac{100\,a}{l \cdot p \cdot d}.$$

For the calculation of specific rotation by these formulæ, the percentage weight and the density of the solution must be known. But since $p\,d$ = the concentration c—that is, the number of grammes of active substance in 100 cubic centimetres of the solution—we need not determine p and d separately. We have simply to dissolve a certain weight K (grammes) of the active substance in a graduated measure of known capacity V (cubic centimetres), and dilute the solution to the mark. Then we have

$$[a] = \frac{a\,V}{l\,K},$$

or,

III.
$$[a] = \frac{100\,a}{l\,c}.$$

The specific rotation of a large number of active substances has been thus determined, without regard to the densities of the solutions employed. But the method in many cases is insufficient. As will presently be seen, specific rotation calculated from solutions is not constant, but varies with the proportion of inactive substance present in each case; and although this can be allowed for when the weight per cent. composition of the solution is known, it is not possible to do so when only the concentration is stated. In preparing solutions, therefore, by means of a graduated flask, it is always necessary, for the sake of determining p and d, that the weight of the contents should be ascertained after the vessel has been filled up to the mark.

In cases where the length of tube is given in millimetres, indicated by L, the foregoing formulæ appear as below :—

I.
$$[a] = \frac{100\,a}{L \cdot d}, \quad \textit{for pure active liqu}$$

II.
$$[a] = \frac{10^4 \cdot a}{L \cdot p \cdot d}, \quad \textit{for solution of a}$$

III.
$$[a] = \frac{10^4 \cdot a}{L \cdot c}.$$

§ 22. *Influence of Temperature on Specific Rotation.*—A decrease in the angle of rotation is observed in active substances when the

E 2

temperature is increased.[1] This is primarily a direct consequence
of the decrease of density; whilst, on the other hand, there is a
simultaneous increase in the length of the tube, producing an opposite
effect, but in much feebler degree. In the calculation of specific
rotation, this element of variation is eliminated when the density
and length of tube[2] are determined for the same temperature as the
angle of rotation; and were such the only element of variation the
value [a] would be the same for all temperatures. Such, for example,
is the case with aqueous solutions of cane-sugar according to Tuchschmid.[3] In most substances, however, the angle of rotation and
the density do not vary together equally when the temperature is
raised, since increase of rotation is observed as a result in some substances as well as decrease. Hence there must be some other influence
at work independently of altered density and consequent on intra-
molecular changes produced by the action of heat. So far as at
present known, the most usual result is a decrease; this has been
observed in invert-sugar (Clerget, Tuchschmid), quinine, quinidine,
cinchonine, cinchonidine, quinine disulphate, quinidine disulphate,
thebain, santonate of soda (Hesse), gelatine (De Bary). The com-
pletest observations on the influence of increase of temperature are
those of Gernez[4] on the specific rotation of certain ethereal oils,
the decrease in which may be expressed for temperatures between
0° and 150° Cent., as follows:—

Oil of turpentine $\quad [a]_D = 36.61 - 0.004437\,t$.
Oil of orange $\quad [a]_D = 115.31 - 0.1237\,t - 0.000016\,t^2$.
Bigarade essence $\quad [a]_D = 118.55 - 0.1175\,t - 0.00216\,t^2$.

The decrease, as Gernez found, still goes on after the boiling-
point is exceeded and the substance takes the gaseous form. The
dispersive powers, on the other hand, were not influenced by heat.

[1] For example, within ordinary temperatures, an increase of 1° Cent. causes a
reduction of the angle of rotation in a tube 2 decimetres long, as follows:—

In Nicotine	. . . . .	by 0·1°
,, Oil of turpentine	. . . . .	,, 0·06°
,, Oil of bitter orange	. . . .	·, 0·3°
,, Oil of orange .	. . . . .	,, 0·38°
,, Aqueous solution of invert-sugar, with 17 grammes per 100 cub. cent. solution		,, 0·22°

The variations are, therefore, somewhat considerable in amount, and hence it is necessary
that the solutions under observation should be kept at an uniform known temperature.
[2] In glass tubes this may be found from the length as measured at ordinary
temperatures, by means of the coefficient of linear expansion 0·0000085 for 1° Cent.
[3] Tuchschmid: *Journ. für prakt. Chem.* [2], 2, 235.
[4] Gernez: *Ann. de l'école norm.* 1, 1.

In tartaric acid, on the contrary, solid as well as in solution (Biot,[1] Tuchschmid[2]) an increase of specific rotation is observed with increase of temperature for all rays of the spectrum, and in solutions of all degrees of concentration, but in unequal amounts. How variable the values obtained may become in this substance is shown by the observations of Krecke[3] in the table annexed :—

Temperature.	Tartaric Acid in 100 parts by Weight of Solution.		
	40 per cent.	20 per cent.	10 per cent.
0°	$[a^1]_D = 5 \cdot 53°$	$[a]_D = 8 \cdot 66$	$[a]_D = 9 \cdot 95°$
10°	7·49°	9·96°	10·94°
20°	8·32°	11·57°	12·25°
30°	9·62°	12·49°	13·93°
40°	11·03°	13·65°	15·68°
50°	12·27°	15·01°	17·11°
60°	12·63°	16·18°	18·31°
70°	13·38°	17·16°	19·42°
80°	14·27°	18·40°	20·72°
90°	15·91°	19·99°	22·22°
100°	17·66°	21·48°	23·79°

Among tartrates, Krecke found that neutral sodium and sodium-potassium tartrate showed a slight increase ; tartar emetic, on the contrary, a decrease. Lastly, in malic acid (Pasteur) and nicotine (see § 32) an increase also of specific rotation with increase of temperature has been observed.

Hence in all specific rotation data the temperature at which the angle of rotation and the specific gravity (or concentration) of the solutions have been observed should be given.

B. Dependence of Specific Rotation on the Nature and Amount of the Solvent.

§ 23. The first substance of which the specific rotation was determined by Biot[4] (1819) was cane-sugar. He found that with tubes of

---

[1] Biot : *Mém. de l'Acad.* 16, 229.   [2] Tuchschmid : *ut supra*.
[3] Krecke : *Arch. Néerland*, 7, 97 (1872).
[4] Biot : *Mém. de l'Acad.* 2, 41 (1819), 13, 39 (1832).

equal length the angles of rotation were in direct proportion to the amounts of sugar in solution, so that the value for [a] was constant whatever the degree of concentration of the solution. The same result was obtained with mixtures of oil of turpentine and ether. Biot therefore assumed that the amount of deviation is simply proportional to the number of optically active molecules encountered by the ray in its passage through the solution, and thence formulated the following axiom :—

" When an optically active substance is dissolved in an inactive liquid, which exerts no influence upon it chemically, the deviation is in direct proportion to the quantity of active substance in the unit-volume of solution, and thus specific rotation [a] is determinable from the equation $[a] = \dfrac{a}{l \cdot \epsilon \cdot d}$." (See § 21.)

Subsequently (1838), from experiments with aqueous solutions of tartaric acid, Biot[1] found that the specific rotation of this substance increases in proportion as the solutions are more dilute. This was long regarded as an exceptional case, until (in 1852) Biot[2], with the aid of improved polariscopic apparatus, found that similar phenomena are exhibited by other substances. Alcoholic and acetic solutions of camphor, for example, were found to show a decrease of specific rotation in proportion to the diluteness of the solutions. With oil of turpentine, on the contrary, an increase was observed on the addition of successive quantities of alcohol or olive-oil; and lastly, even in the case of sugar, a feeble increase with the amount of water in the solution could be detected. Moreover, the influence of the nature of the solvent medium was brought to light, as it was found that different values of [a] for camphor were obtained, according as solution was effected in alcohol or in equal weights of acetic acid. Hence Biot concluded generally that the values of specific rotation deduced from solutions are more or less variable; so that the phenomena cannot, as at one time, be regarded as the result of mere mechanical diffusion of active molecules in an optically indifferent medium.

This fuller and more correct view of specific rotation has, however, remained in a great measure unheeded. The fact that in solutions of cane-sugar the angle of rotation is almost exactly proportional to the degree of concentration, so that the saccharine strength of a solution can be deduced from the deviation observed therein, has

[1] Biot: *Mém. de l'Acad.* 15, 93; *Ann. Chim. Phys.* [3], 10, 385.
[2] Biot: *Ann. Chim. Phys.* [3], 36, 257.

led to the construction of the optical saccharimeter, which has been extensively adopted and employed in the analysis of other substances besides sugar. Notwithstanding that Biot,[1] in 1860, in a comprehensive paper containing a *résumé* of all his previous researches in connection with the subject, again called attention to the facts of the case, the belief is still prevalent that all optically active substances behave essentially like sugar, and that the optical analysis of a substance is complete when the deviation caused by any solution of it has been observed, and the specific rotation calculated therefrom by the formulæ

$$[a] = \frac{a \cdot 100}{l \cdot d \cdot p}, \quad \text{or} \quad = \frac{a \cdot 100}{l \cdot c},$$

the resulting value being regarded as constant.[2] In this manner the specific rotations of a very large number of substances have been calculated, and still appear in chemical and physical text-books without reference to the concentration or nature of the solvent media employed.

A few years ago, Oudemans, jun.,[3] contributed fresh proofs that the specific rotation of substances is susceptible of considerable variations, according as different inactive liquids are employed for solution. Since then, Hesse,[4] in 1875, published a large number of determinations of rotatory power, extending over fifty different active solid substances in solutions of different degrees of concentration. Even for small differences (between 1 and 10 grammes of substance in 100 cubic centimetres of solution), nearly all these substances displayed appreciable variations in the amount of their specific rotation, and nearly always a decrease for increased proportion of active substance. Still greater differences, in some instances exceeding 50^c, resulted from the use of different solvents.

§ 24. Observations like those of Hesse have thus shown conclusively that, as a rule, no value can be attached to specific rotations deduced from an isolated observation of an individual solution. Biot,[5] however, in his investigations of the rotatory power of tartaric acid

---

[1] Biot: *Ann. Chim. Phys.* [3], **59**, 206.

[2] See, by way of illustration, Buff, Kopp, and Zamminer's *Lehrbuch d. phys. u. theoret. Chem.*, p. 387.

[3] Oudemans: *Pogg. Ann.* **148**, 337; *Liebig's Ann.* **166**, 65.

[4] Hesse: *Liebig's Ann.* **176**, 89, 189.

[5] Biot: *Mém. de l'Acad.* **15**, 205 (1838); **16**, 254; *Ann. Chim. Phys.* [3], **10**, 385; **28**, 215; **36**, 257; **59**, 219.

long ago showed the significance attaching to these variations in value, and the following considerations point to the same conclusion :—

The specific rotation of any active liquid can be determined directly, and is constant for any given temperature. But, when such a liquid (*e.g.*, oil of turpentine) is mixed, in different proportions, with an indifferent liquid (as alcohol), and from the composition, density, and angle of rotation of the resulting solution the specific rotation is computed, the values so obtained will differ in a greater or less degree from that of the pure substance. Hence it follows that the specific rotatory power of a substance is in some way influenced by the presence of inactive molecules, so that its value is altered, in the majority of cases suffering an increase, and in rarer cases a decrease for increased proportions of the volume of solvent employed.[1]

If, however, the active substance be a solid, its rotatory power can only be examined in solution, and then different values for [a] will be obtained according to the character of the solvent, none of which is the actual specific rotation of the pure substance, but a value modified by the presence of the inactive liquid, and differing from the real value by a quantity unknown.

When a pure homogeneous liquid is employed as solvent, so that only inactive molecules of one kind are allowed to influence those of the active substance, the variations in specific rotation are best shown by the graphic method, the percentages of inactive solvent (*q*) being taken as abscissæ, and the corresponding values of [a] as ordinates. The increase or decrease of specific rotation will then in many cases appear as a straight line, increasing therefore in direct proportion to *q*. Hence it may be expressed by the formula

I. $$[a] = A + B\,q$$

in which the constants *A* and *B* must be ascertained from direct experiment. In other cases, it will appear as a curve, usually a

---

[1] This may be easily shown with a polariscopic tube set vertically with the upper end open. Let oil of turpentine be poured in to a height of 1 centimetre, and the deviation observed. Then on adding successive quantities of alcohol a continuous increase of rotatory power will be found to take place. The number of active molecules here remains the same, but their action is distributed over a greater length of column. On the other hand, if nicotine be used and diluted with water, a continuous decrease will be observed in the rotatory power on successive additions.

portion of a parabola or hyperbola, when the relation between the specific rotation and q is represented by an expression of the form—

II. $\qquad [a] = A + Bq + Cq^2$, or $[a] = A + \dfrac{Bq}{C+q}$[1],

or by some other equation with several constants.

In these formulæ, A denotes the specific rotation of the pure substance. The values B (I.) and B and C (II.) represent the increase or decrease of A for 1 per cent. of inactive solvent.

If $q = 0$, the specific rotation is that of the pure substance. On the other hand, if in equation I. or II., $q = 100$, we get for $[a]$ a value which may be taken as the specific rotation of the active substance when infinitely diluted. Assuming that, when $q = 100$, the active substance vanishes and the solution consists of the inactive solvent alone, the rotatory power will then necessarily be *nil*. As Biot[2] has pointed out, this may likewise be deduced from the foregoing expressions by equating them with the formula $[a] = \dfrac{a \cdot 100}{l \cdot d \cdot p}$, which is the specific rotation calculated from the directly observed angle of

---

[1] The three constants A, B, and C of the formula $[a] = A + \dfrac{Bq}{C+q}$ may, according to Biot (*Ann. Chim. Phys.* [3], 11, 96, § 69), be calculated in the following manner:— Given three separate solutions with q_1 q_2 q_3 per cent. of active substance, and three specific rotation values $[a]_1$ $[a]_2$ $[a]_3$ respectively,

then putting $\qquad A + B = a, \quad BC = b, \quad C = c,$

the values a and c may be obtained from the equations :—

$$\{[a]_2 \, q_2 - [a]_1 \, q_1\} + \{[a]_2 - [a]_1\} \, c - \{q_2 - \ldots$$
$$\{[a]_3 \, q_3 - [a]_1 \, q_1\} + \{[a]_3 - [a]_1\} \, c - \{q_3 - \ldots$$

and then b may be found from any of the following equations

$$\{[a]_1 - a\} \quad \{q_1 + c\} = -b$$
$$\{[a]_2 - a\} \quad \{q_2 + c\} = -b$$
$$\{[a]_3 - a\} \quad \{q_3 + c\} = -b$$

Lastly,

$$a - \frac{b}{c} = A, \qquad \frac{b}{c} = B, \qquad c = C.$$

Biot also brings the equation

$$[a] = A + \frac{Bq}{C+q}$$

into the form,

$$[a] = A + \frac{B'q}{1 + C'q}, \text{ wherein } B' = \frac{B}{C} \text{ and } C' = \frac{1}{C}.$$

Of course p can be substituted for q in the above formulæ.

[2] Biot: *Ann. Chim. Phys.* [3], 10, 399, § 59; 59, 224, § 15.

REESE LIBRARY UNIVERSITY OF CALIFORNIA.

rotation. Introducing in the latter in place of p, since $p + q = 100$, the value $100 - q$, we get by equating with I.,

$$\frac{a \cdot 100}{l \cdot d (100 - q)} = A + B q,$$

whence,

$$a = l \cdot d \left[A + \left(B - \frac{A}{100}\right)q - \frac{B}{100} q^2 \right].$$

Putting in this equation $q = 100$, we get $a = 0$; that is, the rotatory power vanishes. If $q = 0$, then $a = l \cdot d \cdot A$; that is, the angle of rotation for a length of 1 decimetre of pure substance of density d. And, as in that case $\frac{a}{l \cdot d} = [a]$, we get $[a] = A$, the specific rotation of the active substance in a state of purity.

In such active liquids as are miscible in all proportions with some other indifferent liquid the variations in the specific rotation up to extreme degrees of dilution can be determined by direct experiment, and the complete curve be drawn from $q = 0$ to approximately $q = 100$. In such cases, if the constant A be deduced from observations of a number of solutions, a value will be obtained more closely approximating to the true specific rotation of the pure substance in proportion as the observations extend over a greater length of curve, and the nearer they approach the point where abscissa $q = 0$—that is to say, the greater the concentration of the solutions employed.

When the active substance is a solid, the true specific rotation cannot be determined directly, and we can only construct a portion of the curve, larger or smaller according to the solubility of the substance, but always commencing at some distance from the origin of co-ordinates. If from these observations the constants A and B of formula I. or II. be calculated, the values obtained will only be strictly available for interpolation within the dilution-limits of the solutions employed.

The question then arises to what extent in such cases we are justified in regarding the value obtained for constant A as the specific rotation of the pure substance. The extrapolation, which is here presupposed, is admissible indeed when the variation in the specific rotation takes the form of a straight line—*i.e.*, can be represented by the formula $[a] = A + B q$. But if, on the other hand, it takes the shape of a curve, the value of A, as calculated from the formula $[a] = A + Bq + Cq^2$ or some other similar expression, will represent the true specific rotation of the pure substance, the more imperfectly the shorter the length of the experimental curve. How far a correct determination is attainable in such cases will depend on the solubility of the active substance itself. If indeed only dilute solutions can be prepared, and if, moreover, the increase or decrease in the values of $[a]$ is not constantly proportional to q, an estimation of the true specific rotation of the pure substance is quite impracticable.

§ 25. If we replace q in formulæ I. and II. by p—*i.e.*, the proportion of active substance in 100 parts by weight of solution—the constant A will then represent the specific rotation of the active substance in a state of infinite dilution, and that of the pure substance will be obtained when $p = 100$. But the employment of q (or, adopting Biot's notation e,—*i.e.*, the proportion of inactive substance per unit-weight of solution) as above is preferable.

In estimating rotatory power by the formula $[a] = \dfrac{aV}{l \cdot K} = \dfrac{a \cdot 100}{l \cdot c}$

(see § 21), no determination of the specific gravity of the solutions is required but merely the concentration c, determined by means of a flask of known capacity. Modifying the foregoing equations I. and II. accordingly, we have $[a] = A + Bc$, and $[a] = A + Bc + Cc^2$, and putting $c = 100$, we get the specific rotation of a solution containing 100 grammes of active substance in 100 cubic centimetres of solution. But this would only represent the pure substance if it had a density $d = 1$. If, as is always the case, d has some other value, say δ, we must give c the value $100\,\delta$. This, however, is a condition which can be but rarely satisfied, and never with certainty, as it presupposes a knowledge of the specific gravity of the active substance in an unknown amorphous condition. Hence, no specific rotation data where only the concentration of the solutions is taken account of, and specific gravity or weight per cent. composition is neglected, can be employed for the determination of the rotatory powers of the pure substances.

§ 26. When an active substance is influenced by inactive molecules of two different kinds, as when some other substance is dissolved along with it, or the active substance is dissolved in a mixture of two different liquids, the case becomes much more complicated. Each of the inactive substances exerts its own influence on the true specific rotation, which thus undergoes modifications which differ with every change in the relative proportions of the three components. In such cases several series of solutions must be prepared, such, that whilst in each series the active substance bears some one constant proportion by weight to one of the inactive substances, the proportion of the other inactive substance undergoes variation.

In this way Biot[1] determined the combined action of water and boracic acid on dextro-tartaric acid. In presence of each of these substances singly, as by fusion along with increasing quantities of boracic acid on the one hand, or solution in increasing volumes of water on the other, the rotatory power of tartaric acid increases. The same is the case when a mixture of the two acids is brought into the presence of water, and then the amount of increase is found to depend upon two conditions:—(1) *On the ratio between tartaric acid and boracic acid in the mixture.* The property of increasing the rotation of tartaric acid possessed by boracic acid depends on the formation of an unstable compound (represented, according to Dubrunfaut,[2] by $H_3BO_3 + 2C_4H_6O_6$) of higher rotatory power. The ratio between the tartaric acid and water remaining constant, the increase of rotation caused by the addition of different quantities of boracic acid (β parts by weight per unit-weight of solution) may be represented by the formula

$$[a] = A + \frac{B\beta}{\beta + C},$$

of which the three constants $A\ B\ C$ must be determined by a series of observations. (2) *On the amount of water.* The latter exerts its inherent property of increasing the rotatory power of the tartaric acid; on the other hand, it tends to break up the combination between the two acids, thereby causing a decrease of rotatory power. Experiment shows that so long as the mixture contains less than 0·088 part of boracic acid to 1 part tartaric acid, the rotation undergoes a continuous increase on the addition of increasing proportions of water. When the ratio between the tartaric and boracic acid is exactly 1 : 0·088, the specific rotation remains unchanged at each dilution, the decrease of optical activity due to the progressive breaking-up of the compound being exactly counterbalanced by the increase caused by the action of increasing volumes of water upon the acid set free. Lastly, should the mixture contain more than 0·088 part of boracic

[1] Biot: *Ann. Chim. Phys.* [3], 11, 82; 29, 341, 430; 59, 229.
[2] Dubrunfaut: *Comptes Rend.* 42, 112.

acid to 1 part tartaric acid, the influence of the first of the two factors predominates, and the observed rotation declines with successive additions of water.   In such cases in general the variations may be expressed by the formula $[a] = A + B\,q$, in which q represents the proportion of water per unit-weight of mixture.   If all three ingredients vary, certain weight proportions are found to give a maximum of rotation.   The extensive series of researches by Biot show what great difficulty such ternary systems oppose to a correct appreciation of the rotation-phenomena.

Similar phenomena are exhibited in other cases, as, for example, when alkaloids are dissolved in varying proportions of acids and water, or in mixtures of alcohol and chloroform in different proportions.[1]   The variations in specific rotation in these cases are perfectly obscure, and we may meet with curves showing a maximum at any point whatever.

In polariscopic investigations it is therefore necessary to employk solvents in the purest possible state, or, at any rate, in using mixtures like aqueous alcohol, to see that the composition of the solvent is kept perfectly constant.

§ 27.   A peculiar case of variation of specific rotation occurs in crystallized milk-sugar, $C_{12}H_{22}O_{11} + H_2O$, and also in glucoses having the formula $C_6H_{12}O_6 + H_2O$ (honey-sugar, grape-sugar, starch-sugar, and sugars from glucosides, as salicin, amygdalin, and phlorhizin). When a freshly prepared aqueous solution of one of these substances is placed under observation, it is found to exhibit a gradually decreasing rotatory power, after a certain length of time, however, becoming constant.   The decrease begins immediately after solution, so that observations must be made with all speed if we wish to determine the maximum of rotation with anything like accuracy.   At ordinary temperatures the decrease stops after the lapse of about twenty-four hours. Heat accelerates the change, and boiling brings it to a conclusion in the space of a few minutes ; cold, on the contrary, delays it, as well as addition of alcohol, wood-spirit, or acids.

The extent to which such variations may occur in certain cases is shown by the subjoined results of experiments made by Hesse[2] :—

[1] Oudemans : *Liebig's Ann.* 182, 51 ; 166, 70.
[2] Hesse : *Ann. d. Chem.* 176, 99 and 105.

	Weight per 100 cub.centim. Solution.	Length of Tube.	Angle of Rotation.	Specific Rotation [a]_D.
Milk-sugar (dextro-rotatory) :—				
Immediately after solution . . .	2 gr.	200 mm.	3·55°	80·68°
After the rotation had become constant	2 ,,	220 ,,	2·36°	53·63°
Honey-sugar (dextro-rotatory) :—				
Immediately after solution . . .	6 ,,	200 ,,	10·92°	91·00°
After the rotation had become constant	6 ,,	200 ,,	5·59°	46·58°

As the initial rotation is nearly twice as great as the constant rotation, the name of *bi-rotation* has been applied to this peculiarity.

As Dubrunfaut, Erdmann, and Béchamp[1] have shown, the phenomenon is connected with the fact that these sugars can assume two distinct modifications, the one crystalline, the other amorphous ; the latter being produced by fusion. The crystalline form alone gives the initial maximum of rotatory power; the amorphous gives the minimum at once, and hence it is inferred that when the former is brought into solution, it undergoes a gradual conversion into the latter modification. The higher rotatory power observed at first is probably due to the presence in the liquid of certain groups of molecules (so-called crystal-molecules) possessing optically-active structures, still remaining intact after the solution of the crystals has taken place, and so superadding the rotation due to crystalline form, to that which is possessed by the individual molecules themselves. In course of time the perfect separation of these crystalline combinations into their constituent chemical molecules takes place, and then we have simply the rotation due to the latter.

Another instance of bi-rotation just after solution is found in the case of the crystals belonging to the rhomboidal system, formed by the union of grape-sugar and sodium chloride, $2 C_6 H_{12} O_6 . Na Cl + H_2 O$ (Pasteur[2]). But no other instances of the phenomenon have hitherto been observed.

§ 28. From this inconstancy of specific rotation of substances in solution, it follows that we must no longer, as formerly, assume the complete indifference of the active to the presence of inactive

[1] Dubrunfaut, Erdmann, and Béchamp : *Jahresb. für Chem.* 1855, 671 ; 1856, 639.
[2] Pasteur : *Ann. Chim. Phys.* [3], 31, 95.

molecules, since it is clear that the latter do exert some influence on the former, and it becomes a question what the nature of this influence is.

The statement by Biot in this connection is that the rotatory power of a substance may be modified either by changes in its chemical constitution, or, where this remains unaltered, by the formation of molecule-groups of variable composition, arising from the perfect distribution of the solvent through the active substance, the latter imparting to them that amount of rotatory power, endowed with which they constitute the active elements of the mass.

It is a conceivable view that when between the molecules of some active substance, such as oil of turpentine, in which the molecular attractions throughout the mass are in equilibrium, we introduce the molecules of some other substance, as alcohol, which set up mutual attractions different in degree from the former, some modification in the structure of the active molecule ensues—some change in the mutual distances of the atoms composing it, their arrangement in space, and mode of vibration. In this way would be produced a modification of that non-symmetry in the density of the ether on which the phenomenon of optical activity depends, and the result would be more marked in proportion to the number of inactive molecules taking part in the reaction. The solutions of an active substance in different inactive liquids ought in this way to give different specific rotations, inasmuch as each species of molecule exerts an attraction peculiar to itself.

Of the theories proposed by Biot, the second alternative which supposes a transfer of rotatory power from the active molecules to a number of inactive molecules combined with them, is quite inconceivable. Nor can we possibly suppose that in dissolving such bodies as oil of turpentine in alcohol, sugar in water, &c., the mere solution produces any real alteration of chemical constitution. If we purposely bring about such alteration by the addition of some strong re-agent, we shall at once see how different the two cases are. In the first case we were dealing with modifications of rotatory power of a temporary character; and it was possible by removing the disturbing agent to restore the rotation to its original intensity; but in the latter case this is no longer possible—the modification is permanent. Whatever the true explanation may be, these variations of optical activity point to some mobile condition of the arrangement of atoms in the molecule. How considerable in amount they may at times become, is evident from the fact, as we have already seen (§ 16, towards the end), that in some cases the alteration is so great as to cause complete reversal of the original direction of rotation, lasting only so long as the disturbing substance is present.

C. Determination of the True Specific Rotation of Active Substances from the Rotatory Power of their Solutions.

§ 29. As we have seen (§ 24), the value $[a]$, calculated from a solution of an active substance, is never the true specific rotation of the substance itself, in a state of purity. It is necessary first to ascertain the law of the variation brought about by the inactive liquid, by examining a number of solutions of different strengths, and then the true specific rotation may be approximately obtained by calculating the value of constant A in the formula $[a] = A + Bq$, or $[a] = A + Bq + Cq^2$, &c. (see § 24).

Whether this method is really reliable or not can only be proved by experiments on substances, the real specific rotation of which can be determined independently. A series of solutions of such substances of different strengths must be prepared, from the observed specific rotations of which the value of constant A must be determined, and the result compared with that obtained by direct observation. Up to the present, the only investigations of this kind have been made by Biot with cane-sugar[1] and tartaric acid[2], the latter of which, when cast in solid plates, was found to give nearly the same specific rotation as had been deduced from its solutions by application of the formula $[a] = A + Bq$.

It was, therefore, of importance that further researches of the same kind should be instituted, and more particularly upon liquids, since it is only such substances that admit of the specific rotation of the absolute substance being accurately determined, as well as of a complete examination of the changes produced by the addition of successive quantities of various inactive solvents.

In the following pages are given the results of the author's investigations of the rotatory powers of right- and left-handed oil of turpentine, nicotine, and tartrate of ethyl, taken first alone, and afterwards in

---

[1] Biot: *Mem de l' Acad.*, 13, 39 ; *Ann. Chim. Phys.* [3], 10, 175.
[2] *Ann. Chim. Phys.* [3], 28, 351.
[3] *Liebig's Ann.* 189, 241.

admixture with different inactive liquids.[1] In these, the value [a] exhibits very different rates of increase or decrease for increasing percentages of solvent q, the law of variation, when shown graphically, appearing in some cases as a straight line, in others as a curve.   In the first place a formula was deduced for each, embracing all the curve lying between the most concentrated and the most dilute of the solutions used ; and, as the observations for the most part started with solutions containing about 90 per cent. of active substance, it was to be expected that at this point the value deduced for the constant *A* would approximate very closely to the true specific rotation of the absolute substance.   Next it was sought to determine to what extent the calculated differs from the true specific rotation when only dilute solutions are used, as would be the case with substances but sparingly soluble.

As already stated (§ 23), we find that when an active substance is dissolved in different liquids, different values of [a] are always obtained, even when the degrees of concentration are the same, whence it has been often inferred that for each solvent any substance has a different constant of specific rotation.   Experiment in this direction must decide whether the true rotatory power of a substance does indeed suffer an immediate definite alteration under the influence of small proportions of inactive solvent, which subsequently proceeds at a different rate on further dilution, or whether the values obtained for constant *A* with different solvents do sufficiently agree with each other.   Biot[2] has recorded a solitary experiment of this kind, in which he worked with solutions of camphor in acetic acid and in alcohol.   The results for the red ray (§ 18) gave the formulæ—

$$\begin{array}{cc} A & B \end{array}$$

Solutions in acetic acid $[a]_r = 42 \cdot 54 - 0 \cdot 14236\,q,$
    ,,   alcohol $[a]_r = 45 \cdot 25 - 0 \cdot 13688\,q.$

The two values for A, instead of agreeing as expected, here exhibit a by no means insignificant difference, and Biot assumes, in explanation, that the law of variation of the specific rotation is not expressed with sufficient exactness by the foregoing interpolation-formulæ.

The observations appended were made by the methods described at length in Chapter V.

[1] *Liebig's Ann.*, 189, 241.
[2] Biot : *Ann. Chim. Phys.* [3], 36, 301, 307.

F

The following abbreviations have been used :—

p, percentage of active substance in 100 parts by weight of solution.

q ,, inactive ,, ,, ,, ,, ,,

d, specific gravity of the solution at 20° Cent., that of water at 4° being taken as unity.

$c = dp$, the concentration, *i.e.*, the number of grammes of active substance in 100 cubic centimetres of solution.

L, the length of tube in millimetres.

a, observed angle of deviation of ray D, at a temperature of 20° Cent., expressed in degrees and decimals (circular measure).

$[a]_D = \dfrac{a \cdot 10^4}{L \cdot c}$, the specific rotation for ray D, at a temperature of 20° Cent.

I. LEFT-HANDED OIL OF TURPENTINE.

§ 30. Two kilogrammes of Bordeaux oil, after standing for several weeks over calcium chloride, were submitted to distillation. Nearly the whole came over between 160° and 162° Cent. Height of barometer = 737 millimetres.

Specific gravity 0·86290.

The rotation was observed with a Wild's instrument, having two bath-tubes of different lengths.

Expt.	L.	a.	$[a]_D$.
I.	99·92	31·905°	37·004°,
II.	219·79	70·204°	37·016°.

Mean: $[a]_D = 37·010°$.

It should be remembered that when oil of turpentine is kept in vessels containing air, it undergoes a process of oxidation, whereby the specific gravity of the oil is increased and the rotatory power diminished. A portion of the above oil, after standing thus for four weeks, gave a specific gravity of 0·86779, and with a 219·79 millimetre tube showed a deviation of 68·144, whence $[a]_D = 35·728°$.

(a.) *Mixtures with Alcohol.*

Alcohol, as nearly as possible anhydrous, with a specific gravity 0·7957, was used. The following mixtures were examined with Wild's polariscope :—

Number of Mixture.	Oil of Turpentine p.	Alcohol q.	d.	c.	a for $L = 219·79$.	$[a]_D$.
I.	90·0530	9·9470	0·85558	77·0474	62·716°	37·035°
II.	69·9416	30·0584	0·83923	58·6971	48·05?°	37·247°
III.	49·9658	50·0342	0·82542	41·2428	34·036°	37·548°
IV.	29·9715	70·0285	0·81273	24·3588	20·293°	37·904°
V.	10·0078	89·9922	0·80108	8·0170	6·782°	38·486°

Like the oil itself its alcoholic solutions suffer a gradual decrease of rotatory power when kept in incompletely filled vessels. Mixture II. at the end of eight days gave $[a]_D = 37\cdot164°$.

(b.) *Mixtures with Benzene.*

Crystallizable benzene with a boiling point 80·4° Cent. (barometer, 755 millimetres), and a specific gravity of 0·88029 was used. Observations were taken with Wild's instrument:—

Number of Mixture.	Oil of Turpentine p.	Benzene q.	d.	c.	a for L = 219·79.	[a]_D
I.	89·9185	10·0815	0·86340	77·6356	63·466°	37·194°
II.	77·9272	22·0728	0·86439	67·3595	55·500°	37·487°
III.	65·0553	34·9447	0·86562	56·3131	46·789°	37·803°
IV.	51·0499	48·9501	0·86769	44·2955	37·175°	38·184°
V.	36·8987	63·1013	0·87050	32·1204	27·196°	38·523°
VI.	22·0557	77·0443	0·87377	20·0580	17·207°	39·031°
VII.	9·9839	90·0161	0·87713	8·7572	7·593°	39·449°

(c.) *Mixtures with Acetic Acid.*

The glacial acetic acid used was rectified by two fractional crystallizations, and had a specific gravity 1·0502 at 20° Cent., corresponding to 99·8 to 99·9 per cent. of acetic acid. Wild's polariscope was used:—

Number of Mixture.	Oil of Turpentine p.	Acetic Acid q.	d.	c.	a for L = 219·79.	[a]_D.
I.	90·1636	9·8364	0·87565	78·9520	64·462°	37·148°
II.	78·0658	21·9342	0·89166	69·6082	57·228°	37·406°
III.	64·8610	35·1390	0·91163	59·1293	49·235°	37·885°
IV.	50·9737	49·0263	0·93530	47·6757	40·266°	38·427°
V.	22·9616	77·0384	0·99183	22·7740	19·858°	39·672°
VI.	9·8414	90·1586	1·02330	10·0707	8·903°	40·222°

In these mixtures also an increase of density along with diminution of rotatory power could be detected by keeping. After standing for three days, the following results were obtained :—

Mixture I. $d = 0.87630$ $\alpha = 63.987°$ $[\alpha] = 36.817$'   (when fresh 37.148°).
   ,,   IV. $d = 0.93540$ $\alpha = 39.859°$ $[\alpha]$ 38.034'   (when fresh 38.127).

By reason of this liability to change under the influence of oxidation, oil of turpentine is not altogether a suitable substance for experiments of the above kind, and inattention to this point at the outset entailed the necessity of repeating several of our experiments.

Fig. 15.

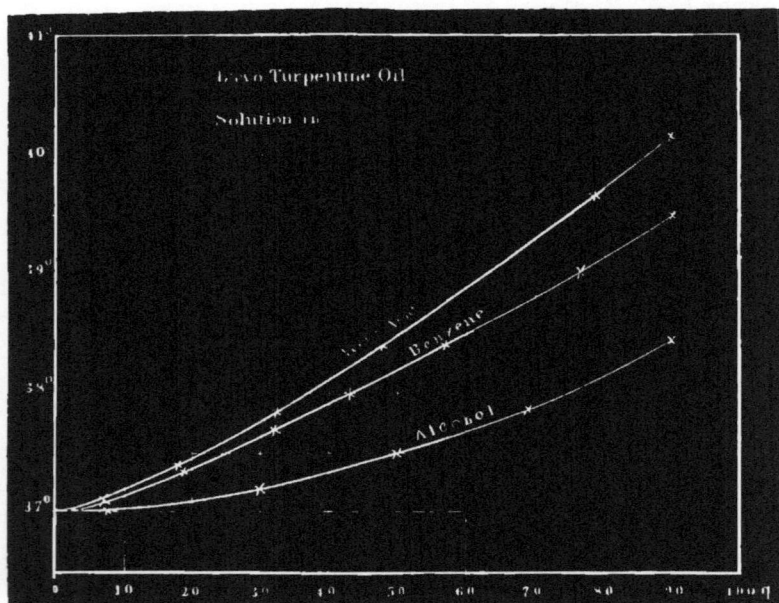

As the observations show, the specific rotation of the oil of turpentine rises with the addition of increasing quantities q of all three solvents, the graphic representations taking the form of curves, of which that for acetic acid appears steepest, that for benzene less so, and for alcohol least (see Fig. 15). Although the curvature in each case is slight, the deviation from a straight line is too great to admit of application of the formula $[\alpha] = A + B\,q$.

If, taking this equation, we proceed to determine the constant A from two mixtures, values will be obtained, which will always be less than the true specific rotation of pure oil of turpentine (37·01), differing from the latter in proportion to the diluteness of the solutions employed in its determination. Thus, with alcohol as the inactive solvent, we get—

					Divergence from 37·01	Extrapolation
From mixtures	I. and	II.	$A = 36·93°$		−0·80°	10 per cent.
„	II. „	III.	$A = 36·79°$		−0·22°	30 „
„	III. „	IV.	$A = 36·66°$		−0·35°	50 „
„	IV. „	V.	$A = 35·87°$		−1·14°	70 „

On the other hand, if we adopt the formula $[a] = A + B q + C q^2$, and compute the values of the constants A, B, and C from the solutions containing the smallest, mean, and largest proportions respectively of the inactive constituent, we get for A a value approaching very near to that of the specific rotation of pure oil of turpentine. Moreover, the formula agrees sufficiently well with the' experimental curve throughout (from $q = 10$ to 90). The mixtures specified furnished the results shown below :—

1. *Alcohol* (computed from solutions I., III., and V.)
   $[a]_D = 36·974 + 0·0048164 q + 0·00013310 q^2.$

2. *Benzene* (computed from solutions I., IV., and VII.)
   $[a]_D = 36·970 + 0·021531q + 0·000066727 q^2.$

3. *Acetic acid* (computed from solutions I., IV., and VI.)
   $[a]_D = 36·894 + 0·024553 q + 0·00013689 q^2.$

These formulæ give the following interpolation values :—

Solvent Medium.	No. of Mixture.	q.	$[a]_D$ Observed.	$[a]_D$ Calculated.	Difference.
Alcohol	II.	30·0584	37·247°	37·239°	− 0·008°
	IV.	70·0285	37·904°	37·964°	+ 0·060°
Benzene	II.	22·0728	37·487°	37·478°	− 0·009°
	III.	34·9447	37·803°	37·804°	+ 0·001°
	V.	63·1013	38·523°	38·594°	+ 0·071°
	VI.	77·0443	39·031°	39·025°	− 0·006°
Acetic Acid	II.	21·9342	37·406°	37·498°	+ 0·092°
	III.	35·1390	37·885°	37·926°	+ 0·041°
	V.	77·0384	39·672°	39·598°	− 0·074°

,If now we seek to determine the specific rotation of pure oil of turpentine from the more dilute solutions only, we get for A in the formula $[a] = A + Bq + Cq^2$ values showing the following divergences :—

Solvent.	From Solutions			We obtain $A =$:	Divergence from 37·01	Extrapolation
Alcohol ..	II.,	IV.,	V.	37·20°	+ 0·19°	30 per cent.
	III.,	IV.,	V.	35·13°	− 1·88°	50 ,,
Benzene ..	III.,	V.,	VI.	37·26°	+ 0·25°	35 ,,
	V.,	VI.,	VII.	35·42°	− 1·59°	63 ,,
Acetic Acid	II.,	IV.,	VI.	36·65°	− 0·36°	22 ,,
	IV.,	V.,	VI.	36·00°	− 1·01°	49 ,,

Hence the divergence from the real value may amount, in some cases, to a whole degree, when the solutions contain more than 50 per cent. of solvent.

II. Right-handed Oil of Turpentine.

§ 31. The American oil here employed had a specific gravity of 0·91083.

For determining the rotation of the pure oil and its mixtures Wild's and Mitscherlich's instruments were used, but at the time these observations were made the experimental tubes had not been furnished with water-baths. The temperature of the liquid was ascertained at the conclusion of the measurements by a thermometer inserted into the tube.

Instrument.	Temperature.	a for $L = 219·90$ millims.	$[a]_D$.
Wild	21·2°	28·354°	14·156°
Mitscherlich	21·0°	28·315°	14·137°

Mean: $[a]_D = 14·147°$.

Mixtures with Alcohol.

The following mixtures were prepared :—

Number of Mixture.	Oil of Turpentine p.	Alcohol q.	d.	c.
I.	73·0927	26·9073	0·87648	64·0643
II.	47·5124	52·4876	0·84642	40·2154
III.	22·2443	77·7557	0·81864	18·2101

The angles of rotation obtained were:—

Number of Mixture.	Instrument.	Tempera-ture.	α for L = 219·90.	[α].	[α] Mean.
I.	Wild	22·5°	20·416°	14·492°	14·496°
	Mitscherlich	22·0°	20·426°	14·499°	
II.	Wild	22·0°	13·096°	14·809°	14·788°
	Mitscherlich	22·0°	13·059°	14·767°	
III.	Wild	24·0°	6·021°	15·036°	15·095°
	Mitscherlich	21·5°	6·068°	15·153°	

Here also the specific rotation undergoes a slight increase with increasing dilution. The graphic representation (Fig. 16) shows that

Fig. 16.

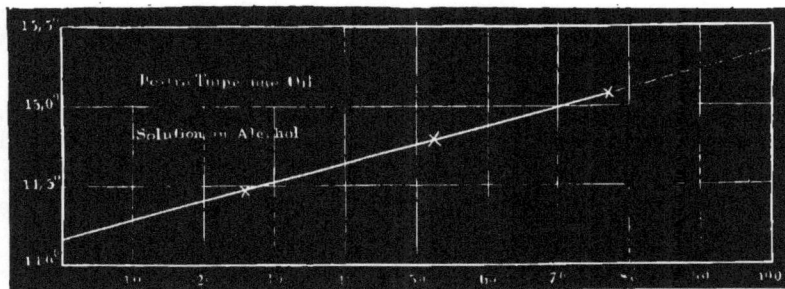

the three points lie almost exactly in a straight line. Introducing mean value of [α] into the formula [α] = A + B q, we obtain—

1. Calculated from mixtures I. and II.   A = 14·189   B = + 0·011415
2.   ,,   ,,   II. ,, III.   = 14·150   = + 0·012150
3.   ,,   ,,   I. ,, III.   = 14·179   = + 0·011780

The values here obtained for A agree very closely with the directly observed specific rotation of the oil, = 14·147°.

Taking the mean of the above values for A and B, respectively,

$$[α]_D = 14·173 + 0·11782\ q,$$

which, of course, corresponds closely with the observations. We have—

Mixture.	[α] Observed.	[α] Calculated.	Difference.
I.	14·496	14·490	− 0·006
II.	14·788	14·791	+ 0·003
III.	15·095	15·089	− 0·006

A further series of experiments with benzene as solvent led to no result, as the oil of turpentine used in preparing the mixtures was not uniform, from not all having been kept an equal length of time exposed to oxidizing influences.

III. NICOTINE (LÆVO-ROTATORY).

§ 32. The pure substance here employed was prepared from 400 grammes commercial nicotine (supplied by H. Trommsdorff, of Erfurt). To remove any small impurities of ether, alcohol, and water, the liquid was heated in a retort, whilst a stream of hydrogen passed over it, for eight hours, at a temperature of 150°, which was finally raised to 180° Cent., whereby a total distillate of 15 cubic centimetres, consisting chiefly of alcohol, passed over. The residue was then distilled in a current of hydrogen, in successive portions of 200 grammes, from a small retort heated by a sand-bath.

At 225° Cent. the nicotine, at first still retaining some water, began to pass over, but the thermometer immersed in the liquid quickly rose to 244° (corrected, 249° Cent.) when the boiling proper began. Raised into the vapour, the thermometer fell to 241·5° to ·242° (corrected, 246·6° to 246·8° Cent.), which temperature remained constant during the rest of the distillation. Height of barometer, 745 millimetres. The hydrogen was admitted in a very slow stream, and no decomposition of the substance resulted. Altogether 350 grammes of pure substance, in the form of a colourless liquid, faintly tinged with yellow, were obtained. It was sealed up in glass tubes. An analysis of this nicotine, the nitrogen being determined in the gaseous form, gave the following numbers :—

	Required for $C_{10}H_{14}N_2$.	Found.
C	74·02	73·95
H	8·66	8·92
N	17·32	17·55

The preparation was submitted to further examination by titration. The nicotine was weighed in thin glass bulbs, which were then broken under water, and, after solution, tincture of litmus added, and the liquid neutralized with a standard hydrochloric acid (containing 0·055013 gramme H Cl in 1 cubic centimetre).

Analysis.	Weighed Substance.	Acid used in Titration.	Hence 1 molecule H Cl (= 36·37) neutralized.
I.	4·1303 grm.	16·9 c.c. = 0·92972 grm. H Cl	161·57 grm. substance
II.	1·0002 ,,	4·1 ,, = 0·22555 ,, ,,	161·28 ,, ,,
III.	7·3048 ,,	29·7 ,, = 1·63390 ,, ,,	162·60 ,, ,,
		Mean	161·82 grm. of substance

But one molecule nicotine, $C_{10} H_{14} N_2 = 161·72$.

Again, in a rough titration with dilute sulphuric acid, containing 0·030369 gramme $H_2 S O_4$ per cubic centimetre, 37·1 cubic centimetres served to neutralize 3·8310 grammes of substance, according to which one molecule $H_2 S O_4$ (97·82) combines with 332·6 parts of nicotine ; but two molecules $C_{10} H_{14} N_2 = 323·4$.

The specific rotation of the nicotine was observed with the Wild's instrument, and observations were taken at three separate temperatures (10°, 20°, 30° Cent.), so as to ascertain the effect of heat. The experimental tube was provided with a water-jacket, and, measured at the mean temperature (20° Cent.), had a length of 99·923 millimetres, from which its true lengths at 10° and 30° Cent. were calculated, assuming the coefficient of expansion of the glass to be 0·0000086. The specific gravity of the nicotine, referred to that of water at 4° Cent., was also observed with the pycnometer at each separate temperature.

Temp.	d.	L.	a.	$[a]_D$.
10·2°	1·01837	99·914 millims.	163·776°	160·96°
20·0°	1·01101	99·923 ,,	163·204°	161·55°
30·0°	1·00373	99·932 ,,	162·450°	161·96°

Another analysis, made with Mitscherlich's instrument, using a tube 49·82 millimetres long, gave $a = 81·283°$. The temperature of the nicotine was about 21·0°, and taking $d = 1·01101$, $[a] = 161·38°$.

Accordingly the specific rotation of nicotine at 20° Cent. has been taken in what follows as

$$[a]_D = 161·55°.$$

(a.) *Mixtures with Alcohol.*

The alcohol used had a specific gravity of 0·7957 at 20° Cent. The following mixtures were observed with the Wild's instrument:—

Number of Mixture.	Nicotine p.	Alcohol q.	d.	c.	a for $L = 99·923$.	$[a]_D$.
I.	90·0945	9·9055	0·98839	89·0488	141·163°	158·65°
II.	74·9336	25·0664	0·95358	71·4553	110·616°	154·92°
III.	59·9345	40·0655	0·92001	55·1405	83·626°	151·78°
IV.	45·0846	54·9154	0·88747	40·0113	59·494°	148·81°
V.	30·0268	69·9732	0·85536	25·6836	37·319°	145·42°
VI.	14·9567	85·0433	0·82506	12·3401	17·460°	141·60°

To ascertain whether the solutions were affected by keeping, mixtures I. and V. were left for a couple of days, and then examined again. The results gave for I. $[a] = 158·63°$, and for V. $[a] = 145·45°$, which values agree almost perfectly with those above obtained from freshly-prepared solutions.

As the table shows, the specific rotation of nicotine undergoes a pretty considerable decrease for successive additions of alcohol. Represented graphically, it takes the form of a straight line, with but small divergences either way, and consequently the values of the constants in the formula $[a] = A - Bq$ do not materially differ, whichever of the solutions we take.

From Mixtures	We obtain $A = :$	Divergence from 161·55°	Extra-polation	$B =$
I. and III.	160·90°	− 0·65°	10 per cent.	− 0·22805
II. „ IV.	160·06°	− 1·49°	25 „	− 0·20490
III. „ V.	160·31°	− 1·24°	40 „	− 0·21272
IV. „ VI.	161·96°	+ 0·41°	55 „	− 0·23928
I. „ VI.	160·90°	− 0·65°	10 „	− 0·22686

Here even dilute solutions yield values for A, which, considering the large rotation-angle of nicotine, come pretty close to the real value.

Taking for A and B the mean of the above values, we get
$$[a]_D = 160·83 - 0·22236 \, q,$$
which gives the following interpolation values:—

Number of Mixture.	$q.$	$[a]$ Observed.	$[a]$ Calculated.	Difference.
I.	9·9055	158·65°	158·63°	− 0·02
II.	25·0664	154·92°	155·26°	+ 0·34
III.	40·0655	151·78°	151·92°	+ 0·14
IV.	54·9154	148·81°	148·62°	− 0·19
V.	69·9732	145·42°	145·27°	− 0·15
VI.	85·0433	141·60°	141·92°	+ 0·32

(b.) *Mixtures with Water.*

Nicotine mixes with water in all proportions, the more dilute solutions exhibiting a faint opalescent clouding, which necessitated a larger number of polariscopic observations. With the addition of water a considerable amount of heat is developed, so that in mixing 24 grammes nicotine with 6 grammes water, the thermometer rose from 20° to 35° Cent. The rotations were all observed with the Wild's instrument, the 99·923 millimetre tube being used for mixtures I., II., III., IV., VIII., and a short tube 49·82 millimetres in length for mixtures V., VI., and VII.

Number of Mixture.	Nicotine $p.$	Water $q.$	$d^1.$	$c.$	$a.$	$[a]_D.$
I.	89·9155	10·0845	1·02671	92·3170	123·467°	133·85°
II.	78·3920	21·6080	1·03528	81·1578	88·823°	109·53°
III.	65·8972	34·1028	1·04010	68·5397	64·543°	94·24°
IV.	53·4750	46·5250	1·03649	55·4256	47·952°	86·58°
V.	34·2854	65·7146	1·02282	35·0680	14·113°	80·78°
VI.	17·6793	82·3207	1·01158	17·8840	6·855°	76·94°
VII.	16·3356	83·6644	1·00957	16·4920	6·317°	76·88°
VIII.	8·9731	91·0269	1·00469	9·0152	6·804°	75·53°

Hence it appears that the specific rotation of nicotine goes on diminishing on successive additions of water, at first very markedly, but by-and-bye at a much smaller rate. The strongly curved line

[1] The specific gravity of the mixtures at first increased by the addition of water, attaining the maximum with mixture III., and then declining with each fresh addition.

(Fig. 17) forms part of a hyperbola, and it is impossible by means of the formula $[a] = A + Bq + Cq^2$, even when extended to four or

Fig. 17.

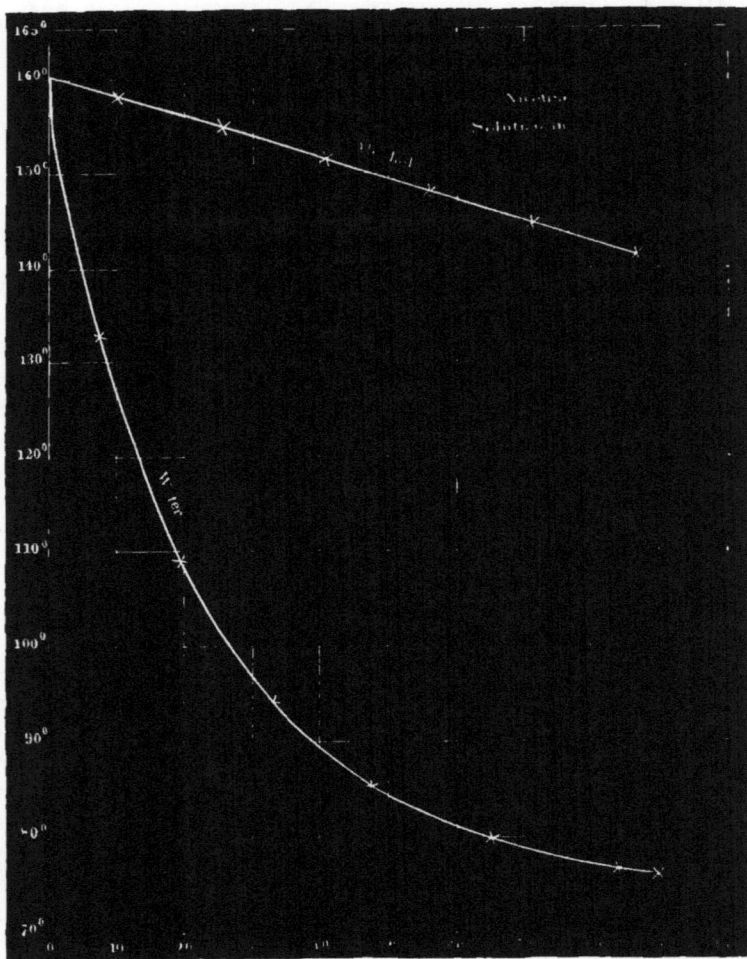

five terms, to obtain sufficient agreement with the results furnished by observation. In this case the value for pure nicotine can only be ascertained approximately by basing the calculations on the values of A for the most concentrated solutions. Thus, for mixtures I., II., III., $A = 163\cdot17°$; whereas from I., IV., VII. we get $A = 153\cdot00°$, and from IV., V., VIII., $A = 141\cdot16°$. For a complete expression of the curve an equation of different form must be used, such as that calculated by Dr. Vogler, of Aachen, viz :—

$$[a]_D = 115\cdot019 - 1\cdot70607\,q + \sqrt{2140\cdot8} - 108\cdot867\,q + 2\cdot5572\,q^2,$$

which affords values comparable with the results of observation as below :—

Number of Mixture.	q.	$[a]$ Observed.	$[a]$ Calculated.	Difference.
I.	10·0845	133·85°	133·92°	+ 0·07
II.	21·6080	109·53°	109·49°	– 0·04
III.	34·1028	94·24°	94·28°	+ 0·04
IV.	46·5250	86·58°	86·74°	+ 0·16
V.	65·7146	80·78°	80·56°	– 0·22
VI.	82·3027	76·94°	77·08°	+ 0·14
VII.	83·6644	76·88°	76·84°	– 0·04
VIII.	91·0269	75·53°	75·56°	+ 0·03

For pure nicotine ($q = 0$), the above formula gives $[a]_D = 161\cdot29°$, instead of the value found $= 161\cdot55°$; for $q = 100$, the value of $[a] = 74\cdot13°$, so that nicotine, when diluted largely with water, has its specific rotation reduced to less than one-half of the original amount.

IV.   TARTRATE OF ETHYL (DEXTRO-ROTATORY).

§ 33. To prepare this substance, an alcoholic solution of tartaric acid was heated for some days in a water-bath along with one-tenth its volume of concentrated sulphuric acid. The mixture was then diluted plentifully with water, saturated with barium carbonate, and the filtered liquor agitated with ether. From the ethereal extract the ether was removed by distillation, and the residue in the retort heated to a temperature of 110° to 120° Cent., whilst a current of dry air was led through it so long as any traces of ether, alcohol, or aqueous vapour were observable. In this way was obtained an ethereal liquid of a pale yellow colour and syrupy consistency, which, when heated on a platinum plate, volatilized without leaving any carbonaceous residue. An attempt to distil it *in vacuo* was frustrated by the violence of the shocks.

As a test of its purity, 2·6079 grammes of the substance were boiled with 30 cubic centimetres of potash solution (containing

0·06336 gramme KOH per cubic centimetre), and the excess of alkali titrated with dilute hydrochloric acid, of which 37·4 cubic centimetres were equivalent to 50 cubic centimetres of the potash. For this purpose 5·6 cubic centimetres were required, so that the amount of KOH required to decompose the ether was 1·4262 grammes, corresponding to 2·6171 grammes ethyl tartrate, or 100·35 per cent.

The specific gravity of the ether was 1·1989 at 20° Cent. The rotation was observed with the Mitscherlich's instrument :—

Observation.	L.	Temp.	α.	$[\alpha]_D$.
I.	99·92 millims.	20·0°	9·932	8·291°
II.	49·82 ,,	20·3°	4·974	8·328°

Mean : $[\alpha]_D = 8·309°$.

(a.) *Mixtures with Alcohol.*

Specific gravity of alcohol employed = 0·7962 at 20° Cent.

Three solutions were examined in the Mitscherlich, with a jacketed tube 219·90 millimetres long :—

No. of Mixture.	Tartrate of Ethyl p.	Alcohol. q.	d.	c.	α.	$[\alpha]_D$.
I.	77·9774	22·0226	1·08373	84·5064	16·315°	8·780°
II.	35·7366	64·2634	0·90892	32·4818	6·870°	9·618°
III.	22·3297	77·6703	0·86337	19·2787	4·174°	9·846°

The specific rotation rises therefore slowly with successive additions of alcohol, the law of variation taking the form of a slight curve departing but little from a straight line. The formula $[\alpha] = A + Bq$ gives—

From Mixtures I. and II.	$A = 8·343$	$B = + 0·019839$
,, ,, II. ,, III.	$= 8·525$	$= + 0·017006$
,, ,, I. ,, III.	$= 8·358$	$= + 0·019156$

Whence, taking the mean,

$$[\alpha]_D = 8·409 + 0·018667\, q.$$

Adopting the formula $[\alpha] = A + Bq + Cq^2$, we get

$$[\alpha]_D = 8·271 + 0·024216\, q - 0·000050648\, q^2.$$

Thus the values for the constant A agree very nearly with the specific rotation of tartrate of ethyl (8·31).

(b.) *Mixtures with Wood-Spirit.*

Specific gravity of the pure wood-spirit at 20° Cent. = 0·80915. Observations taken with the Wild's polariscope, with a jacketed tube 219·79 millimetres in length :—

No. of Mixture.	Tartrate of Ethyl p.	Wood-Spirit q.	d.	c.	a.	$[a]_D$.
I.	77·4567	22·5433	1·08820	84·2888	17·876°	9·649°
II.	56·6527	43·3473	1·00066	56·6900	12·973°	10·411°
III.	39·9196	60·0804	0·93808	37·4476	8·984°	10·915°
IV.	26·9681	73·0319	0·89462	24·1261	5·870°	11·070°
V.	15·3065	84·6935	0·85675	13·1139	3·232°	11·213°

In this case also the increase of specific rotation experienced is small; but the rate of increase on the addition of successive quantities of wood-spirit is not uniform, being represented by a curve strongly marked at first, but afterwards much less so. Mixtures I., III., V. give the equation

$$[a]_D = 8·418 + 0·062466 \, q - 0·00034786 \, q^2,$$

in which constant A agrees fairly well with the specific rotation of pure ethyl tartrate. Taking only the more dilute solutions III., IV., V. into account, $A = 10·25$, showing at once a marked deviation from the true value, 8·31.

(c.) *Mixtures with Water.*

Three mixtures were prepared, of which I. and III. were examined in the Wild's polariscope, with the 219·79 millimetre tube, and II. in Mitscherlich's instrument, with the 219·90 millimetre tube :—

No. of Mixture.	Tartrate of Ethyl p.	Water q.	d.	c.	a.	$[a]_D$.
I.	69·6867	30·3133	1·15079	80·1948	24·673°	14·001°
II.	39·8205	60·1795	1·08841	43·3412	19·271°	20·220°
III.	13·8864	86·1136	1·02921	14·2921	7·916°	25·200°

The rise in specific rotation here observed in ethyl tartrate on the addition of increasing volumes of water, is almost proportional to the amounts added. Adopting, therefore, the formula $[a] = A + Bq$, we have,

From Mixtures I. and II. $A = 7·689$ $B = + 0·20823$
    ,,   ,,   II. ,, III. $= 8·664$ $= + 0·19203$
    ,,   ,,   I. ,, III. $= 7·917$ $= + 0·20070$

and taking the means,

$$[a]_D = 8·090 + 0·20032\, q.$$

The departure here shown by constant A from the specific rotation of pure ethyl tartrate (8·31) is explicable by the fact that water causes a gradual decomposition of the substance, and thereby reduces the rotatory power. This effect was apparent in the marked decrease in the angles of rotation when the same solutions were again observed after standing for forty-eight hours. Solution I. now showed a decrease of 0·028°, II. a decrease of 0·113°, and III. a decrease of 0·166°. But even smaller differences would affect the calculation of the formulæ very considerably.

Fig. 18 is a graphic representation of the foregoing results.

Fig. 18.

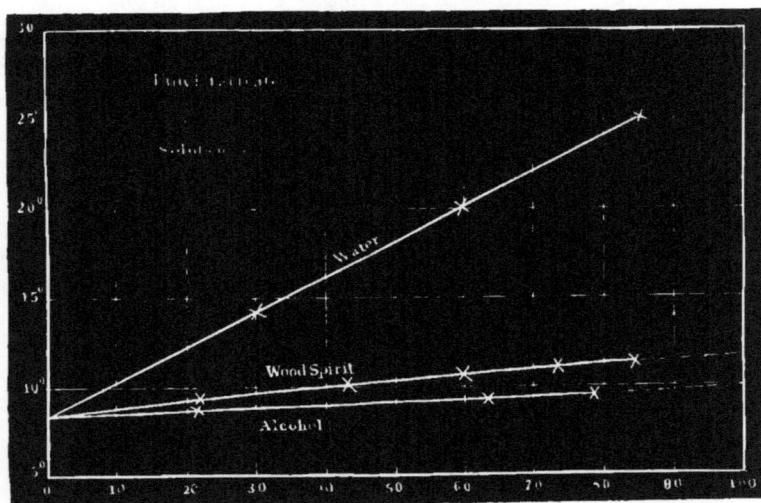

The extent to which the observational results agree with the results furnished by the interpolation formulæ of the three solvents viz. :—

For Alcohol $\quad[a]_D = 8 \cdot 409 + 0 \cdot 018667\, q$
,, Wood-Spirit $[a]_D = 8 \cdot 418 + 0 \cdot 062466\, q - 0 \cdot 00034786\, q^2$
,, Water $\quad[a]_D = 8 \cdot 090 + 0 \cdot 20032\, q$

is shown in the table annexed :—

Solvent.	No. of Mixture.	$q.$	$[a]_D$ Observed.	$[a]$ Calculated.	Difference.
Alcohol	I.	22·0226	8·780°	8·820°	+ 0·040
	II.	64·2634	9·618°	9·609°	− 0·009
	III.	77·6703	9·846°	9·859°	+ 0·013
Wood Spirit	II.	43·3473	10·411°	10·472°	+ 0·061
	IV.	73·0319	11·070°	11·124°	+ 0·054
Water	I.	30·3133	14·001°	14·162°	+ 0·161
	II.	60·1795	20·220°	20·145°	− 0·075
	III.	86·1136	25·200°	25·340°	+ 0·140

INFERENCES.

§ 34. The results of the investigations detailed in the previous sections may be stated as follows :—

1. *The Rate of Change in the Specific Rotation of an Active Substance, when progressively diluted with some Inactive Liquid, is gradual throughout.*—The nature of the change, whether an increase or decrease, will depend on the nature of the active substance : thus, oil of turpentine and ethyl tartrate, in whatever liquids dissolved, always exhibit an increase, whilst nicotine and camphor (for which see § 36), on the other hand, exhibit a decrease in the amount of their specific rotation. Moreover, the rates of variation produced by increasing quantities of different solvents with the same active substance are very different, so that if we represent them graphically we shall have a series of curves radiating from the common point which represents the rotation of the pure substance.

Thus the more dilute any solution of an active substance, the more will its specific rotation differ from that of the absolute substance; and the total amount of possible variation may be shown by putting in the interpolation-formulæ the limiting values $q = 0$ (absolute substance), and $q = 100$ (maximum of dilution). Taking the substances we have been investigating and applying this process, we obtain the following results :—

Active Substance.	Solvent.	$[a]_D$ for Pure Substance $q = 0$.	$[a]_D$ for Maximum of Dilution $q = 100$.	Difference.
Left-handed Oil of Turpentine	Alcohol	36·97°	38·79°	+ 1·82°
	Benzene	36·97°	39·79°	+ 2·82°
	Acetic Acid	36·89°	40·72°	+ 3·83°
Right-handed Oil of Turpentine	Alcohol	14·17°	15·35°	+ 1·18°
Nicotine (Lævo-rotatory)	Alcohol	160·83°	138·59°	− 22·24°
	Water	161·29°	74·13°	− 87·16°
Tartrate of Ethyl (Dextro-rotatory)	Alcohol	8·27°	10·19°	+ 1·92°
	Wood-Spirit	8·42°	11·19°	+ 2·77°
	Water	8·09°	28·12°	+20·03°

From this it will be seen that the amounts of variation of the specific rotation of an active substance differ widely for different solvent media.

2. *The True Rotatory Power of an Active Substance can be calculated from Observations on a Number of its Solutions.*—The degree of exactitude attainable varies for each substance, and is dependent on the following conditions :—(a) On the general extent of the variation produced by the inactive liquid upon the rotation of the substance. The larger the scale, the more unfavourable will be the elements of the calculation (as in the case of nicotine). (b) On the law of dependence of the variations on the increasing percentages of the solvent present, in accordance with which such variations must be represented by a straight or a more or less curved line. (c) On the strength of the solutions employed. The higher the degree of concentration, the greater the possible exactitude of the calculation. The above examples show that where the formula $[a] = A + B q$ applies, the constant A approximates with sufficient closeness (i.e., within a few tenths of a degree) to the true specific rotation of the absolute substance even if the most concentrated solution

contains only about 50 per cent. of the active substance. On the other hand, in cases where it is necessary to use the formula $[a] = A + B\,q + C\,q^2$, divergences of over 1° will occur whenever solutions containing less than about 80 per cent. of the active substance are employed.

3. *In calculating the True Specific Rotation of a Substance, the same Value is obtained, whatever Inactive Liquid may have been used as Solvent.* —Collecting the various values obtained for A in the substances we have already had under investigation, they appear as follows :—

		Divergence.
I. *Left-handed Oil of Turpentine :*—		
By direct observation	$[a]_D = 37\cdot01°$	—
Calculated from mixtures with Alcohol	$[a]_D = 36\cdot97°$	$-\ 0\cdot04°$
,, ,, ,, ,, Benzene	$[a]_D = 36\cdot97°$	$-\ 0\cdot12°$
,, ,, ,, Acetic Acid	$[a]_D = 36\cdot89°$	$-\ 0\cdot12°$
II. *Right-handed Oil of Turpentine :*—		
By direct observation	$[a]_D = 14\cdot15°$	—
Calculated from mixtures with Alcohol	$[a]_D = 14\cdot17°$	$+\ 0\cdot02°$
III. *Nicotine* (left-rotating) :—		
By direct observation	$[a]_D = 161\cdot55°$	—
Calculated from mixtures with Alcohol	$[a]_{D|} = 160\cdot83°$	$-\ 0\cdot72°$
,, ,, ,, ,, Water	$[a]_D = 161\cdot29°$	$-\ 0\cdot26°$
IV. *Ethyl tartrate* (right-rotating) :—		
By direct observation	$[a]_D = 8\cdot31°$	—
Calculated from mixtures with Alcohol ..	$[a]_D = 8\cdot27°$	$-\ 0\cdot04°$
,, ,, ,, ,, Wood-spirit ..	$[a]_D = 8\cdot42°$	$+\ 0\cdot11°$
,, ,, ,, ,, Water	$[a]_D = 8\cdot09°$	$-\ 0\cdot22°$

The differences between the values are so small as to be clearly within the limits of experimental error.

4. *In making Comparisons between Solid Bodies, in respect to their Rotatory Powers, only those Values should be used which hold good for the Absolute Substances, that is, only the Constants A.*—If we employ the modified values obtained from solutions in inactive liquids, we shall find that any relations will become less apparent the more dilute the solutions are from which the values have been derived.

§ 35. The foregoing results point to the method to be employed in determining the true specific rotation of active solids. The most important point to be observed in the process is the employment of the most concentrated solutions attainable, so that, since it is of no consequence otherwise what inactive liquid be employed as solvent, we ought to select that one which best satisfies this condition. Having fixed upon the proper liquid, we must then prepare at least three solutions of different strengths and ascertain their respective rotatory powers. If we now proceed to represent graphically the relation between the values obtained for specific rotation [a], and the percentage of solvent present q, we shall obtain either a straight line or a curve. In the former case, where the three points, representing the three observations, lie in a straight line, [a] being simply proportional to q, the equation [a] = A + B q applies, and the value of the constant A calculated from this equation, will be the specific rotation of the absolute substance. Should the middle one of the three points, however, diverge to either side of the line joining the other two, we have then to deal with a curve, and in this case must proceed to extend our observations over a whole series of solutions, so as to make the data for the construction of the curve as full as possible, using some appropriate interpolation-formula ([a] = A + B q + C q², or some other such) for calculation. By the graphic method alone, indeed, an approximate value for the specific rotation of the absolute substance may be arrived at, by simply prolonging the straight or curved line so obtained till the abscissa q = 0.

That values obtained by such extrapolations must be used with caution is self-evident. For greater accuracy, we should never omit to repeat the experiments with other solvents, and, should the values obtained for constant A agree sufficiently well, the mean of the whole may then be taken as the true value of the specific rotation; if otherwise, the results should be rejected altogether.

The imperfect solubility of many active substances is a serious obstacle to the determination of their true specific rotation. As the foregoing experiments show, it is only in cases where solutions containing at least 50 per cent. of active substance can be prepared, and where the rotation-curve does not depart too considerably from a straight line, that anything like trustworthy numbers can be obtained; so that when we have to do with sparingly soluble bodies, we can have no hope of arriving at any knowledge of the rotatory powers of the absolute substances.

§ 36. By the method above described, the author has endeavoured to determine the rotatory power of a solid substance, choosing for the purpose common camphor.

The camphor employed for this purpose was first purified by distillation from a retort with short, wide neck. In this way, on the first application of heat, oily drops, which did not solidify, separated out, and were put aside. The melting-point of the purified material was 175° Cent., the boiling-point 204° Cent. As solvents, a number of liquids in the purest possible condition were employed, viz., acetic acid, acetic ether, monochlor-acetic ether, benzene, dimethyl-aniline, wood-spirit, and alcohol. The observations were all made with the Wild's polariscope.

Solvent.	No. of Solution.	In 100 parts of Solution.		Specific gravity at 20° d.	Rotation for $l = 2\cdot1979$ dcm at 20°= a.	$[a]_D$.
		Camphor p.	Solvent q.			
Acetic Acid	I.	65·2519	34·7481	0·98983	72·117°	50·801°
	II.	39·7183	60·2817	1·01128	41·652°	47·181°
	III.	15·8819	84·1181	1·03389	15·887°	44·021°
Acetic Ether	I.	53·7260	46·2740	0·93269	58·492°	53·109°
	II.	34·5489	65·4511	0·91987	36·520°	52·283°
	III.	14·9221	85·0779	0·90686	15·290°	51·408°
Monochlor-acetic Ether	I.	54·2184	45·7816	1·04206	65·356°	52·631°
	II.	31·3990	68·6010	1·08670	38·340°	51·123°
	III.	14·2332	85·7668	1·12243	17·543°	49·961°
Benzene	I.	63·1250	36·8750	0·93067	63·575°	49·236°
	II.	49·6359	50·3641	0·91920	47·097°	46·966°
	III.	24·3169	75·6831	0·89910	20·638°	42·948°
Dimethyl-aniline	I.	57·1519	42·8481	0·95997	59·533°	49·370°
	II.	36·0428	63·9572	0·95914	35·151°	46·263°
	III.	15·1028	84·8972	0·95813	13·708°	43·101°
Wood-spirit	I.	49·3866	50·6134	0·88093	46·840°	48·996°
	II.	30·3154	69·6846	0·85318	26·820°	47·179°
	III.	11·2590	88·7410	0·82700	9·382°	45·844°
Alcohol	I.	54·7281	45·2719	0·88021	50·634°	47·823°
	II.	49·8142	50·1858	0·87194	44·806°	46·934°
	III.	30·1620	69·8380	0·84031	25·013°	44·901°
	IV.	15·0920	84·9080	0·81752	11·840°	43·661°
	V.	9·6883	90·3117	0·80943	7·378°	42·806°

In all the solutions the specific rotation experienced a decrease with the increase of inactive substance q, but at a rate varying widely, and depending on the nature of the latter.

As the diagram, Fig. 19, shows, the rate of decrease in each case

Fig. 19.

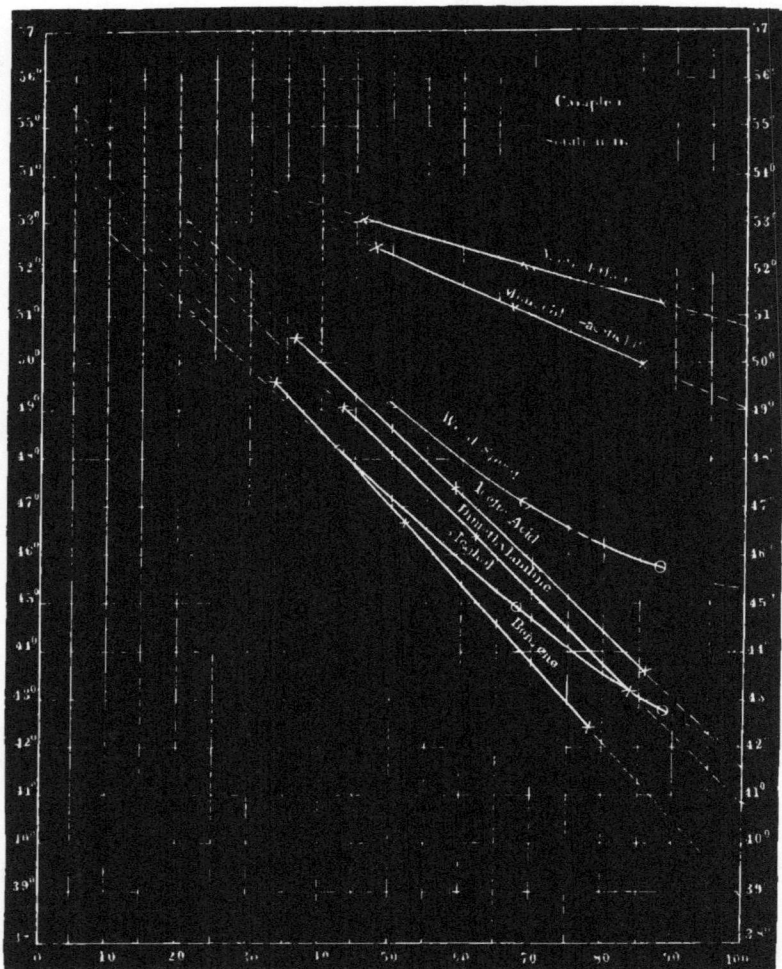

is expressed pretty closely by a straight line when acetic acid, acetic ether, monochlor-acetic ether, benzene, or dimethyl-aniline is used as solvent, and hence for these substances the formula $[a] = A + Bq$

may be used. With wood-spirit and alcohol, on the contrary, the deviations from a straight line are too considerable, and in these cases the

Solvent.	1 $[\alpha] = A - Bq$			2	3		
	Calculated from Solution	$A.$	$B.$	Mean.	Solution.	$[\alpha]$ as Calculated.	Diff. from $[\alpha]$ as Observed
Acetic Acid	I. and II.	55·73	0·1418	$[\alpha]_D = 55·49 - 0·1372\,q$	I.	50·72	− 0·08
	II. ,, III.	55·17	0·1326		II.	47·22	+ 0·04
	I. ,, III.	55·58	0·1373		III.	43·95	− 0·07
Acetic Ether	I. and II.	55·11	0·04307	$[\alpha]_D = 55·15 - 0·04383\,q$	I.	53·12	+ 0·01
	II. ,, III.	55·21	0·04458		II.	52·28	0·00
	I. ,, III.	55·14	0·04384		III.	51·41	0·00
Mono-chlor-acetic Ether	I. and II.	55·65	0·06608	$[\alpha]_D = 55·70 - 0·06685\,q$	I.	52·64	+ 0·01
	II. ,, III.	55·77	0·06769		II.	51·12	0·00
	I. ,, III.	55·69	0·06677		III.	49·97	+ 0·01
Benzene	I. and II.	55·45	0·1683	$[\alpha]_D = 55·21 - 0·1630\,q$	I.	49·19	− 0·05
	II. ,, III.	54·96	0·1587		II.	47·00	+ 0·03
	I. ,, III.	55·21	0·1620		III.	42·87	− 0·08
Dimethyl-aniline	I. and II.	55·68	0·1472	$[\alpha]_D = 55·78 - 0·1491\,q$	I.	49·40	+ 0·03
	II. ,, III.	55·92	0·1510		II.	46·25	− 0·01
	I. ,, III.	55·76	0·1491		III.	43·13	+ 0·03
Wood-spirit	I. II. III.			$[\alpha]_D = 56·15 - 0·1749\,q + 0·0006617q^2$			
Alcohol	I. III. V.			$[\alpha]_D = 54·38 - 0·1614\,q + 0·0003690\,q^{2\,a}$	II.	47·21	+ 0·28
					IV.	43·33	− 0·33

formula $[\alpha] = A + Bq + Cq^2$ must be taken as the basis of calculation. The table annexed shows (1) the values of constants A and B,

[1] The formula $[\alpha] = A + \dfrac{Bq}{C + q}$ gives $[\alpha]_D = 54·83 - \dfrac{42·879\,q}{231·82 + q}$.

calculated from the several solutions; (2) the derived interpolation-formulæ obtained by putting in the mean values ; (3) the specific rotation of the solutions employed calculated from these equations, and the differences between these and the observed values, as given in the preceding table (p. 85).

Comparing now with each other the values for constant A derived from different solvents, we find an agreement which, in view of the large amount of extrapolation from $q = 0$ onwards, varying from 35 to 50 per cent., must be regarded as very close, and the mean of their values may accordingly be taken as the true specific rotation of pure camphor. The values for constant B, on the contrary, exhibit very marked variations. Calculating the specific rotation from the same formulæ, by putting in limiting values $q = 0$ and $q = 100$, we obtain the following as the range of variation which the rotatory power of camphor may undergo under the influence of various inactive liquids employed as solvents.

Solvent.	$[a]_D$ for $q = 0$ Pure Substance.	$[a]_D$ for $q = 100$ Infinite Dilution.	Total Variation.
Acetic Acid	55·5°	41·8°	13·7°
Acetic Ether	55·2°	50·8°	4·4°
Monochlor-acetic Ether	55·7°	49·0°	6·7°
Benzene	55·2°	38·9°	16·3°
Dimethyl-aniline ..	55·8°	40·9°	14·9°
Wood-spirit	56·2°	45·3°	10·9°
Alcohol	54·4°	41·9°	12·5°

Lastly, taking the mean of the values obtained for the pure substance, we have as the true specific rotation A_D of camphor, at a temperature of 20° Cent.,

$$A_D = 55\cdot4°,$$

with a mean variation of $\pm$ 0·4°.

§ 37. In the same way the true rotation-constant of cane-sugar was determined by Tollens,[1] and simultaneously by Schmitz,[2] water being the only solvent used. In the case of sugar, the specific rotation

---

[1] Tollens : *Ber. d. deutsch. chem. Gesell.* 1877, 1403.
[2] Schmitz : *Idem.*, 1877, 1414 ; also, *Zeitsch. d. Ver. für Rübenzuckerind.* 1878, 48.

increases with dilution, or, conversely, decreases with increase of concentration, but the variations are small. Tollens examined seventeen solutions, of which the most concentrated, with 69·2144 per cent. by weight of sugar, gave a specific rotation $[a]_D = 65·490°$, and the most dilute with 3·8202 per cent., gave $[a]_D = 66·803°$. From the experimental results were derived the following interpolation-formulæ for the calculation of the specific rotation of any given solution, by putting in values for p, percentage of sugar, and q, percentage of water respectively:—

(*a*) *For strong Solutions, containing from 18 to 69 per cent. of Sugar.*

I. $[a]_D = 66·386 + 0·015035\,p - 0·0003986\,p^2.$
II. $[a]_D = 63·904 + 0·064686\,q - 0·0003986\,q^2.$

(*b*) *For weak Solutions containing less than 18 per cent.*

III. $[a]_D = 66·810 - 0·015553\,p - 0·000052462\,p^2.$
IV. $[a]_D = 64·730 + 0·026045\,q - 0·000052462\,q^2.$

Putting $p = 100$ in formula I., or $q = 0$ in II., we obtain the specific rotation of anhydrous sugar,

$$A_D = 63·90°.$$

Schmitz examined the following eight solutions:—

No. of Solution.	In 100 parts by Weight of Solution.		Sp. Gr. at 20° Cent. d.	Concentration c = p d.	Rotation, a for 100 mm. at 20° Cent.	$[a]_D$.
	Sugar p.	Water q.				
I.	64·9775	35·0225	1·31650	85·5432	56·134°	65·620
II.	54·9643	45·0357	1·25732	69·1076	45·533°	65·919
III.	39·9777	60·0223	1·17664	47·0392	31·174°	66·272
IV.	25·0019	74·9981	1·10367	27·5938	18·335°	66·441
V.	16·9926	83·0074	1·06777	18·1442	12·064°	66·488
VI.	9·9997	90·0003	1·03820	10·3817	6·912°	66·574
VII.	4·9975	95·0025	1·01787	5·0868	3·388°	66·609
VIII.	1·9986	98·0014	1·00607	2·0107	1·343°	66·802

The equation with reference to q, derived from these observations, stands—

$$[a]_D = 64·156 + 0·051596\,q - 0·00028052\,q^2;$$

according to which, the rotation-constant for pure sugar at a temperature of 20° Cent. is

$$A_D = 64·16°,$$

which only differs from that obtained by Tollens by 0·26°.[1]
Tollens[2] has attempted, as Biot[3] had already done, to determine the
rotatory power of anhydrous sugar directly, by employing plates
cast from the melted substance. In this way he obtained a value
considerably below that of the calculated specific rotation, viz.,
$[a]_D = 46·9°$. This is not surprising, as under the influence of
heat sugar undergoes various important changes, as indicated by the
assumption of a yellow coloration, as well as a strong reducing action
on cupric salts. Even after solution in water, such a sugar exhibits a
notably smaller rotatory power than before fusion, and the decrease
is greater in proportion to the length of time during which it
was kept fused (Hesse[4]). Probably, in such cases a formation of
inactive glucose takes place.

§ 38. The true specific rotation of right-handed glucose (grape-
sugar) has been determined by Tollens[5] both for the hydrate $C_6H_{12}O_6$
+ H_2O, and the anhydrous substance.

In the subjoined table are given the solutions employed (with
p per cent. by weight of glucose), along with the values of $[a]_D$
observed in each case, and side by side the values calculated from the
interpolation-formulæ given below.

No. of Solution.	Hydrate of Glucose $C_6H_{12}O_6 + H_2O$.			Anhydrous Glucose $C_6H_{12}O_6$.		
	p.	$[a]_D$ Observed at 20° C.	$[a]_D$ Calculated.	p.	$[a]_D$ Observed at 20° C.	$[a]_D$ Calculated.
I.	8·4501	48·50°	48·08°	7·6819	53·35°	52·89°
II.	10·2216	48·18°	48·12°	9·2924	53·00°	52·94°
III.	10·3083	47·99°	48·13°	9·3712	52·79°	52·94°
IV.	11·0675	48·20°	48·14°	10·0614	53·02°	52·96°
V.	11·6907	48·16°	48·16°	10·6279	52·97°	52·98°
VI.	14·2459	48·34°	48·23°	12·9508	53·17°	53·05°
VII.	20·4832	48·55°	48·41°	18·6211	53·40°	53·25°
VIII.	34·7753	48·76°	48·94°	31·6139	53·64°	53·83°
IX.	44·8175	49·41°	49·40°	40·7432	54·35°	54·34°
X.	48·3870	49·70°	49·59°	43·9883	54·67°	54·54°
XI.	53·7534	49·66°	49·88°	48·8667	54·62°	54·87°
XII.	58·3254	50·15°	50·15°	53·0231	55·16°	55·17°
XIII.	90·8722	52·45°	52·54°	82·6111	57·70°	57·80°

[1] This trifling difference is partly explained by Tollens having taken the specific
gravity of the sugar solutions at a temperature of 17·5 Cent., whilst Schmitz took it at
20° Cent. The angles of rotation were observed by both at a temperature of 20° Cent.
[2] Tollens : *Ber. der deutsch. chem. Gesell.* 1877, 1413.
[3] Biot : *Mém. de l'Acad.* 13, 130.   [4] Hesse: *Liebig's Ann.* 192, 167.
[5] Tollens : *Ber. der deutsch. chem. Gesell.* 1876, 1531.

(As the molecular weights of $C_6 H_{12} O_6 + H_2 O$ and $C_6 H_{12} O_6$, and, therefore, also the corresponding values of p for the two substances, are in the ratio 198 : 180, or 11 : 10, the specific rotation of the hydrate and the anhydride must stand to each other in the inverse ratio 10 : 11.)

As will be seen, the specific rotation increases with increase of concentration, or, conversely, decreases with increased dilution. Glucose thus exhibits a behaviour the reverse of that of cane-sugar.

For glucose hydrate, the experiments gave the formula
$$[a]_D = 47\cdot 925 + 0\cdot 015534\, p + 0\cdot 0003883\, p^2;$$
or, with reference to q, the percentage of water present in the solution :
$$[a]_D = 53\cdot 362 - 0\cdot 093194\, q + 0\cdot 0003883\, q^2.$$

For anhydrous glucose, by raising the preceding values by one-tenth, we get the equations
$$[a]_D = 52\cdot 718 + 0\cdot 017087\, p + 0\cdot 0004271\, p^2,$$
$$[a]_D = 58\cdot 698 - 0\cdot 10251\, q \quad + 0\cdot 0004271\, q^2.$$

Lastly, from the foregoing formulæ we get, as the true values of the rotation-constants,

for $C_6 H_{12} O_6 + H_2 O$ $A_D = 53\cdot 36°,$

,, $C_6 H_{12} O_6$ $A_D = 58\cdot 70°.$

§ 39. Camphor, cane-sugar, and glucose[1] are the only solids, up to the present time, the direct specific rotations of which have been accurately determined. Numerous investigations, indeed, have been published as to the variation of the specific rotation in a large number of substances, but the observers have, as a rule, employed only solutions containing small percentages of active substance, so that only a few points have been determined, and that at the outer end of the respective curves, where the variation of rotatory power is at its maximum. From results of this kind the value for A cannot be determined. But, indeed, this were impossible at any rate, from the fact that neither the percentage composition by weight nor the density of the solutions is stated, but merely their concentration, i.e., the number of grammes of substance in 100 cubic centimetres.

For the determination of individual values of $[a]$ this is enough; but, as before explained (§ 25), it is altogether insufficient for determining the value of A.

Nevertheless, in chemical writings we still find many specific

[1] Arndtsen (see § 19) has determined for tartaric acid the relation between specific rotation and percentage of water, and hence deduced the formula $[a]_D = 1\cdot 95 + 0\cdot 1303\, q$, which gives the value for the pure substance $A_D = 1\cdot 95$. However, as the number of solutions observed was but small, a verification of the constants given is desirable.

rotation-data based on the old view that the value is constant, and may be obtained by observation of a single solution of any optically-active substance. Accordingly, neither the weight-percentage of active substance nor the concentration is stated, and in most cases no reference is made to the ray with which the observations were made. For example, in many text-books the specific rotation of tartaric acid is given briefly thus: $[a] = + 9.6°$; whereas, as we have seen from the table already given (§ 19), the specific rotation in solutions containing 10 to 90 per cent. of this substance varies for the yellow ray D from $3.25°$ to $13.68°$; for the green ray b, from $1.08°$ to $16.40°$; and for the blue ray e, as much as from $- 6.51°$ to $+ 18.64°$. Again, we find the specific rotation of cane-sugar given as $[a] = + 73°$ to $74°$, without mention of the fact that this is the value for the *transition tint*, although for the neighbouring yellow ray D the value for solutions holding, say 25 per cent. of sugar, is, as shown in § 37, only $[a]_D = 66.44°$, and the value ranges, moreover, for the same ray from $64°$ to $67°$, according to the degree of concentration of the solutions employed. That data of this sort, as remarked in § 23, are utterly worthless, must now be obvious after what has been said.

§ **40.** *The Specific Rotation exhibited by an Active Substance in a Solution of given Composition is Constant, and hence can be employed as a Distinguishing Characteristic of the Substance.* But that it may possess this value, it is indispensably necessary that, along with the value of $[a]$, the following data should be stated:—

1. The ray with which the observations have been made—the index-letter being placed after the bracket.

2. The description of solvent used (as water, alcohol, &c.; in the case of the latter, either the per cent. composition or specific gravity being stated).

3. The proportion of active substance in 100 parts by weight of solution (per cent. composition p), or else the number of grammes in 100 cubic centimetres of solution (the concentration c).

4. The temperature t, of the solution when the angle of rotation was observed. The determination of the specific gravity of the solution, or the adjustment of the volume in a graduated measure, must be done at this same temperature.

5. The direction of rotation (dextro-rotatory $+$, lævo-rotatory $-$).

These data may be recorded as follows :—

Cane-sugar (solution in water, $p = 16\cdot993$, $t = 2C^\circ$), $[a]_D = + 66\cdot49^\circ$.

Ordinary camphor (solution in alcohol of specific gravity $0\cdot796$, at 20°, $p = 15\cdot092$, $t = 20^\circ$), $[a]_D = + 43\cdot66^\circ$.

Santonin (solution in alcohol of 97 per cent. by volume, $c = 2$, $t = 15^\circ$), $[a]_D = - 174\cdot00$.

Quinine hydrate, $C_{20} H_{24} N_2 O_2 + 3 H_2 O$ (solution in alcohol of 80 per cent. by volume, $c = 1$, $t = 15^\circ$), $[a]_D = - 158\cdot63^\circ$.

,,   (solution in alcohol of 80 per cent. by volume, $c = 6$, $t = 15^\circ$), $[a]_D = - 114\cdot92^\circ$.

,,   (solution in a mixture of 2 volumes chloroform + 1 volume alcohol of 97 per cent. by volume, $c = 5$, $t = 15^\circ$), $[a]_D = - 140\cdot50^\circ$.

In this way Hesse[1] has estimated the specific rotation of a great number of optically-active substances dissolved in different liquids, thus supplying data which, as constant marks of the several substances, are of great value in determining the identity or purity of different preparations.

In all cases it is advisable to record the per cent. composition p, rather than the concentration c of the solutions, and so to calculate the specific rotation by the formula $[a] = \dfrac{a \cdot 100}{l \cdot d \cdot p}$, which, moreover, renders it necessary to determine the specific gravity of the solutions. The resulting values, at least in cases where several solutions have been observed, can then be used in determining the specific rotation of the absolute substance. This, as frequently already mentioned, is not the case when only the concentration is determined by means of a graduated vessel, and the specific rotation calculated by the otherwise more convenient formula $[a] = \dfrac{a \cdot 100}{l \cdot c}$.

§ 41. *Molecular Rotation.*—Specific rotation $[a]$ is frequently referred to by Biot under the name of molecular rotation, indicating, as observed (§ 10), that the rotatory power of liquids is a property resident in the molecules.

---

[1] Hesse: *Liebig's Ann.* **176**, 89, 189; **178**, 260; **182**, 128. Hesse indicates the number of grammes of active substance in 100 cubic centimetres solution by p. It is, however, much better to employ c for this purpose (concentration) and let p denote the true per cent. composition (or number of parts by weight of active substance in 100 parts by weight of solution).

But this expression has been applied by Wilhelmy,[1] Hoppe-Seyler,[2] and more recently by Krecke,[3] to a different value, viz., to the number obtained by multiplying the specific rotation of any substance into its molecular weight P. The values thus obtained being inconveniently large, Krecke has proposed to divide them uniformly by 100. The molecular rotation $[M]$ of a given substance then appears as

$$[M] = \frac{P\,[a]}{100},$$

which expresses the angles of rotation produced by passage of the ray through layers 1 millimetre thick of substances when the unit-volumes contain the same number of molecules.

It has been attempted, by means of this formula, to discover relations between an active substance and its derivatives in respect to rotatory power, and the existence of certain multiple relations has been supposed to have been detected (Krecke,[3] Landolt[4]). But the observations on which these comparisons were based were made, as was formerly the practice, with a single solution of each substance, whereas we have seen (§ 34) that the constant A of the pure substance should alone have been employed. Before, therefore, the hypothetical so-called law of multiple rotation is ripe for discussion a much more extensive series of experiments is necessary.

[1] Wilhelmy : *Pogg. Ann.* 81, 527.
[2] Hoppe-Seyler : *Journ. für prakt. Chem.* [1], 89, 273.
[3] Krecke : *Journ. für prakt. Chem.* [2] 5, 6.
[4] Landolt : *Ber. der deutsch. chem. Gesell.* 1873, 1073.

V.

PROCESS OF DETERMINING SPECIFIC ROTATION.

—◦◦◦—

§ 42. In calculating specific rotations by the formulæ given in §§ 20, 21, viz.,

I. (For liquids) $\quad [a] = \dfrac{100 \cdot a}{L \cdot d}$,

II. (For solutions) $\quad [a] = \dfrac{10^{4} \cdot a}{L \cdot p \cdot d}$,

III. (,, ,,) $\quad [a] = \dfrac{10^{4} \cdot a}{L \cdot c}$,

the following data must be obtained by direct experiment:—

1. The measurement of the angle of rotation a for a given ray.

2. The measurement of the length of the experimental-tube, in millimetres.

3. The weight p of active substance in 100 parts by weight of solution.

4. The specific gravity d of the active liquid.

5. The concentration c—i.e., the number of grammes of active substance in 100 cubic centimetres of solution.

If the object of determining the specific rotation of a solution of a solid substance, is merely to obtain a characteristic of its presence in solution, formula III., based on the knowledge of its concentration c, will suffice. But if, on the contrary, it is desired to ascertain the actual specific rotation of the substance itself, from observations on a number of different solutions, it is necessary (see § 25) to employ formula II., involving a knowledge of the percentage composition, and specific gravity of the several solutions.

A. Determination of the Angle of Rotation.

POLARISCOPIC APPARATUS.

§ 43. *Apparatus for the Qualitative Examination of Rotatory Power.*—To determine merely whether a given substance is or is not optically-active, and, if active, the direction in which the

Fig. 20.

rotation takes place, the instrument here represented (Fig. 20[1]), which is delicate enough to detect even feeble degrees of rotatory power, may be used.[2]

A brass trough a b, of semi-circular section, fitted with a cover c so as to form a tube, carries at the extremity a, in a fixed case, a polarizing Nicol d. In front of the latter is placed the convex lens e, and on the other side of the polarizer at f, a so-called *Soleil double-plate*, formed of two plates, one of dextro-rotatory, the other of lævo-rotatory quartz, fitted vertically together and ground to a uniform thickness of either 3·75 or 7·5 millimetres.

The opposite end of the brass tube holds the movable Nicol g,

---

[1] [f is given apparently out of proper section, representing a front view, whilst the rest of the figure shows a longitudinal section.—D.C.R.]

[2] The instrument shown above is manufactured by F. Schmidt and Haensch, Stallschreiberstrasse 4, Berlin. Instruments on the same optical principle, but of simpler construction (described by C. Neubauer in Fresenius' *Zeitschr. für analyt. Chem.* 16, 213) intended for determining grape-sugar in wine, but equally applicable for all other active substances, are procurable from the Optical Instrument Works of Dr. Steeg and Reuter, Homburg v. d. Höhe.

besides a small Galilean telescope, consisting of an object glass h, and an ocular i. The Nicol is turned by the handle k, which moves round the face of a small graduated disc l, so as to allow the amount of rotation to be determined, at least approximately. The brass trough receives the glass tube p p (the ends of which may be closed by glass plates fixed with brass screw-caps) containing the liquid to be examined. The whole rests on a stand o. As a considerable depth of liquid is requisite for the detection of feeble rotatory power, the brass case is so constructed as to take glass tubes 5 or 6 decimetres in length. It is to this, and the introduction of the Soleil double-plate, that the sensitiveness of this instrument is due.

In using the instrument, the glass tube is at first left out whilst the extremity is directed towards a bright flame, for which purpose the gas-lamp, shown in Fig. 25, will be found best. The eyepiece of the telescope is then adjusted so that the vertical division of the double-plate appears sharply defined. By turning the analyzer g, a certain position will readily be found in which the two halves of the field of vision exhibit a perfectly uniform purplish tint, which the least turn of the Nicol to the right or left changes, one half becoming red, the other blue. Further particulars of this so-called *sensitive tint* will be given later on (§ 78) in speaking of the Soleil saccharimeter. Having thus established perfect uniformity of colour in the two halves of the field of vision, with the index standing at the zero-point on the scale of the analyzer, the glass tube containing the liquid to be tested is laid in the trough, when its optical activity will at once be declared by inequality of tint in the field of vision. To know whether the rotation be right-handed or left-handed, it is requisite, in the first place, to determine in the instrument, once for all, what relative positions the red and blue take up when some substance of known rotatory power, such as a (dextro-rotatory) solution of cane-sugar, is inserted. If the substance under examination shows the colours in the same relative order in which they are shown by the sugar, it likewise is dextro-rotatory ; if the positions are interchanged, it must be lævo-rotatory. Further, with dextro-rotatory substances uniformity of tint in both halves of the field of vision is restored by turning the analyzer to the right, or in the direction of the hands of a watch, and with lævo-rotatory substances, to the left. The position of the index on the graduated disc of the analyzer shows the angle of rotation in each case.

- H

Instead of this instrument any of the forms of polariscope described further on may be used. The advantage of the above instrument lies in its sensitiveness and the facility with which with it the direction of the rotation can be determined.

§ 44. For the accurate measurement of the angle of rotation, a variety of instruments have been devised, which may be divided into two classes, according to their objects :—

1. The so-called *polaristrobometers*, what in England are known as polariscopes,[1] which indicate the amount of rotation in angular measure, and are applicable to *all* optically-active substances.

2. The saccharimeters, which are specially intended for the analysis of solutions of cane-sugar, the angular measurement being replaced in them by an empirical scale.

(a.) *Mitscherlich's Instrument.*[2]

§ 45. This simplest of all forms of polariscope consists, as already stated (§ 5), of a pair of Nicol prisms, placed one at each end of a brass or wooden rod or bar *d*, Fig. 21. The polarizer *a* is provided with a brass case, by means of which it can be turned if required, and then clamped with the small screw *e*. The circular movement of the analyzer *b* is effected with the handle *c*, the angle through which it is revolved being read off on a fixed graduated disc by means of opposite index arms, with or without verniers. The

---

[1] [At this point something requires to be said as to the nomenclature of various polarizing instruments. The word *polariscope*, commonly used in English for any instrument the essential parts of which consist of a polarizer and analyzer, and which may or may not be applicable for showing the rotation phenomena with which this work deals, has no such range of meaning in German, as will be seen by reference to the use of the word in describing a special contrivance, the Savart polariscope forming part of the instrument described in § 49. On the other hand, we have in English no word limited to describe what the Germans call rather clumsily a *polaristrobometer*, and the French a *polarimètre*, a polariscope that is of special form, suited to observe and to measure the rotatory power of substances. If we deemed it advisable to introduce an expression for the purpose, it seems that "rotation polarimeter" would, as nearly as possible, represent the German polaristrobometer ; but since the word polariscope has become so familiar to practical people it seemed better to retain it and make the needed explanation.—D.C.R.]

[2] Mitscherlich : *Lehrbuch der Chem.* 4 Aufl. Bd. 3, 361 (1844).

graduation is in degrees, and the reading is taken in degrees and tenths. The experimental tube f is laid between the prisms, and usually has a length of 2 decimetres (7·6 in.). For increasing the illuminating power and giving a circular field of vision, a small convex lens is inserted in the case of the polarizer.

Fig. 21.

Fig. 22.

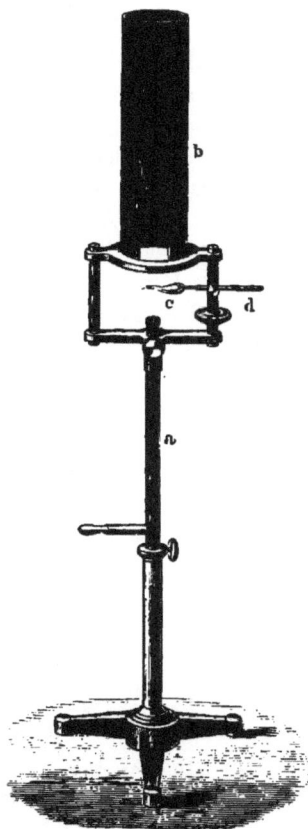

§ 46. In using Mitscherlich's instrument, it will be found best to employ homogeneous yellow sodium light, thus determining the angle of rotation for ray D. To obtain a sodium flame which shall last for some time, the lamp,[1] shown in Fig. 22, may be used, which consists of a vertically adjustable Bunsen burner a, with a sheet-metal chimney, having a side aperture, b. In the movable pillar d is inserted horizontally a small cross-bar, carrying at its extremity a bundle of fine platinum wires, arranged so as to form a small pointed spoon c, the hollow being filled with well-dried common salt; the

[1] Laurent: *Dingler's Polyt. Journ.* 223, 608. This lamp may be obtained of Schmidt and Haensch, Berlin.

spoon is moved forward to the front edge of the flame, and the salt fusing and running to the point, volatilizes, and produces an intense yellow colour. Or the stem *d* may be provided with a small brass revolving collar, having several arms with holes, into which can be fitted platinum wires with beads of salt fused on. When one bead is consumed the next arm is brought round to the flame, and so on. Instead of common salt calcined soda may be used, but this, whilst volatilizing more slowly, has less illuminating power.

In making observations, the instrument should be set up at a distance of an inch or so from the flame, and a black screen placed behind the latter, so as to shut off extraneous light. The room should be darkened too, at least partially, as the observations in general are more satisfactory the darker the place is. The zero-point must first be determined. For this purpose the tube is put in its place either empty or filled with water, and the analyzer

Fig. 23.

set to the position of greatest darkness. If the circular field of vision is at all large, there will not be perfect obscuration over the whole, but merely a vertical dark band, getting lighter towards the sides, as in Fig. 23, and this band must be brought, by backward and forward motions of the analyzer, as nearly as possible into the middle of the field. Repeating the adjustment several times, and taking the mean of the readings on the disc, we get the true zero-point of the instrument. To make the zero of the scale agree, at least approximately, therewith, we set the index against the mark, loosen the clamp *e* (Fig. 21), and rotate the polarizer until the dark band appears in the middle. Usually this correction is made by the instrument-maker himself. As before stated (§§ 4 and 5), there are two positions, 180° apart, at which the analyzing prism gives maximum darkness, and the zero-point of the second, which must lie somewhere about 180° on the scale, should similarly be accurately determined by a few observations.

If the tube, filled with active liquid, be now laid in the instrument, the analyzer having been previously set to zero, the field of vision will again appear bright, and in order to restore the black band it will be necessary to rotate the analyzer to the right in the case of a dextro-rotatory substance—that is, in the direction of the hands of a watch—and in the opposite direction, to the left, when the substance is lævo-rotatory. The same order of phenomena will be observed if we start from a position 180° from the first. In case, however, we do not know beforehand the direction of rotation peculiar to the substance, the following considerations must be borne in mind :—Suppose that the plane of polarization, having originally the direction $A B$ (Fig. 24), is diverted to $C D$ (at an angle of 30° from $A B$) by passing through an optically-active medium, the dark band

Fig. 24.

will then appear when the index stands at 30° or 210°, and the substance may either be dextro-rotatory, 30°, or lœvo-rotatory, 360° − 210° = 150°. In most cases the side on which the smaller amount of deviation occurs is the true direction of the rotation. We cannot, indeed, be so guided when the smaller of the two angles exceeds 90°, which, however, only happens with substances having very high rotatory power, or in using tubes more than 2 decimetres in length. In these cases, the question can easily be decided by examining the liquid in a tube only half the length of that originally used, or by diluting the solution to half its strength. The deviation should then be only half of the original amount, and it is thus easy to discover which is the direction in which this holds. If, for example, darkness now occurs when the analyzer lies in the direction $E F$—that is, at 15° and 195°—the decrease shows the rotation to be dextro-rotatory, since the position measured for lœvo-rotatory power would indicate an increase of rotation from 150° to 165°, which is absurd.

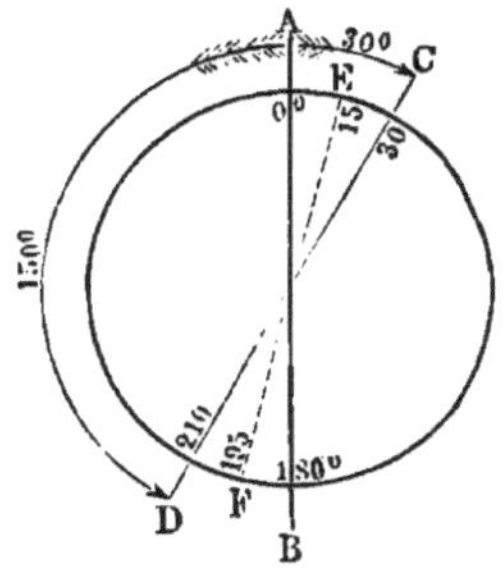

It is well to take observations on the filled tube at both positions, 180° degrees apart, as, owing to defective construction, the Nicol prisms may be somewhat eccentric, causing the observed angles to differ appreciably from each other. Any such source of error is accordingly eliminated by taking the mean of the two readings. Mitscherlich's instruments frequently have two opposite index arms, but as with fairly good graduation the difference between their readings does not amount to $\frac{1}{10}$th degree, it is generally sufficient to use one. The differences between the angles in successive observations usually amount to several tenths of a degree, and the accuracy of the final result will, of course, be greater in proportion as the observations are more numerous. As an example we may give the following results :—

	Half-circle I.		Half-circle II.	
	Empty Tube. (Zero-point).	Filled Tube.	Empty Tube. (Zero-point).	Filled Tube.
	0·2°	16·4°	180·4°	196·3°
	5°	16·2°	6°	3°
	4°	15·9°	3°	6°
	2°	16·0°	5°	5°
	3°	16·2°	6°	6°
Mean	0·32°	16·14°	180·48°	196·46°
Rotation Angle a_D =	15·82°		15·98°	
Mean		15·90°		

Instead of the sodium flame, monochromatic light, obtained by placing a red glass slide in front of an ordinary gas lamp, was used by Biot and Mitscherlich. But in this way observation is rendered much more difficult through defect of brilliancy, besides which, the red light so produced does not correspond to any one distinct ray (see p. 46).

§ 47. When white day or lamp-light is used with Mitscherlich's instrument, the angle of rotation observed is that for mean yellow rays, and, as already stated (§ 18), is denoted by a_j. For this purpose

Fig. 25.

the most suitable form is a gas or petroleum lamp, fitted outside its glass chimney with a metal screen coated inside with white porcelain, and having a side opening (see Fig. 25).

When with an empty tube placed in the instrument the analyzer is set to zero there appears, exactly as with the sodium flame, a dark band with fainter margins, which, as before, must be brought into the middle of the field of vision. If the active liquid is now placed in the tube, the different coloured rays composing this white light will experience different degrees of rotation, so that we shall have the phenomenon of rotatory dispersion. And here we may have two cases :—

1. When the liquid possesses feeble rotatory power and the dispersion is therefore trifling, by turning the analyzer the dark band may be made to reappear with a border of blue on one side and red on the other. Now, if the blue border is to the left of the observer and the red to his right the substance is dextro-rotatory; if *vice versâ*, it is lævo-rotatory ;

and this independently altogether of whether the proper position has been found by turning the analyzer to the right or to the left. It is to be observed, however, that when a Mitscherlich instrument is provided with any kind of Galilean telescope, the foregoing conditions are reversed. The position of the dark band indicates the point of extinction of the yellow rays.

2. When, on the other hand, the rotatory power of the active liquid is high, the dark band appears broad and undefined, or else cannot be brought back by any movement of the analyzer at all. By turning the latter we get merely a succession of colours, produced by the analyzer extinguishing, according to the position of its principal section, certain of the unequally-rotated coloured rays, and allowing the rest to pass on with different intensities, thus producing a succession of colour-mixtures. With solutions in which the angle of rotation for any ray is less than 90°, the sequence of coloured tints, when the analyzer is turned from the initial zero-point, is as follows:—

In Lævo-rotatory Liquids		In Dextro-rotatory Liquids	
By turning the analyzer to		By turning the analyzer to	
The Left.	The Right.	The Left.	The Right.
Yellow	Yellow	Yellow	Yellow
Green	Orange	Orange	Green
Blue	Red	Red	Blue
Red	Blue	Blue	Red
Orange	Green	Green	Orange
Yellow	Yellow	Yellow	Yellow

As a point of reference for the analyzer, that position is chosen where the transition from blue to red stands exactly in the middle of the field of vision. This being the point of extinction of the yellow rays, the observed angles will be a_j.

With substances of still higher rotatory and dispersive powers, intermediate tints make their appearance; for example, between the blue and the red a reddish-violet, which, on the least touch of the analyzer, passes into one or the other. This is known as the *sensitive* or *transition tint*, and appears when the position of the analyzing Nicol is exactly such as to bar the passage of the mean yellow rays. This position also gives the angle a_j.

For the relation between the angles of rotation a_j and a_n see § 18.

Observations taken with white light cannot be made so exactly as those with the sodium flame; the former light is therefore only employed when the latter is not available.

. §48. *Mitscherlich's larger Instrument for Observations at Constant Temperature.*—In exact researches it is requisite, as already stated (§ 22), that the temperature should not only be known, but be constant during the period of observation. This can only be effected by surrounding the tube with water. Moreover, in examining substances of feeble rotatory power, it is necessary to employ tubes of considerable length, sometimes a whole metre long, to obtain rotation-angles of sufficient magnitude. Fig. 26 represents an instrument fulfilling these conditions, constructed at the works of Dr. Meyerstein, of

Fig. 26.

Göttingen, and in use in the chemical laboratory of the Polytechnic School at Aachen.

Starting from the end next the light, the instrument consists of the following parts, resting loosely upon a frame formed of two strong iron bars $Q\,Q$:—

1. A fixed tube A, containing the polarizing Nicol, a convex lens of long focus, and a diaphragm with a square aperture of 5 millimetres side. Affixed to the same support as the tube is a circular dark screen, a, to shut off extraneous light.

2. A glass bottle with parallel walls B,[1] filled with bichromate

_____
[1] May be obtained of Dr. J. G. Hofmann, 29, Rue Bertrand, Paris.

of potash solution, the object of which is to free the transmitted sodium rays from any admixture of blue or green light. This is important in the case of solutions of high dispersive power, as, without it, other tints make their appearance when the Nicols are crossed, and interfere with the sharp recognition of the dark band.

3. A sheet-metal case, CC, through which the solution tube passes, the ends passing water-tight through india-rubber corks. The tube in Fig. 26 is one metre long ; shorter tubes, of course, need cases of proportionate length. The case is filled with water, which is then raised to the desired temperature, usually 20° Cent., by moving about in it a hot bar, K. For higher temperatures a Bunsen lamp with a row of burners, JJ, must be used.

4. A support, D, carrying a tube containing the analyzing Nicol, which, together with the graduated disc attached to it, is susceptible of movement round a common axis. This movement is communicated by the screw G, working in the toothed rim of the disc E. A small Galilean telescope, F, is fitted to the tube, the eye-piece of which must be so adjusted that the aperture in the diaphragm of the polarizer appears sharply defined. The support also carries two fixed verniers, and the divisions can be read off by the light of a small gas-jet, H.

The sodium flame for the observations is obtained by means of the blow-pipe L, arranged vertically with chimney, M, over it, and connected by means of india-rubber tubing with the bellows P. Over the nozzle of the burner, and projecting from the support N, is fixed a ring of platinum wire, which, when dipped in fused soda, imparts to the whole mass of flame an intense yellow, thus producing a strong light, essential for observations with great lengths of liquid, since the slightest opacity will, in such cases, often obscure the field so much as entirely to frustrate the experiment.

The rotation-angles are determined in the same way as with Mitscherlich's smaller instrument. The analyzer is turned, by means of the milled head G, until the dark band appears exactly in the middle of the field of vision formed by the square aperture of the diaphragm, or, in other words, until the light spaces on each side of the dark band appear of equal width. The tube is first introduced empty, and the two zero-points determined, after which it is filled with the active liquid, the metal case being turned up on one end for the purpose. When shorter tubes are used, the intervention of extraneous light must be prevented by enclosing the course of the rays after they

leave the tube with a paste-board cylinder, always taking care to
have the polarizer and analyzer properly placed at the ends. The
supports A and D can be slid along and screwed to the cross-bars
in other positions as required.

In all exact observations it is necessary to determine the zero-
point afresh with each observation, as changes in the temperature of
the place as well as differences of tension in the metallic screw-joints
have an appreciable affect on the readings.

The following values, obtained with a 10 per cent. solution of
cane-sugar, are given as a working example:—The zero-points were
found at about 20° and 200° on the right and left sides respectively
of 0° and 180°. The vernier could be read accurately to 0·1° and
approximately to 0·01°. The temperature of the solution was 20° Cent.

OBSERVATION-SERIES a.				OBSERVATION-SERIES b.			
Length of Tube 219·90 millims.				Length of Tube 1000·60 millims.			
Half-circle I.		Half-circle II.		Half-circle I.		Half-circle II.	
Empty Tube.	Full Tube.	Empty Tube.	Full Tube.	Empty Tube.	Full Tube.	Empty Tube.	Full Tube.
20·55°	5·25°	200·45°	185·38°	20·40°	311·55°	200·60°	131·75°
·54°	·25°	·49°	·50°	·45°	·65°	·40°	·80°
·38°	·23°	·55°	·38°	·40°	·72°	·42°	·85°
·48°	·33°	·50°	·28°	·55°	·78°	·50°	·75°
·40°	·25°	·59°	·53°	·42°	·60°	·55°	·70°
·39°	·20°	·55°	·50°	·50°	·65°	·57°	·72°
·42°	·30°	·44°	·35°	·55°	·55°	·40°	·85°
·55°	·20°	·55°	·30°	·40°	·70°	·43°	·83°
·43°	·30°	·45°	·30°	·50°	·80°	·60°	·75°
·47°	·33°	·57°	·40°	·40°	·77°	·60°	·85°
20·461°	5·264°	200·514°	185·392°	20·457°	311·677°	200·507°	131·785°
$a = 15·197°$		15·122°		68·780°		68·722°	
15·160°				68·751°			
For 1 decim. $a = 6·894°$				For 1 decim. $a = 6·871°$			

(b.) *Wild's Polariscope.*

§ **49.** The polariscope invented by Wild[1] in 1864, which has already come largely into use, affords results considerably more accordant than those obtained by the apparatus of Mitscherlich.

Fig. 27.

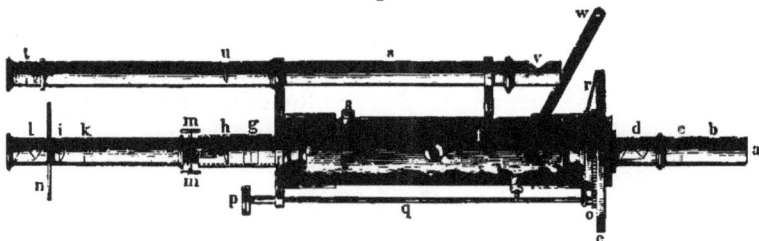

Fig. 28.

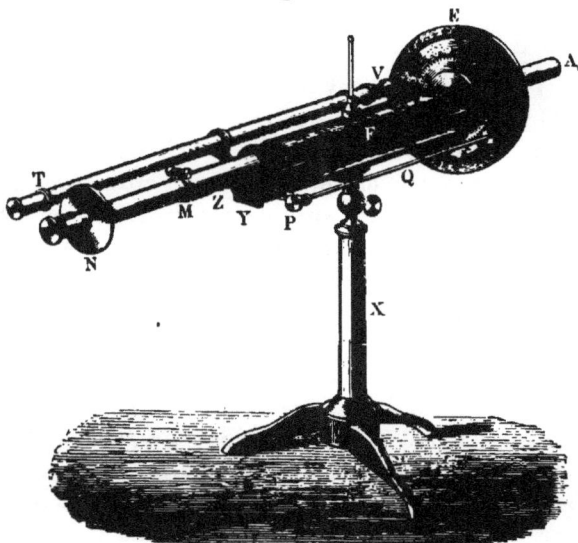

Its novelty consists in the introduction of a *Savart-prism*[2] between the polarizer and analyzer (the former of which has the rotatory move-ment), whereby a number of parallel interference-bands are brought into the field of vision, which vanish in certain positions of the polarizer. These positions, which can be determined with great accuracy, furnish the reference marks of the instrument. A sodium flame is used as the source of light.

[1] H. Wild: *Ueber ein neues Polaristrobometer*, Berne, 1865.
[2] ["*Savart'sches Polariskop*," see footnote, page 98.—D.C.R.]

The details of the instrument, as constructed by Hermann and Pfister, mechanicians, Berne,[1] are shown in Figs. 27 and 28, the same parts being indicated in Fig. 27 by small and in Fig. 28 by large letters.

A metal stand X, Fig. 28, supports a brass cradle Y, which is capable of vertical and horizontal movement, and carries at its extremities the polarizing and analyzing arrangements of the instrument. Entering at a, Fig. 27, into the dark chamber b, the light passes through a circular diaphragm c (10 millimetres in diameter), thus reaching the Nicol d. The latter is fixed to the graduated disc e, with which it turns on a common axis. Thence, the polarized ray, after traversing the solution-tube f, passes on to the analyzer. Here it meets, first, the so-called *Savart polariscope*, a prism g, composed of two plates of calc-spar 3 millimetres thick, cut at an angle of 45° to their optic axes, and cemented together with their principal sections crossing each other at right angles. To this succeed two lenses forming a telescope of low power (about 5 times), the one, h, having a focal length of 120 millimetres, the other, i, of infinite length. Between the two, and in the focus of the objective h, is a circular diaphragm, k, of about 4 millimetres diameter, provided with cross-threads. Lastly comes the Nicol prism l, fixed with its principal section horizontal. The latter will therefore form an angle of 45° with the principal sections of the double-plate g. That this relative position of the parts g and l may remain unaltered, the draw-tube, containing the Nicol and the lens i, is furnished with a guide-pin. The whole front part is set in the tube Z, which projects from the arm Y, and which allows it only a small movement about its own axis. For this purpose the tube is provided with a slot and adjusting screws $m\,m$, which clamp a projection on the inner tube. The object of the arrangement is for fixing the zero-point. Lastly, at n is placed a circular screen to shade the observer's eyes from extraneous light. The mode of effecting rotation of the polarizing Nicol is as follows :—The circular disc and the Nicol move in a piece within a fixed ring projecting from the arm Y. The disc is provided on the side next the observer with a toothed wheel driven by the pinion o, worked by the rod q, with milled-head p. The graduation is close to the edge of the disc, and in front of it is a fixed vernier or simply an index arm r. To read off the divisions a telescope, s, is used, consisting of the movable eye-

[1] Dr. J. G. Hofmann, 29, Rue Bertrand, Paris, and Schmidt and Haensch, Stall-schreiberstrasse 4, Berlin, also supply the instrument.

piece t and objective n, and having at the farther end, v, an inclined metallic reflector with round hole in the centre, by means of which the light from a small gas-flame on a movable arm w, is thrown upon the vernier. In conclusion, it should be stated that the instrument is generally constructed for tubes 220 millimetres (8·6 in.) long.[1]

Fig. 29.

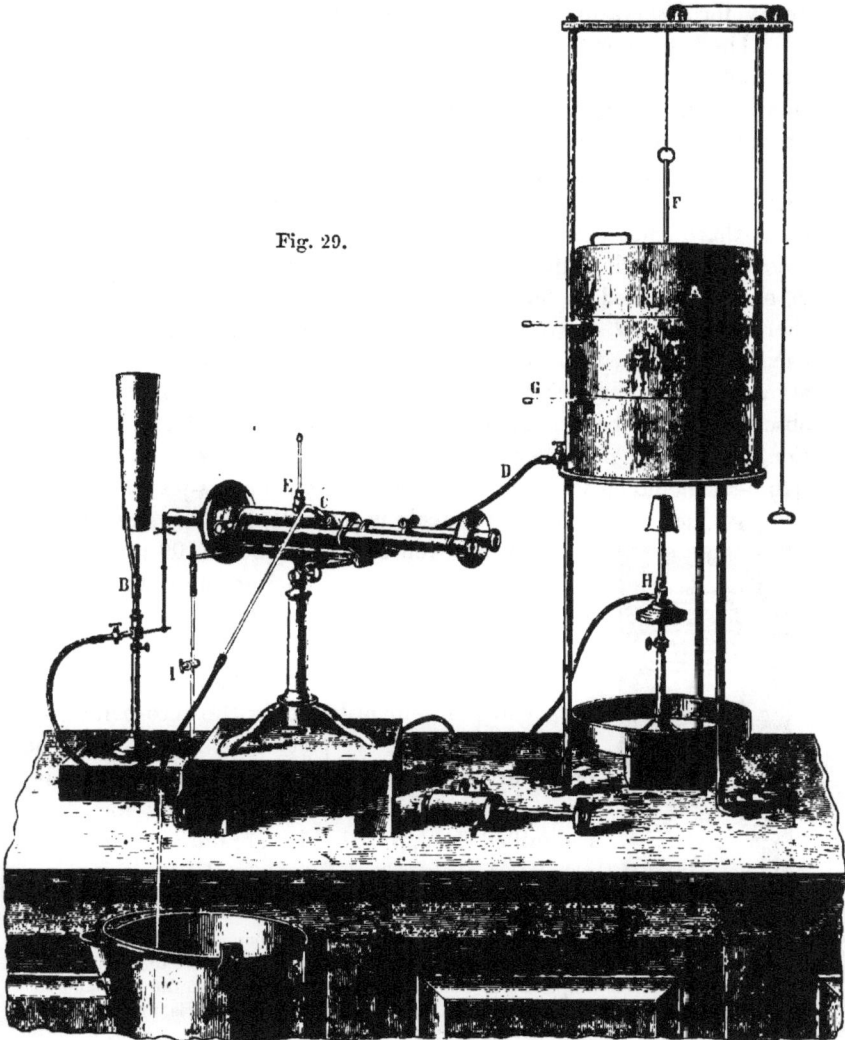

[1] Should the instrument be wanted for general scientific work and not merely as a saccharimeter, it is necessary to take care that the disc be graduated all round, and not

§ 50. That the liquids may possess a fixed temperature maintained constant, the experimental tube requires to be laid in a water-bath. For this purpose it is enclosed in a metal jacket of considerably larger diameter, so as to allow a current of water to flow between (see § 65, Fig. 44). The complete arrangement of the apparatus ready for observation, is shown in Fig. 29.

The instrument is set up opposite a Bunsen lamp B, which, with its bead of common salt, gives the sodium flame (or the lamp shown in § 22, Fig. 46, may be used instead). The outer case of the tube is provided with two side pieces, of which the lowermost is connected by india-rubber tubing, D, with the reservoir A, while the other at C serves as an outflow-pipe. A third opening, E, is for the insertion of a thermometer in the water as it flows through. The zinc reservoir A, resting upon a tall iron stand, is provided with a stirrer F, and is swathed in flannel to prevent loss of heat. One or two thermometers, G, serve to indicate the temperature of the water, usually maintained at 20°, by means of the lamp H. The water is allowed to flow through the tube for about twenty minutes before the observations are begun, the stream being interrupted during observation, while the thermometer E must remain steady throughout. The instrument should be set up in a darkened room ; the gas-jet required for reading the scale is controlled by the stop-cock I.

Fig. 30.

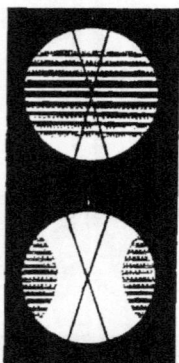

At K is represented a short (100 millimetre) tube, provided with a junction-piece to allow of its insertion in the instrument.

§ 51. In carrying out observations a tube, empty at first—in order to fix the zero-points—is introduced and the eye-piece drawn out, so that the cross-threads appear sharply defined. The polarizing Nicol must then be turned by means of the milled-head, p. 107, Fig. 27, until a number of parallel dark bands or fringes make their appearance in the field of vision (Fig. 30, a). As the move-

in one or two quadrants only. The instruments made by Hermann and Pfister have their discs divided to the third of a degree (20 minutes), and either a vernier to give readings to 5 minutes, or else a simple index-point which suffices to read to the same amount approximately. The polariscopes made by Dr. Hofmann, of Paris, read to single minutes... It would be more convenient if the divisions on the scales were not minutes but decimals of a degree (as 0·02°), as it is always in this form that the rotation angle of active substances is expressed. In reading as minutes one has to convert into decimals of a degree by dividing by 0·6.

ment continues these become fainter, until at last a position is reached at which a luminous space, devoid of lines, occupies the field. By a slight movement of the milled-head to and fro this luminous space is brought, as nearly as possible, into the middle of the field of vision, so that the remains of the fringes appear to stand at equal distances to the right and left of the cross-threads (see Fig. 30, *b*). This position serves as reference-point for the angular measurement.

If the Nicol be turned further, the dark lines will grow darker till they attain a certain maximum intensity, then become fainter again, and again vanish; these maxima recurring at intervals of 90° in the course of a complete revolution.[1] Generally the dark lines exhibit certain peculiarities of form in each position which can be recognized.[2]

Their disappearance indicates positions of the movable Nicol, in which its principal section either coincides with, or is perpendicular to the plane of the principal section of the first of the calc-spar plates of the Savart, while they occur with maximum intensity when these

[1] For the theory of these interference-bands, see Wild, *Polaristrobometer*, Berne, 1865; or Wüllner, *Lehrbuch der Physik*, 3 Aufl. Bd. 2, 604.

[2] In many Wild's polariscopes the luminous space is too wide to allow any remains of the dark lines to show on the right and left of it. Some other reference position must then be chosen. Perhaps the best method is when the dark lines have nearly passed out of the field to fix the eye on one side of the field of vision, say, the right, and continue turning the analyzer slowly until the last traces of the lines disappear at that side. This position will necessarily be the same in every observation—that is to say, the milled-head *p* will always have communicated to the polarizer precisely the same amount of adjustment in each case. The cross-threads are not required here. With many people, however, this method does not afford the same amount of accuracy as by bringing the fringes into a position at equal distances from the cross-threads on each side.

In the case of an instrument in which the luminous space is too wide, Tollens (*Ber. der deutsch. chem. Gesell.* 10, 1405) adopts the plan of loosening the adjustment-screws and turning the ocular draw-tube, containing the stationary Nicol *l*, Fig. 27, through an angle of 20° to 40° on its own axis. In this way the phenomena presented by a complete revolution of the analyzer are altered, the field becoming more or less darkened at two points 180° apart, and acquiring at two intermediate points a maximum luminosity. The dark lines vanish at each position as before, but while in the illumined quadrants the luminous space is now broader, in the darkened quadrants it is narrower than before. The former positions are entirely unsuitable for purposes of observation; but the two latter permit of a very sharp adjustment of the fringes in regard to the cross-threads. The observations must then be made in these two quadrants only. See, also, Tollens (*Ber. der deutsch. chem. Gesell.* 11, 1801).

Some instruments are so constructed that the interference-bands appear vertical, in which case the cross-threads are placed horizontally.

In Wild's instruments, should other lines appear crossing the field of vision obliquely, it is a sign that the principal sections of the two calc-spar plates of the Savart are not truly perpendicular to each other, and the instrument should be returned to the maker for readjustment.

two planes form angles of 45°. The parts are generally so regulated by the maker that the position of the index at the absorption-points is approximately at the readings 0°, 90°, 180°, 270° on the scale, admitting, however, of a certain amount of adjustment by means of the screws *m m*, Fig. 27.

If the Nicol is set to one of the four zero-points, and the empty tube replaced by one filled with some optically-active liquid, the interference-bands reappear. In its passage through the active medium the plane of polarization of the transmitted ray is made to rotate through a certain angle, and to restore it to a position either parallel or perpendicular to the principal section of the first calc-spar plate of the Savart, the Nicol must be turned in the opposite direction to that in which the rotation has taken place, when the fringes will again disappear. The graduated disc must therefore be turned to the left if the substance is dextro-rotatory, and to the right if lævo-rotatory. But, as regards the movement of the milled-head *P p*, Figs. 28 and 27, inasmuch as a change of direction is involved in the wheel-and-pinion movement, the direction in which the milled-head is turned must be *the same* as that of the rotation. When, as usually happens, the graduation follows the same direction as the figures on a watch-dial, the readings for a dextro-rotatory substance will be greater, and for a lævo-rotatory substance less than the number on the disc at the zero-point.

§ 52. If the direction of the rotatory power of an active liquid be unknown, it will be best to begin observations with a weak solution in the tube, so as to get a feeble amount of rotation. It will then be easy to see whether the direction is to the right or the left of the zero-point. On the contrary, when the rotation is considerable, a doubt may remain as to the direction, in the same way as with Mitscher-lich's instrument (§ 46). For instance, let the four zero-points be

0°	90°	180°	270°

and after the insertion of the tube with its liquid, the vanishing points of the dark lines,

30°	120°	210°	300°.

Here the medium may be, as shown in Figs. 31 and 32, either dextro-rotatory with an angle of 30°, or lævo-rotatory with an angle of 60°.

To decide the question, a second observation is necessary with a shorter tube, or a more dilute solution. If the length of tube or strength of the solution be half that in the former experiment, then also the angle of rotation will be the half only of that first observed. The vanishing points of the dark lines will then appear either at

| 15° | 105° | 195° | 285°, |

as in Fig. 34, in which case the substance is dextro-rotatory, or at

| 60° | 150° | 240° | 330°, |

as in Fig. 33, when it is lævo-rotatory.

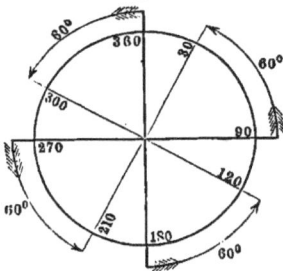

Fig. 31.

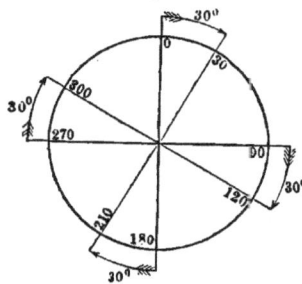

Fig. 32.

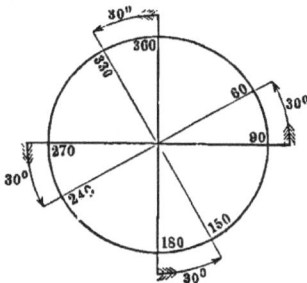

Fig. 33.

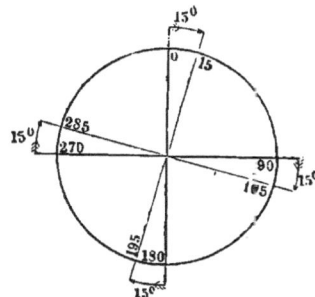

Fig. 34.

Accordingly, if observations with the shorter tube or weaker solution give lower readings than the original, the rotation is right-handed, whilst if the readings are higher than at first, the rotation is left-handed. The conditions are, of course, reversed when the graduation of the instrument is towards the left.

§ 53. In examining solutions of very high rotatory power it may happen that the angle of rotation exceeds 90°, so that the readings are always found in the quadrant beyond. In such cases, to avoid error, the observations should be made with two tubes of

I

different lengths.   For example, the following were, in round numbers, the values obtained with lævo-rotatory nicotine, in a 100
millimetre tube :—

	Quad. I.	Quad. II.	Quad. III.	Quad. IV.
Empty tube	90°	180°	270°	360°
Full ,,	18°	108°	198°	288°
Angle of rotation	72°	72°	72°	72°

Apparently, therefore, a layer of nicotine 100 millimetres in depth
rotates through an angle of 72°.   A second observation was now
taken with a tube 50 millimetres long, when the following results
were obtained :—

	Quad. I.	Quad. II.	Quad. III.	Quad. IV.
Empty Tube	90°	180°	270°	360°
Full ,,	9°	99°	189°	279°
Angle of rotation	81°	81°	81°	81°

Here the angle, instead of being reduced to half, as we should expect,
is larger even than that given by twice the thickness of medium.   If
50 millimetres of nicotine rotate through 81°, 100 millimetres should
rotate through 162°.   Now this is found to be the case when the zero-
point of the observations with a tube of the last-named length is
moved a quadrant to the right.   We then get :—

	Quad.	Quad.	Quad.	Quad.
Empty Tube	II. 180°	III. 270°	IV. 360°	I. 90°
Full ,, ·	I. 18°	II. 108°	III. 198°	IV. + 72°(= 360 − 288)
Angle of rotation	162°	162°	162°	162°

The foregoing conditions are shown in Figs. 35 and 36, the former

Fig. 35.   Fig. 36.

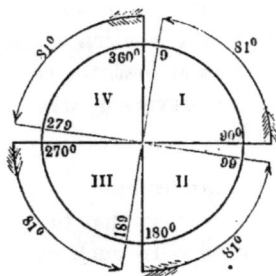

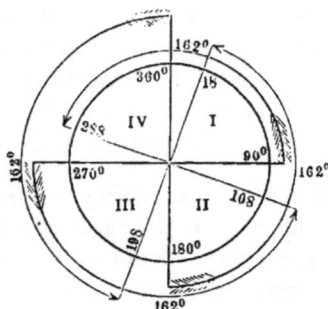

of which shows the rotation for a length of 50 millimetres, the latter for 100 millimetres.

§ 54. Wherever exactness is required, the observations should be repeated in each of the four quadrants of the circle. It will be found that appreciable differences exist between the angles of rotation thus obtained, differences which van de Sande Bakhuyzen[1] has shown originate in defective construction of the Nicol, as well as in improper placing of the two calc-spar plates of the Savart. These errors, however, disappear altogether when the mean of the four values for the angle of rotation is taken. When the observations are repeated in two opposite quadrants only, and the mean of the readings is taken, the compensation of errors is not indeed complete, but the degree of accuracy attained is usually enough for all ordinary purposes. The deviations from the true value do not exceed $0.03°$ at the most, and are generally less than $0.01°$. To take observations in two adjoining quadrants has not, of course, the same compensatory effect.

To obtain very precise results, it is obviously requisite that a still larger number of observations should be taken. As a rule, five observations in each quadrant will be enough, so that allowing for the verification of the zero-points, which should be repeated at least once each day on taking observations, the angle of rotation finally obtained will be the result of forty readings. Where considerable differences are found in the readings, the number of observations must be increased.[2] This will occur when the solutions are not absolutely clear; slight colorations, on the other hand, do not materially affect the observations.

The degree of accuracy attainable is shown in the two series of observations appended. These were taken with an instrument of Hermann and Pfister's manufacture, graduated to divisions of 5 minutes, and allowing of approximate reading to single minutes. The liquid employed was an aqueous solution of cane-sugar containing 19·45 grammes in 100 cubic centimetres. The length of tube was 219·79 millimetres.

[1] van de Sande Bakhuyzen : *Pogg. Ann.* **145**, 259.
[2] Differences of 20 minutes in the readings may easily occur with unpractised observers, but with care these can soon be much reduced in amount, so that a hasty opinion should not be formed of a newly-purchased Wild polariscope. When the observer has become accustomed to the instrument, and the latter is properly constructed, the difference in the readings will seldom exceed 5 minutes.

OBSERVATION-SERIES A.

Quadrant I.		Quadrant II.		Quadrant III.		Quadrant IV.	
Empty Tube.	Full Tube.	Empty Tube.	Full Tube.	Empty Tube.	Full Tube.	Empty Tube.	Full Tube.
0° 20′	28° 40′	90° 15′	118° 40′	180° 10′	208° 35′	270° 18′	298° 50′
24′	46′	20′	45′	17′	35′	15′	45′
18′	43′	20′	45′	15′	45′	20′	45′
20′	45′	18′	38′	15′	43′	18′	48
18′	45′	22′	32′	12′	35′	18′	46′
0° 20·0′	28° 43·8′	90° 19·0′	118° 40·0′	180° 13·8′	208° 38·6′	270° 17·8′	298° 46·8′

α = 28° 23·8′   28° 21·0′   28° 24·8′   28° 29·0′
or = 28·397°   28·350°   28·413°   28·483°

Mean of Quadrants I. and III. 28° 24·3′ = 28·405°.
  „   „   II.   „   IV. 28° 25·0′ = 28·417°.
  „   „   I., II., III., IV. 28° 24·65′ = 28·411°.

OBSERVATION-SERIES B.

Quadrant I.		Quadrant II.		Quadrant III.		Quadrant IV.	
0° 18′	28° 50′	90° 18′	118° 45′	180° 18′	208° 38′	270° 20′	298° 42′
22′	42′	15′	35′	15′	34′	18′	45′
25′	45′	15′	32′	12′	40′	20′	47′
17′	42′	20′	40′	12′	40′	19′	54′
20′	44′	22′	42′	15′	40′	16′	48′
0° 20·4′	28° 44·6′	90° 18·0′	118° 38·8′	180° 14·4′	208° 38·4′	270° 18·6′	298° 47·2′

α = 28° 24·2   28° 20·8′   28° 24·0′   28° 28·6′
or = 28·403°   28·347°   28·400°   28·477°

Mean of Quadrants I. and III. 28° 24·1′ = 28·402°.
  „   „   II.   „   IV. 28° 24·7′ = 28·412°.
  „   „   I., II., III., IV. 28° 24·4′ = 28·407°.

(c.) *Half-Shade Instruments (Polarimètres à Pénombre) of Jellett, Cornu, and Laurent.*

In these instruments the mechanism for sensitiveness is arranged to produce a circular field of vision divided into halves, which in certain positions of the analyzing Nicol are unequally illuminated, but in one particular position exhibit a uniformly faint shade. This position, which can be fixed with great accuracy, is taken as the point of reference. The use of monochromatic sodium light is pre-supposed.

§ 55. The earliest instrument of this kind was constructed by Jellett in 1860.[1] In this, between the polarizing and analyzing Nicols, and close behind the former, is placed a prism of peculiar form. An elongated rhombohedron of calc-spar which, by grinding the ends, has been converted into a right prism, is divided longitudinally into halves by a plane nearly, but not quite, perpendicular to its principal section, and the two halves then reunited, but in reversed positions. The prism is mounted in a case, furnished at the extremities with diaphragms having circular apertures. The circular field so obtained appears divided diametrically by the section into equal halves, in which the planes of polarization are slightly inclined to each other. A plane polarized ray passing through can, by turning the analyzer, be extinguished by either half of the prism, these points of extinction lying very close together, whilst between them lies the position of uniform shade. The appearance of uniform shade can also be made to vanish by the introduction of an active liquid, and, to bring it once more into view, the analyzer must be turned on its axis through a certain angle, which can be taken as measure of the deviation of the ray produced by the active substance.

§ 56. Cornu's instrument[2] consists of an ordinary Nicol as analyzer, with a polarizer of peculiar construction. The latter is formed out of a Nicol prism, by bisecting it in the direction of the plane passing through the two shorter longitudinal diagonals, cutting down the sectional faces $2\frac{1}{2}°$ and reuniting the halves. In this way we have a double Nicol prism, having its two principal sections forming an angle of $5°$ with each other. When, therefore, by turning

[1] Jellett : *Reports of the British Association*, 1860, **2**, 13.
[2] Cornu : *Bull. Soc. Chim.* [2], **14**, 140.

the analyzer, we bring its principal section exactly perpendicular
to one of the two principal sections of the polarizer, perfect obscuration follows in the corresponding half of the field of vision, the other
half remaining illumined.   A rotation of 5° reverses these conditions,
the dark half then becoming bright and *vice versâ*, while midway
between these two positions lies a point where the halves exhibit
equal degrees of incipient shadow.

Fig. 37.

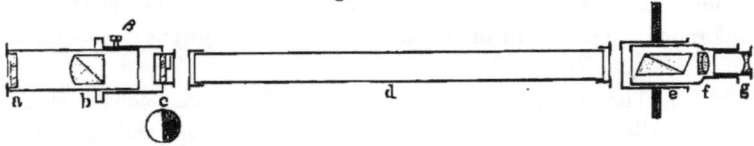

Fig. 38.

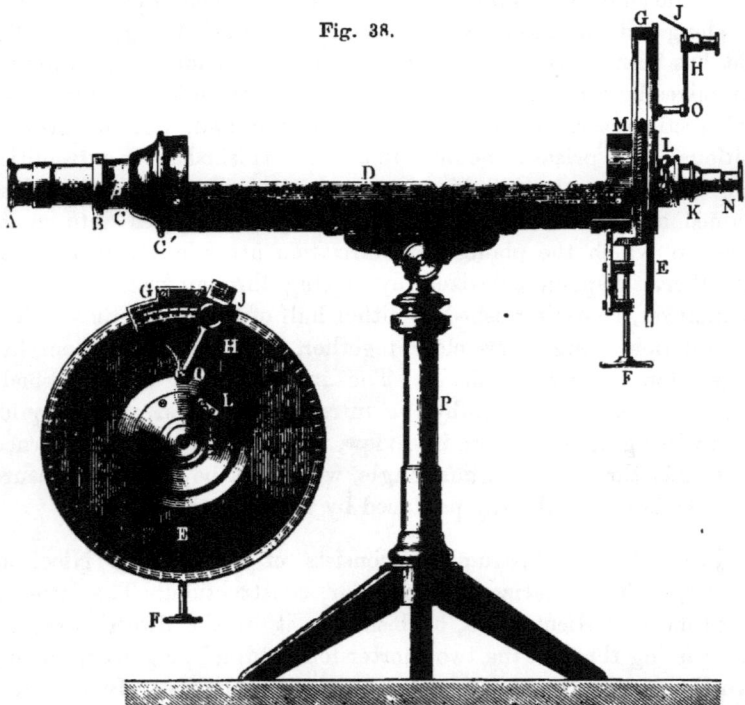

§ 57. The half-shade instrument, however, which has come into
most general use is that of Laurent,[1] of which a representation is
given in Figs. 37, 38.

' Laurent: *Dingler's Polyt. Journ.* **223**, 608.

In this,[1] the light from a sodium flame passes through the following optical apparatus :—

1. A thin plate, a (Fig. 37), cut from a crystal of bichromate of potash, serving to free the yellow ray from intermixture of green, blue, and violet light. This is enclosed between a couple of glass plates and fixed in a movable diaphragm.

2. A double refracting calc-spar prism, b, as polarizer.

These two pieces are placed one at each end of the tube A B, Fig. 38, which is inserted in the fixed portion C C', of the instrument, within which it is capable of rotation through a small angle. The amount of this movement is regulated by means of the screw-stop β, passing through the slot at C.

3. A circular diaphragm c, containing a glass plate, to which is affixed a thin plate of quartz, cut parallel to the axis, and just large enough to cover exactly one half of the circle. The thickness of the quartz plate must be so regulated that the yellow rays polarized parallel and perpendicularly to the axis may in their transmission undergo a retardation of half a wave-length. (In an instrument manufactured by Dr. Hofmann, of Paris, the thickness of this quartz plate is 0·11 millimetre.)

4. The solution-tube d.

5. An analyzing Nicol e, furnished with rotatory movement.

6. The lenses f and g, forming a small Galilean telescope.

The analyzer rotates in a piece with the divided disc E, within the stout ring, M. For this purpose, the back of the disc is furnished with a bevelled toothed wheel, driven by a small pinion worked by the milled-head F. The vernier is screwed firmly to the arm G, hanging down over the graduated edge of the disc. In reading, a magnifier, H, is used, which has a motion round the point O, and is provided at the top with a metal reflector, J. The latter can be made to reflect light on the divisions either from the sodium flame or some other convenient source. The Nicol can be turned in its case slightly by means of the screw L, so as to alter the zero-point. The telescopic lenses are mounted in tubes, K N, the latter of which has a draw motion. The graduated circle, a front view of which, with the parts pertaining thereto, is given in the figure, has a diameter of 250 millimetres, and the vernier reads to single minutes. The optical

[1] May be obtained of Dr. Hofmann, 29, Rue Bertrand, Paris; Schmidt and Hacusch, Berlin; J. Duboscq, 21, Rue de l'Odéon, Paris; Bartels and Diederichs, mechanicians, Göttingen. Fig. 38 is drawn from one of Hofmann's instruments.

arrangements are fixed at the ends of a brass trough, of semi-circular section, D, resting on the stand P. The size of the trough should be such as easily to take tubes 3 decimetres long with their water jackets.

§ 58. The peculiar feature in Laurent's instrument is the thin plate of quartz PQ, cut parallel to the axis, Fig. 39. Let the polarizer

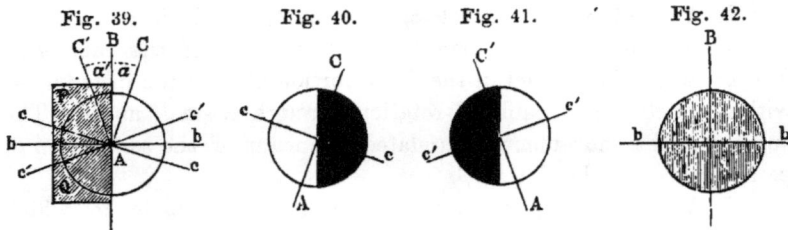

Fig. 39.   Fig. 40.   Fig. 41.   Fig. 42.

be first so adjusted that the plane of polarization of the transmitted pencil of light is parallel to the axis of the plate—that is, lies in the direction $A B$—the two halves of the field of vision will then appear equally dark or equally bright in every position of the analyzer. But if the polarizer be inclined to $A B$, at an angle a, the plane of polarization of the rays passing through the quartz plate will undergo deviation through an equal angle, a', in the opposite direction. Therefore, when in the uncovered half the plane of polarization has the direction $A C$, in the covered half it will have the direction $A C'$. If now we turn the analyzer, then, according as its plane of polarization lies in the direction $c c$ or $c'c'$, so will either the rays polarized parallel to $A C$ or to $A C'$ be extinguished, and the corresponding half of the field of vision will appear completely dark, while the other half merely suffers a partial decrease of brightness (Figs. 40, 41). In the middle position, $b b$, Fig. 42, there is a uniform shading over the two halves, but a very slight movement to and fro of the analyzer will at once destroy the equality. These phenomena repeat themselves when the analyzer has been moved through an angle of 180°.[1]

The degree of uniform shade obtained by bringing the analyzer into the middle position will be greater, the smaller the angle a (Fig. 39), which the plane of the polarizer makes with the axis of the quartz plate. The parts are set by the instrument-maker so that these two

[1] For the theory of the phenomena produced by polarized light with plates cut parallel to the axes of uni-axial crystals, see Wüllner's *Lehrbuch der Physik*, 3 Aufl. Bd. II. S. 568. Laurent, in his earlier instruments, employed a thin plate of gypsum instead of the quartz, which gave the same results (*Comptes Rend.* 78, 349).

directions are parallel to one another, but, as before observed, § 57, adjustment is arranged for by allowing the polarizer a slight amount of rotation, by means of the ring B, in the slot β, Fig. 38, whereby the field of vision is brightened. In this way the sensitiveness of the instrument can be altered. This is always greater the smaller the departure from the parallel position, when, consequently, the less will be the amount of movement of the analyzer requisite to produce perfect obscurity of one or other half of the field of vision. The deepest shading suitable should therefore be chosen.

§ 59. In setting up the instrument it is directed towards a sodium flame, and the telescopic eye-piece so adjusted that the edge of the quartz plate appears to divide the diaphragm by a sharply defined vertical line. The analyzing prism must then be turned until both halves of the field of vision appear equally dark, and the polarizer adjusted to that position where the least displacement of the analyzer is required to produce an appreciable change in the appearance of the field.

In determining the zero-point, the analyzer is brought into the middle position, where the partition-line becomes invisible. Moreover, it is better to fill the experimental tube with water, so as to equalize the conditions in respect to absorption of light with those holding in observations of active liquids. One can then make the actual zero-position correspond as nearly as possible with the zero-mark by means of the screw L (Fig. 38). Then, introducing the liquid, a rotation of the analyzer with its disc to the right will be necessary to restore the reference position if the substance be dextro-rotatory, and to the left if it be lævo-rotatory.

If now, owing to colouring or any slight turbidity, the field of vision is too dark, greater brightness can be obtained by a slight movement of the polarizer on its axis (see § 58), but this entails the disadvantage that larger movements of the analyzer are then required before any alteration on the uniformity of shade is apparent, and the readings accordingly are more divergent. In clear solutions these do not differ by more than single minutes. For determining the direction of rotation in active substances of high rotatory power, the procedure given under the head of Mitscherlich's instrument (§ 46) is equally appropriate here.

A bright sodium flame is necessary, which is best obtained by using the lamp shown, § 46, Fig. 22. If tubes provided with water-jackets are employed, the complete arrangement of the instrument corresponds exactly with the description given § 50, Fig. 29. To eliminate errors in the Nicols, the observations should be made at two positions 180° apart, and the mean taken. The subjoined table contains, as an example, a series of observations with one of Hofmann's instruments, graduated from 0° both ways to 180°. The zero-point was approximately at 90° in each half circle. The active liquid was a solution of cane-sugar :—

Tube with Water.		Tubes with Sugar Solution.			
Zero-Point.		Observation-Series *a*. Tube Length 200·48 mm.		Observation-Series *b*. Tube Length, 300·08 mm.	
Half Circle I.	Half Circle II.	Half Circle I.	Half Circle II.	Half Circle I.	Half Circle II.
89° 54′	89° 56′	103° 44′	76° 10′	110° 32′	69° 14′
53′	57′	42′	9′	35′	11′
56′	55′	45′	5′	36′	10′
57′	55′	39′	5′	31′	15′
59′	55′	46′	4′	33	12
52′	59′	41′	7′	33′	16′
54′	54′	38′	10′	32′	13′
57′	56′	40′	5′	30′	15′
56′	56′	41′	9′	35′	16′
56′	56′	42′	7′	31′	14′
89° 55·4′	89° 55·9′	103° 41·8′	76° 7·1′	110° 32·8′	69° 13·6′
		103° 41·8′	89° 55·9′	110° 32·8′	89° 55·9′
		89° 55·4′	76° 7·1′	89° 55·4′	69° 13·6′
Observed Angle of Rotation *a* =		13° 46·4′	13° 48·8′	20° 37·4′	20° 42·3′
Mean *a* =		13° 47·6′ = 13·793°		20° 39·85′ = 20·664°	
For 1 decim. *a* =		6·880°		6·886°	

(d.) *Comparison of Mitscherlich's, Wild's, and Laurent's Instruments.*

§ 60. To determine the degree of concordance possible between

the above instruments, observations were made with the same tubes
on two suitable solutions of sugar. In each case, forty readings were
taken, one-half of which were for the determination of the zero-
point. The following were the results :—

Instrument.	Solution I.		Solution II.	
	Tube Length 219·79 mm.	Angle of Rotation for 100 mm.	Tube Length 220·00 mm.	Angle of Rotation for 100 mm.
Mitscherlich	20·227°	9·203°	14·570°	6·623°
Wild	20·285°	9·229°	14·574°	6·625°
Laurent	20·268°	9·222°	14·602°	6·637°

From this it will be seen that the results obtained with these
several instruments agreed to the tenth of a degree, the variations not
appearing till the second, and in some cases not before the third
decimal place. It is therefore immaterial which instrument is used,
any superiority consisting merely in the comparative facility with
which observations can be made. An idea of the amount of difference
that may be expected between the results obtained by different observers is shown by the following table, in which are recorded the angles
of rotation obtained with the same sugar-solution by two observers of
almost equal experience in the use of the three several instruments :—

Polariscope used.	Length of Tube, in millim.	Observed Angle of Rotation.		Angle of Rotation for 1 decim.		Differ- ence, A − B.
		Observer A.	Observer B.	Observer A.	Observer B.	
Mitscherlich	219·89	15·165°	15·160°	6·896°	6·894°	+ 0·002°
	1000·60	68·809°	68·751°	6·876°	6·871°	+ 0·005°
Wild	99·92	6·941°	6·849°	6·947°	6·854°	+ 0·093°
	219·79	15·171°	15·078°	6·903°	6·860°	+ 0·043°
Laurent	200·48	13·786°	13·793°	6·876°	6·880°	− 0·004°
	300·08	20·628°	20·664°	6·874°	6·886°	− 0·012°
Mean				6·695°	6·874°	+ 0·021°
Mean error in a single determination[1]				±0·028°	±0·015°	

Here we see the observations of the two observers varying
within limits of hundredths or even only thousandths of a degree.

[1] Mean error for a single determination $= \pm \sqrt{\dfrac{\delta_1^2 + \delta_2^2 + \ldots \delta_n^2}{n-1}}$, when
$\delta_1, \delta_2 \ldots \delta_n$ represent the differences between individual observations and their
arithmetical mean, and n the number of observations taken.

The variations that occur in observations with different tubes are similar in amount, provided indeed that, as in the above instances was the case, their lengths have been measured correctly to within 0·05 millimetre. (See further § 75.)

(e.) *Determination of the Angle of Rotation for different Rays.*— [METHOD OF BROCH.]

§ 61. The instruments thus far described serve only to measure the rotation of the yellow sodium ray D. It is possible, indeed, at least with Mitscherlich's and Wild's instruments, by introducing into the flame, instead of common salt, some other substance, such as lithium or thallium compounds, to produce monochromatic light of another colour. The red lithium flame has the disadvantage of being too weak in illuminating power to admit of exact observations ; besides which, it retains an admixture of yellow rays, which, however, can be absorbed by placing a red glass slide in front of the flame. The volatility of thallium compounds, on the other hand, renders it difficult to maintain the green-coloured light with sufficient intensity for any length of time.

§ 62. A method which admits of the determination of the rotation for a whole series of rays of known wave-length was proposed by Broch in 1846[1], and about the same time by Fizeau and Foucault,[2] and has since been adopted by various other observers, as Hoppe-Seyler, Wiedemann, &c. For this purpose, solar light is employed, reflected horizontally by means of a heliostat into a darkened chamber. The beam passes in succession through—(a) a polarizing Nicol ; (b) the layer of active substance ; (c) the analyzing Nicol ; (d) a spectroscope, consisting of a collimator, prism, and telescope, which last must be furnished with cross-threads. If, leaving out the liquid, the analyzer is first adjusted to the position of greatest darkness—the zero-point—and the active liquid then introduced, the spectrum with Fraunhofer's lines will at once appear in the telescope. If the analyzer be then rotated, a position will be found at which a vertical black line makes its appearance, and can be made to move across the field of vision by continuing the rotation. This is caused by the Nicol as it revolves, extinguishing in succession the rays whose planes of polarization are perpendicular to its own. If the cross-

---

[1] Broch: *Dove's Repertorium d. Physik.* 7, 113.
[2] Fizeau and Foucault : *Comptes Rend.* 21, 1155.

wires of the telescope have been previously made to coincide with some one of the Fraunhofer lines, and the dark band be now brought to cover it, the reading on the graduated disc will give the angle of rotation for that particular ray. In a similar manner the amounts of rotation can be determined for other portions of the spectrum.

§ 63. Instead of solar light, the use of which is of course very much restricted, artificial light may also be employed for this method of observation. V. von Lang's[1] method of using lithium, sodium, and thallium light, is first to adjust the cross-lines of the telescope upon the lines produced by volatilizing a salt of the respective metal in a Bunsen burner; then, replacing the Bunsen by the luminous flame from an Argand, to bring into view the continuous spectrum necessary for the production of the dark absorption-band. By rotating the Nicol until the band is made to coincide with the cross-threads—the position of which must, of course, remain undisturbed during the experiment—the angle of rotation for the given ray is obtained. In the same way the light from a hydrogen Geissler-tube can also be used. By these means it is possible to determine the angle of rotation for six different rays of known wave-length. In the subjoined table these artificial lines are collated with the Fraunhofer lines so as to indicate their position in the spectrum.[2]

Artificial Lines.		Fraunhofer Lines.	Wave-length, in millim.
		A	0·0007607
		B	0·0006872
Red lithium line	Li_α		0·0006712
Red hydrogen line	H_α	C	0·0006567
Yellow sodium line	Na	D	0·0005893
Green thallium line	Tl		0·0005350
		E	0·0005271
		b	0·0005182
Green-blue hydrogen line	H_β	F	0·0004862
Blue-violet hydrogen line	H_γ		0·0004343
		G	0·0004304
		H	0·0003956

[1] v. Lang: *Pogg. Ann.* 156, 422.
[2] The wave-lengths of the Fraunhofer lines given above (of which C coincides with

The arrangement of the apparatus is shown in Fig. 43. It consists of a Mitscherlich polariscope, with its polarizing Nicol at

Fig. 43.

a, and its analyzing Nicol, with rotatory motion, at c; in front of it a spectroscope, consisting of a collimator, d, a prism, e, and a movable telescope, f. Instead of the usual cross-threads, the telescope is fitted

H_α, D with Na, and F with H_β), and also of line H_γ, are the means of the results obtained by Fraunhofer, van der Willigen, Ditscheiner, Ångström, Stefan, and Mascart (Wüllner's *Lehrbuch der Phys.* 3 Aufl. 2, 136 and 141). For Li_α and Tl Ketteler (*Beobachtungen über die Farbenzerstreuung der Gase*, Bonn, 1865) gives the values 6706 and 5345, that of Na being taken as 5888; so that putting Na = 5893, we obtain the values Li = 6712, and Tl = 5350.

The lines of other metals might probably be made available, as the following:—

Wave-lengths in millimetres.

Red potassium line	0·0007607	(Fraunhofer line A).
Red cadmium line	0·000644	(Hurion).
Green cadmium line	0·000534	(,,).
Green magnesium line	0·0005182	(Fraunhofer line b).
Blue cadmium line	0·000468	(Hurion).
Blue strontium line	0·0004607	(Lecoq de Boisbaudran).
Blue-violet cadmium line	0·000441	(Hurion).
Violet indium line	0·0004101	(Lecoq de Boisbaudran).

Hurion (*Pogg. Beiblätter*, 2, 83). Lecoq de Boisbaudran (*Spectres lumineux*. Paris, 1874).

with two parallel vertical threads, separated from each other by a distance somewhat greater than the breadth of the black absorption-band. The determination of the angle of rotation of a given substance comprises the following operations :—1. The determination of the zero-point of the analyzer. The spectroscope is removed, and the experimental tube left empty or filled with water. The Argand lamp i serves as the source of light. 2. The spectroscope is set up in its place, and in front of the Nicol a, the source of light h, which gives the lines (a Bunsen burner, with a bead of salt or a hydrogen tube) ; the analyzer is then rotated till the light can pass through. After widening pretty considerably the slit of the collimator d, the telescope f is moved horizontally until the bright line under examination lies accurately between the parallel threads, when the clamp-screw g is tightened. This position of the telescope can be more readily found when the Argand lamp i, with a small flame, is placed behind the Bunsen burner h, so that the threads stand out against a bright background. 3. The tube containing the active liquid is then laid in its place, the Argand light turned full on, and the analyzer moved on its axis until a broad dark band, with faint edges, appears in the field of the telescope. This also is brought between the parallel threads, and the angle read off on the graduated disc. Lastly, by reproducing the bright line again, it can be ascertained whether the position of the telescope has been disturbed in the meantime.

In observing the parts of the spectrum which lie near the extremities, as the red lithium-line, the Drummond lime-light should be used for the production of the continuous spectrum, the Argand being too weak in red rays sufficiently to illumine the edges of the dark band.

The above method does not admit of the same degree of concordance of results as when the polariscope is used alone, the position of the dark absorption-band being less definite, owing to the indistinctness of its edges. Thus von Lang, in the paper referred to, gives the following table as showing the varying positions of the analyzing Nicol in a determination of the rotatory powers of a quartz plate.

Observation-series 1 and 2 were obtained with an Argand lamp, 3 and 4 with a Drummond light, the position of the polarizer in the second being different from that in the first two series. The temperatures of the quartz plate are also given in the table; these exhibit certain variations, but too small in amount to account for the differences in the observed positions.

1		2		3		4	
Sodium Line.		Thallium Line.		Sodium Line.		Lithium Line.	
Temp.	Angle.	Temp.	Angle.	Temp.	Angle.	Temp.	Angle.
17·8°	72·23°	20·0°	69·28°	19·0°	91·81°	22·0°	102·65°
.20·1°	71·53°	20·1°	70·20°	17·0°	93·53°	20·5°	103·15°
20·0°	70·29°	19·7°	68·12°	19·4°	93·37°	21·5°	103·57°
20·4°	72·14°	20·0	68·73°	20·0°	92·89°	18·3°	103·71°
20·1°	71·90°	20·5°	69·62°	19·5°	93·33°	19·6°	104·03°
21·0°	71·11°	20·6°	68·82°	20·0°	93·08°	20·0°	103·15°
21·7°	71·20°					19·9°	102·64°
20·2°	71·49°	20·2°	69·19°	19·1°	93·00°	20·3°	103·27°
	± 0·25°		± 0·20°		±0·25°		± 0·20°

B. Measurement of the Length of Tubes and their Adjustment.

§ 64. The experimental tubes of polariscopes should invariably be made of glass. They are generally 2 decimetres (about 8 in.) in length, but shorter ones, 1 decimetre, and longer ones, 3 decimetres and upwards in length, can be used where the construction of the instrument will admit of it. The internal diameter varies from 6 to 10 millimetres ($\frac{1}{4}$-in. to $\frac{2}{5}$-in.), and the thickness of the glass walls should be about 2 millimetres. The extremities are ground flat with great care, the ground surfaces being kept as nearly as possible perpendicular to the axis of the tube, otherwise it will be impossible to determine the length of the tube with any exactness. Moreover, it is convenient to have the internal walls of the tube ground with coarse emery powder, so as, by dulling the surface, to prevent disturbing reflections.

The mode of closing the ends of the tubes usually adopted is, as shown in Fig. 44, by plane parallel glass plates, which can be fixed down by a screw-cap, having a washer of india-rubber or soft leather between. As Scheibler[1] first observed, this mode of closing

[1] Scheibler: *Ber. d. deutsch. chem. Gesell.* 1868, 268.

the tubes may prove a source of error owing to the tendency of the glass plates under pressure to become double-refracting, whereby light passing through them is circularly-polarized. Moreover, when differences of tension exist in the body of the glass, owing to imperfect annealing, the same results will appear altogether apart from pressure, although the latter intensifies them. Indeed, by applying sufficient pressure to the glass plates, the errors thus arising may even amount to several degrees. It will be evident then how essential it is that all new glasses should be carefully tested, before they are taken into use.

For this purpose a series of observations should be made of the zero-point of the instrument, first with a tube open at the ends, then with a glass plate applied to one end under moderate pressure. Moreover, as the glass may be differently affected at different parts, the tube should be turned about on its axis so as to test every part. Generally speaking, if one glass is found to be circular-polarizing, all others of the same lot, cut at the same table, will prove to be so likewise.

§ 65. The tubes supplied with the apparatus are usually fixed within a simple brass tube, or they may be left uncovered; in either case the liquid inside will be exposed to the temperature of the surrounding air. Now as (see § 22) the rotatory power of most substances is materially affected by heat, it is necessary to be able to control the temperature of the solutions during the continuance of the observations. This can be done by enclosing the tube in a jacket of brass, 4 to 5 centimetres wide, and allowing water to flow between, supplied by a reservoir in which it has been previously raised to the desired temperature. Fig. 44 represents a tube enclosed in a

Fig. 44.

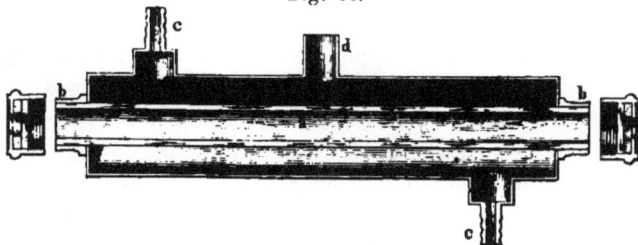

water-bath of this description. The glass tube a is fastened within the brass necks b b with shellac. The openings c c serve for the inflow and outflow of the water ; d is an opening to receive a thermometer. The mode of using the whole apparatus is given in describing Wild's polariscope (§ 49, Fig. 29).

Another arrangement is that adopted in Mitscherlich's large instrument, shown in § 48, Fig. 26, in which the observation-tube is enclosed in a rectangular metal box filled with water.

Or, lastly, the method shown in Fig. 45 may be adopted, in which the tube is laid round with a fine lead spiral through which the water is allowed to flow ; the tube itself being provided with two side openings through which the liquid is filled and emptied, and which serve also to hold the thermometers. The two terminal plates can in this case be fixed permanently in their places, for which purpose a solution of isinglass in acetic acid may be used. To prevent loss of heat the lead tubing should be protected by wrapping round it a good layer of flannel.

Fig. 45.

§ 66. *Measurement of Length of Tube.*—In the formula for the determination of specific rotation, the length of tube appears as an absolute quantity, so that to render results obtained by different observers comparable, it is necessary to have a uniform standard of measurement ; and for this purpose, the first consideration is that the millimetre division on the measure be rigidly accurate.

The determination of the tube-length is frequently left to the instrument-maker. As a means of verifying the ordinary 1 and 2 decimetre tubes, Scheibler[1] recommends round brass rods, made accurately 100 and 200 millimetres respectively in length, with flat ends. Screwing a glass plate on one end of the tube, the rod is pushed inside the open end, taking care that it stands straight, and the other glass plate put on, when, if the tube is of the proper length, it should fit exactly, and no room be left for the rod to move about when the whole is shaken.

In all exact experiments, it is necessary to know the length of tube to within at least 0·1 millimetre, and it is therefore desirable to be able oneself to measure it with this degree of precision. Fig. 46 represents an instrument constructed by Feldhausen, mechanician, Aachen, for this purpose, which can be used for ordinary tubes of any

---

[1] Scheibler: *Zeitschr. des Vereins für Rübenzuckerindustrie*, 1867, 226 ; 1874, 786.

diameter, as well as for those provided with water-bath surroundings. On two supports AA (A' in the side elevation) is fixed, horizontally, the brass bar BB, $3\frac{1}{2}$ decimetres in length, carrying the plate C fixed at one extremity, and the sliding-piece D. The latter consists of the piece a, which can be firmly clamped by the screw c, and is connected by means of the micrometer-screw d and the spring e with the second piece b, so as to communicate to the latter a fine movement. The bar BB is graduated to millimetres, and the sliding-piece b carries a vernier, reading to $\frac{1}{10}$th millimetre. Into the face of the fixed plate C

Fig. 46.

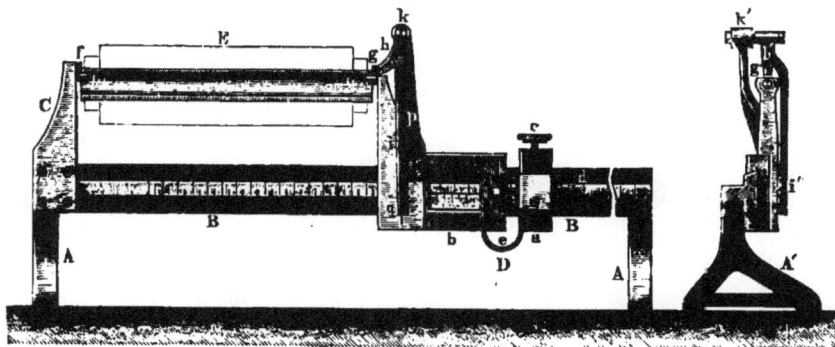

and the opposite face of the sliding-piece b, are screwed horizontally two round steel pins m n, to serve as supports for the tube we wish to measure E (which in the Fig. is shown as jacketed). Above the pin m is a wedge of steel f, fixed with its sharp edge vertical, against which one end of the tube is made to rest. A similar steel wedge g, is affixed to the short arm of the bent lever h i, which works on the pivot k; these parts (indicated in the side-view by g' h' i' k') being in connection with the sliding-piece b. The outside of the longer arm, i, which is directed downwards, has a spring p resting upon it, which tends to make it move towards the left; and at its extremity an index mark q is placed which can be made to coincide with a similar mark on b.

In using the instrument the pins m n are first removed by unscrewing, and the sliding-piece D pushed close up to C, until the two edges of the prisms f and g nearly meet. The part a is then clamped, and with the aid of the micrometer-screw d the part b is moved forward, until by perfect contact of the edges f and g, the index mark q, which at first stood to the left, is made to coincide with the mark on b. This gives the zero-point. The pins m n are then screwed into their place, D being pushed back far enough to allow of the tube to be measured, being suspended freely upon the pins. One end of the tube is pressed against the edge f, while the contact of g with the opposite end is completed with the micrometer-screw, until the marks at q again coincide. The length is then read off with the vernier on the graduated bar. Since the annular end-surfaces of the tube are never truly parallel, the latter should be turned on its axis through a quarter, half, and three-quarters of a circle, and the mean of the measurements in the several positions taken as the true length.

Where a cathetometer is available, the following mode of determining tube-lengths may also be adopted. A piece of glass tubing $a\,b$ (Fig. 47), of such diameter that it will slide easily inside the tube to be measured without shaking about in it, is closed at one end b with the blowpipe, the glass being drawn to a blunt point, which can then be sharpened off a little with the file. Enough is cut off the open end to leave the tube some millimetres shorter than the tube to be measured, after which it is filled with mercury to about one-fourth of its length, the metal being retained in its place by a cork c pressed down on its surface. Into the mouth of the tube is inserted a close-fitting, well-greased piece of india-rubber tube d, through which passes a glass rod e, of diameter just sufficient to move easily within the india-rubber collar. To allow escape of air the india-rubber should be slit down its whole length at one side. The ends of the glass rod are drawn out a little, that at f being brought to

Fig. 47.

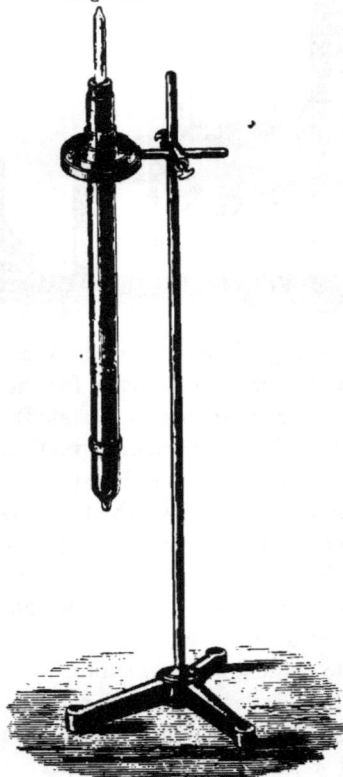

Fig. 48.

a point. The other end is passed through the cork g, so as to ensure straight motion within the tube, and the rod is pushed down so far that the total length of the combination is somewhat greater than that of the tube to be measured. The latter is then closed at one end with glass plate and screw-cap, the contrivance just described slipped into it, and the glass rod pushed down by pressing the point f with the other glass end, and screwing the latter firmly down. The lower screw-cap is now removed, the inner tube withdrawn without disturbing the position of the rod, and fixed in gimbals (Fig. 48), whereby the weight of the mercury in the bottom keeps it truly perpendicular. By means of a cathetometer the distance between f and b can then be determined. Repeated measurements taken in this way are found to agree with a mean variation of only $\pm$ 0·02 millimetre. Where the apparatus is too wide and requires steadying in the tube, india-rubber bands, $h\,h$ (Fig. 47), can be slipped over it.

In determining the angle of rotation of a liquid at different temperatures, the consequent variations in the length of tube must be taken into account. For this purpose it will be found sufficient to take the *mean value* 0·0000085 as the coefficient β of linear expansion of glass for 1° Cent. Representing the length of tube in millimetres by L_t and the temperature of measurement by t, the length at any other temperature t' will be given by the formula

$$L_{t'} = L_t\left[1 + \beta\,(t' - t)\right], \text{ when } t' > t,$$

or

$$L_{t'} = L_t\left[1 - \beta\,(t - t')\right], \text{ when } t' < t.$$

Thus if a tube measures exactly 200 millimetres, at a temperature of 20°, it will measure 200·02 millimetres at 30° and 199·98 millimetres at 10°. The correction is therefore only needed for great variations of temperature and long tubes.

C. Estimation of Percentage Composition of Solutions.

§ 67. *Preparation of Solutions by Weighing the Active and Inactive Constituents.*—For this purpose small blown glass flasks (Fig. 49) of 25 to 100 cubic centimetres capacity, with wide necks, and provided with ground-glass stoppers, will be found most suitable. The active substance is first weighed into a flask of this kind, after which the calculated amount of solvent necessary to give the desired percentage is introduced by means, first, of a wide, and subsequently a narrow-necked pipette. But as in this way a drop or two may be easily added too much, and thus the right percentage not accurately attained, it is best to prepare the solutions roughly at first by weighing in a pair of scales, and afterwards determine accurately by a chemical balance the real amount added.[1]

§ 68. The percentage composition of fresh-prepared solutions can easily be found to the third place of decimals by weighing to milligrammes, but this accuracy vanishes when it becomes necessary to filter from turbidity, the evaporation of the solvent which takes place during the process, increasing the percentage of non-volatile active substance.

Fig. 49.

To estimate the magnitude of the error arising from this source, a few experiments were made partly by placing the filtering apparatus bodily into the balance-pan, and partly by determining the percentage after as well as before filtration. Filters of Swedish paper were invariably used, and both the funnels and the other vessels employed were covered over as much as possible.

Aqueous Solutions.—*a.* 43·131 grammes of water took four minutes to filter, losing 0·019 gramme by evaporation. Temperature of the air 18° Cent. 99·614 grammes of water took eleven minutes to filter, and lost 0·041 gramme by evaporation. Temperature, 20° Cent. At this rate, filtration of 40 to 100 grammes of a 10 per cent. solution would be tantamount to an addition of about 0·004 per cent.

---

[1] Instead of putting the object to be weighed in the left and the weights in the right scale, in the usual way, the following method is convenient :—A certain weight, greater than any likely to be used in the experiments, is put in the left scale. A 50 gramme weight will generally do. In the right scale is placed first the empty flask, together with weights enough to produce equilibrium, and the same process repeated after the substances have been introduced. The advantage of this method is that, as the weight in the balance remains constant, the oscillations remain the same, and, by finding once for all the amount of swing corresponding to 1 milligramme, it will be easy always, when the balance is approaching equilibrium, to fix upon the proper division for the rider from observation of a single oscillation. Thus the process of weighing is facilitated. It is, however, an indispensable condition that the length of beam-arms suffer no alteration during any series of connected weighings. The substitution method (Borda's) of weighing is the only method which meets this difficulty.

of active substance. *b.* Filtration, lasting for three minutes, of 50
cubic centimetres of an aqueous solution of nitrate of silver raised
the proportion of the solid substance from 9·708 to 9·713 per
cent. Temperature, 24·5°. In this case the addition amounts to
0·005 per cent., thus agreeing with the result of the preceding
experiment.

Alcoholic Solutions.—*a.* 31·007 grammes of alcohol, of 94 per
cent. by weight, took four minutes to filter, and lost 0·067 gramme by
evaporation. Temperature, 18° Cent. 71·494 grammes of the same
alcohol filtered in ten minutes, and lost 0·114 gramme. Temperature,
19° Cent. Had these solutions contained, to begin with, 10 per cent.
of active substance, the process of filtration would have raised it to
10·022 per cent. in the first experiment, and to 10·016 per cent. in
the second. *b.* 50 cubic centimetres of a solution of nitrate of silver
in alcohol of 78 per cent. by weight took ten minutes to filter (temperature, 23° Cent.), and the proportion of active substance rose in
one experiment from 9·686 to 9·714, and in another to 9·736, or by
amounts ranging from 0·028 to 0·050 per cent.

Assuming, therefore, that evaporation is independent of the
concentration of the solutions, and, further, is proportional to the
amount of filtrate (both of which postulates are only approximately
true), the result of the foregoing experiments may be taken as
proving that for every 10 per cent. of active substance originally
present, the proportion is raised by filtration—in aqueous solutions
by 0·005, and in alcoholic solutions by from 0·02 to 0·05 per
cent.

Thus it will be seen that in the case of concentrated solutions
this increase of percentage may become very considerable, and where
alcohol is employed as the inactive solvent, may alter the first
decimal by several units. Filtration is therefore to be avoided as far
as practicable.

The degree to which errors of this kind affect the calculation
of specific rotation is stated in § 75.

§ 69. *Reduction of Weighings to Weight in Vacuo.*—If it be desired
to determine with great exactness the percentage composition of the
solutions, the results of the several weighings should be reduced to
their values *in vacuo.* When pains have been taken to weigh accurately to milligrammes, which should, as a rule, be done, it costs but
little extra trouble to apply the trifling correction requisite, which is

the more desirable, as neglect of it may affect the value of the percentage to the second place of decimals. As a rule, the error arising from non-reduction of the weights will be greater in proportion to the difference in density between the active substance itself and the solution of it employed. The effect on the specific rotation value will be greater the more concentrated the solution is, and the smaller the angle of rotation.

The following simple method will suffice for the reduction:—

Let p be the observed weight in air of a given substance,
   d its specific gravity,
then the weight γ, by which the substance, weighed with brass weights, appears too light, owing to the pressure of the atmosphere, is given by the formula[1]

$$\gamma = p \cdot 0 \cdot 0012 \left(\frac{1}{d} - 0 \cdot 12\right).$$

The number $0 \cdot 0012$ is the mean density of atmospheric air, and $0 \cdot 12$ is obtained by dividing unity by the specific gravity of brass, which latter may be taken as $8 \cdot 4$.

The weight *in vacuo* P of the substance will then be[2]

$$P = p + \gamma.$$

These coefficients are amply sufficient for the reduction of all weighings that occur in determining the specific rotation of active substances, and it is unnecessary to allow for changes of density in the air, so that we need not take observations of temperature and pressure at the time of weighing.[3]

To facilitate calculation the following table has been prepared, giving the values of the factor $0 \cdot 0012 \left(\frac{1}{d} - 0 \cdot 12\right)$ for solutions with specific gravities ranging from $0 \cdot 74$ to $3 \cdot 0$. Putting R for this value, as in the table, the reduced weight becomes

$$P = p + p R.$$

---

[1] For the *rationale* of this formula see Kohlrausch, *Leitfaden der praktischen Physik*, 3 Aufl. S. 29. [Translated into English from the second German edition, under the title of *An Introduction to Physical Measurements*, by Messrs. Waller and Proctor. Churchill, London, 1873. 8vo, 12s.—D.C.R.]

[2] If the density of the substance exceeded $8 \cdot 4$, which, however, is never the case with the substances with which we have to deal, $P = p - \gamma$.

[3] Moreover, the circumstance that the smaller weights are of platinum instead of brass has no appreciable effect.

d.	*R.*	*d.*	*R.*	*d.*	*R.*	*d.*	*R.*
0·74	0·00148	0·92	0·00116	1·1	0·00095	1·7	0·00056
0·76	144	0·94	113	1·15	90	1·8	52
0·78	140	0·96	110	1·2	86	1·9	49
0·80	136	0·98	108	1·25	82	2·0	46
0·82	132	1·00	106	1·3	78	2·2	40
0·84	128	1·02	103	1·35	74	2·4	36
0·86	125	1·04	101	1·4	71	2·6	32
0·88	122	1·06	099	1·5	66	2·8	29
0·90	119	1·08	097	1·6	61	3·0	26

In the calculation it will be sufficient to take the weights p to decigrammes and the specific gravity to two places of decimals. The following example of the preparation of a solution of tartrate of ethyl in wood-spirit, will show the mode of the reduction :—

I. Tartrate of ethyl weighed in air $\qquad p = 10·898$ grammes.
   Specific gravity of tartrate of ethyl, $d = 1·2$.
   Value of R for 1·2 from table $= 0·00086$.
   10·898 or (substantially) 10·9 × 0·00086 $= \quad 0·009 \qquad$,,

$$\text{Weight } in\ vacuo = 10·907 \text{ grammes.}$$

II. Weight in air of the solution in wood-spirit, $p = 27·269$ grammes.
   Specific gravity of solution, $d = 0·94$.
   Value of R for 0·94 from table $= 0·00113$.
   27·269 or (substantially) 27·3 × 0·00113 $= \quad 0·031 \qquad$,,

$$\text{Weight } in\ vacuo = 27·300 \text{ grammes.}$$

The percentage composition of the solution therefore stands as follows :—

	Uncorrected.	Corrected.	Difference.
Tartrate of ethyl	39·965	39·952	− 0·013
Wood-spirit	60·035	60·048	+ 0·013

In the process of preparing solutions there are thus two separate weighings requiring to be reduced to weight *in vacuo* :—(1) That of the active substance, for which a knowledge of its specific weight is necessary ; and (2) that of the prepared solution, the density of which must be known at any rate for calculating the specific rotation.   In case the density of an active substance is entirely unknown, which may occur with solids, the difficulty may be got over by weighing first the solvent, whose specific gravity is known, in the flask, and then adding the active substance.

The table below gives the specific gravities of a number of optically active solids and liquids, as also of several substances suitable as solvents. The specific gravities of solutions employed in polariscopic experiments seldom appear outside the limits 0·8 to 1·4.

Active Substances.

	d.		*d.*
Oil of turpentine	0·85 to 0·91	Cane-sugar	0·85 to 1·58
Colophonium	0 85 to 1·07	Sorbin	0·85 to 1·65
Camphor	0·85 to 0·99	Ammonium acid malate	0·85 to 1·55
Camphoric acid	0·85 to 1·18	Tartaric acid	0·85 to 1·75
Nicotine	0·85 to 1·01	Tartrate of ethyl	0·85 to 1·20
Cholesterin	0·85 to 1·07	Sodium-ammonium tartrate	0·85 to 1·59
Santonin	0·85 to 1·25	Potassium-ammonium tartrate	0·85 to 1·70
Salicin and Phlorhizin	0·85 to 1·43	Sodium-potassium tartrate	0·85 to 1·78
Asparagin, crystallized	0·85 to 1·50	Tartrate of sodium	0·85 to 1·79
Aspartic acid	0·85 to 1·66	Potassium acid tartrate	0·85 to 1·96
Mannite	0·85 to 1·52	Tartrate of potassium	0.85 to 1·97
Milk-sugar	0 85 to 1·54	Tartar emetic	0·85 to 2·60

Inactive Solvents.

Ether	0·85 to 0·71	Acetic acid	0·85 to 1·05
Alcohol	0·85 to 0·79	Nitro-benzene	0·85 to 1·20
Wood-spirit	0·85 to 0·80	Formic acid	0·85 to 1·22
Acetone	0·85 to 0·80	Ethylene chloride	0·85 to 1·26
Benzene and Toluene	0·85 to 0·86	Carbon bisulphide	0·85 to 1·27
Acetic ether	0·85 to 0·90	Chloroform	0·85 to 1·48
Aniline	0·85 to 1·04	Carbon tetrachloride	0·85 to 1·60

D. Determination of the Specific Gravity of Liquids.

§ 70. The only method which affords the requisite degree of exactness is that of weighing a determinate volume. For this purpose we may use narrow-necked pycnometers of 10 to 20 cubic centimetres capacity, which can be filled or emptied by means of a pipette with capillary stem and india-rubber ball (Fig. 50.)

The washing and drying of the apparatus is quickest done by rinsing with alcohol, followed by a little anhydrous ether. To bring the level of the liquid to the mark a roll of filter-paper or cigarette-paper may be used.   If the neck of the pycnometer have a width of about 1 millimetre, differences of from 0·5 to 2 milligrammes may be observed in the weight of the flask in successive adjustments of level,

Fig. 50.

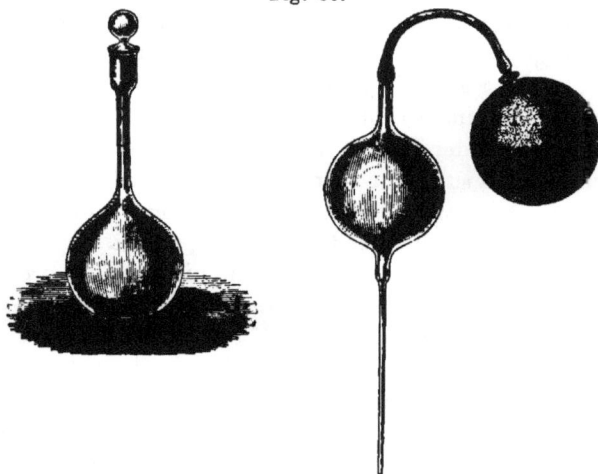

at a constant temperature.   The temperature is maintained constant by keeping the flask in a water-bath.

Fig. 51.

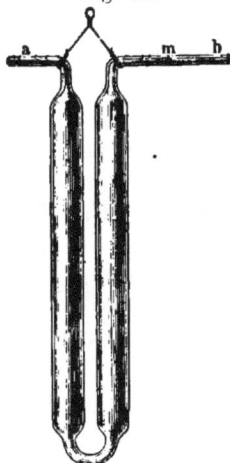

Fig 52.

A greater degree of exactness is attainable with Sprengel's pycnometer[1] (Fig. 51). This consists of a U-shaped tube of thin glass,

---

[1] Sprengel : *Pogg. Ann.* **150**, 459.

the ends of which are drawn out narrow and bent at right angles.
The internal diameters of these two capillary tubes a and b are made
unequal; in one, b, on which is placed a mark m, it measures about
$\frac{1}{2}$ millimetre; in the other it is less, not exceeding $\frac{1}{4}$ millimetre at
most. The filling of the apparatus is performed by the method
shown in Fig. 52, the narrow tube a being fitted by means of a
small cork to the glass bulb g, to the other limb of which is attached
a piece of slender india-rubber tubing.

Dipping the wider tube b into the liquid and sucking at the
india-rubber end, a sufficient vacuum may be produced if the bulb g
be large enough, so that by keeping the tubing pinched with the
finger enough liquid will enter to fill the apparatus.

The operation is complete when the liquid begins to drop out at a.
The bulb is then removed, and the instrument placed nearly up to
the level of the ends in a water-bath of the desired temperature.   In
the resulting changes of volume it will be seen that it is only in the
wider capillary tube b that the level of liquid oscillates—that is to say, in
the line of least resistance.   In the narrower limb, it remains steady
throughout at a. If at the desired temperature the liquid in the tube b
stands outside the mark m, it can be adjusted by applying a piece of
blotting-paper to the end a; if, on the other hand, it does not reach
the mark, an additional drop of the solution may readily be introduced
by applying it at a on the end of a glass rod; the capillary action
of the tube sufficing to absorb it and carry forward the level of the
liquid within b.

This operation is capable of so much exactness that in successive experiments, assuming the temperature to continue perfectly
uniform, the weight of the charged instrument will not vary by more
than 0.1 to 0.2 milligramme.   In removing the apparatus from the
water-bath and wiping it previous to weighing, it is obviously
requisite to avoid touching the point at a. The emptying is done
by again attaching the glass bulb and blowing out the contents of
the tube; then a little alcohol and ether sucked into it will serve to
rinse it out and dry it.

Another form of Sprengel pycnometer is shown in Fig. 53.[1]
In this a thermometer is fused into the body of the instrument,
whereby the temperature of the solution can be known with absolute
certainty.   The apparatus is furnished with capillary tubes as in
the instrument already described; but the end of the wider tube

---

[1] May be obtained of Dr. Geissler, Bonn; or, Heintz, glassblower, Aachen.

is ground and fitted with a bend for immersion in the liquid to be aspirated.

Moreover, both ends can be closed with ground-glass caps to prevent loss by evaporation. This form of pycnometer is exceedingly convenient and accurate to work with, the specific gravities determined from successive observations not varying more than two or three units in the fifth place of decimals.

If the specific gravity is to be determined accurately to the fourth decimal place, neglecting variations in the fifth, the temperature of the water-bath must not be allowed to vary by more than 0·2° Cent. With a pycnometer of 10 cubic centimetres capacity, filled with water at a temperature between 17° and 20° Cent., a variation of this amount will produce a difference of 0·4 milligramme in the weight; whilst in the case of other more expansible liquids the same amount of variation

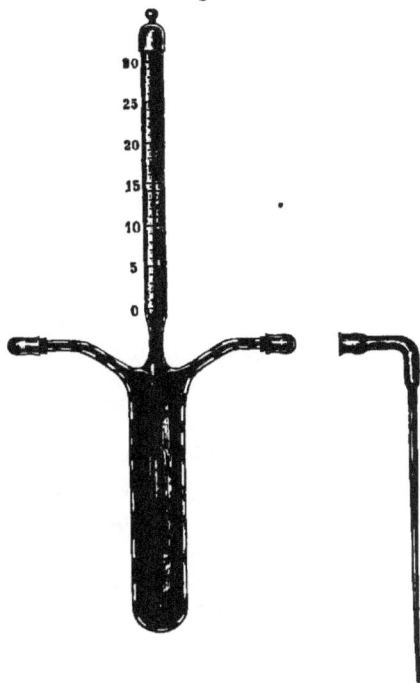

Fig. 53.

of temperature may cause a difference amounting to 2 milligrammes in the weight of 10 cubic centimetres, whereby the specific gravity will be altered by nearly 2 units in the fourth decimal place. In a pycnometer of 20 cubic centimetres capacity, the error will be half this amount.

As in observing the angles of rotation, a normal temperature of 20° Cent. is here also to be preferred. Cylindrical glass jars of several litres capacity, so as to maintain the temperature constant for some time, may be used as water-baths. The immersed thermometer should be graduated to at least fifths of a degree. In a pycnometer of 10 to 20 cubic centimetres capacity complete uniformity of temperature is generally attained in the course of ten minutes.

§ **71.** *To determine the specific gravity or density* we proceed as follows :—The pycnometer is filled at a temperature $t°$, first with distilled water, from which the air has been previously expelled by boiling, and afterwards with the liquid under examination. If now we subtract from each of these weights the weight of the instrument empty, then putting

W for the weight of water,

F for the weight of liquid,

$\frac{F}{W}$ will represent the specific gravity of the solution at a temperature of $t°$ relative to that of water at the same temperature. Multiply this by the density of water at $t° = Q$ (that of water at 4° Cent. being taken as 1), we get the specific gravity of the liquid at $t°$ relative to water at 4° Cent. Lastly, taking into account the influence on the weighings of the pressure of the atmosphere (density δ) we obtain the true specific weight, relative to water at 4° Cent., of the liquid at $t°$ *in vacuo*, which we may designate d_4^t from the formula[1]

$$d_4^t = \frac{F}{W}(Q - \delta) + \delta.$$

The value so obtained expresses the weight in grammes of 1 cubic centimetre of the liquid at $t°$ weighed *in vacuo*.

At the normal temperature of 20° Cent. the density of water $Q = 0\text{·}99826$. Where some other temperature is employed the corresponding value of Q can be found in the table given in § 73.

For the density δ of air—that is, the weight in grammes of 1 cubic centimetre, which varies with temperature and pressure, it will be sufficient to adopt the mean value $0\text{·}0012$, or, reckoning to five places of decimals, $0\text{·}00119$. If the specific gravity of the solution lies between $0\text{·}7$ and $1\text{·}7$, as is the case with nearly all solutions of optically active substances, the temperature of the air at the time of weighing being between 10° and 25°, and the barometric pressure between 720 and 770 millimetres, the above coefficient suffices to correct the influence of atmospheric pressure accurately to within at most 4 units in the fifth place of decimals. But in calculating specific rotations, it is sufficient to know the densities to the fourth decimal place.

Accordingly having ascertained the weight of the pycnometer

---

[1] For the mode of deriving the formula, see F. Kohlrausch, *Leitfaden d. prakt. Phys.*, 3 Aufl. S. 40. [The English reader may consult the translation already referred to on page 136.—D.C.R.]

filled to the mark at 20° Cent. first with water and then with the liquid, the annexed formula will give the specific gravity:—

$$d_4^{20} = \left(\frac{F}{W}\; 0.99707\right) + 0.00119,$$

$$log.\; 0.99707 = 0.998726\; ^{-1}.$$

Example: In determining the specific gravity of an aqueous solution of sugar, the pycnometer filled with water at 20° Cent. weighed $W = 13.6158$ grammes, and filled with the solution $F = 15.4015$ grammes. Hence, by the formula above we get $d_4^{20} = 1.1290$. Neglecting the reduction to vacuum—that is, taking the value simply as $\dfrac{F}{W}\; 0.99826$—we get 1.1292, making the specific gravity 0.0002 too high.   Again, an alcoholic solution of camphor gave $F = 11.4260$ grammes, and, as before, $W = 13.6158$ grammes, whence $d_4^{20} = 0.83863$. By neglecting δ the result would be 0.83763, or 0.001 too small.

Having once determined the weight of water W, contained at the normal temperature by any particular pycnometer, we may reckon the quotient $\dfrac{Q - 0.00119}{W} = C \left(\text{for } 20^\circ : \dfrac{0.99707}{W}\right)$.   And with this constant the specific gravity of any liquid may be found from the value F by the formula

$$d_4^t = F . C + 0.00119.$$

If, as is not unusual, 17·5° Cent. is taken for normal temperature, then since at this temperature $Q = 0.99875$, we get

$$d_4^{17.5} = \left(\frac{F}{W}\; 0.99756\right) + 0.00119.$$

Lastly, if the specific gravity of a solution has to be determined at some temperature other than that at which the weight of water contained by the pycnometer has been ascertained, the change in capacity of the apparatus—that is, the coefficient of cubic expansion of glass—has to be taken into account.   Let

F represent the observed weight of liquid in the pycnometer at the temperature t°;

W the weight of water contained at temperature T°;

Q the density of water at temperature T° (see table § 73);

δ the mean density of the air (0·0012);

γ the coefficient of cubic expansion of glass, which may be taken at 0·000025, then the specific gravity of a solution at the temperature t° referred to water at 4° Cent. and reduced to its value *in vacuo*, may be obtained with sufficient accuracy from the formula

$$d_4^t = \frac{F}{W}\left[1 + (T - t)\,\gamma\right](Q - \delta) + \delta.$$

E. Estimation of the Concentration of Solutions.

(*Preparation of Solutions in Graduated Flasks.*)

§ 72. The *concentration*—by which term we mean the number of grammes of active substance in 100 cubic centimetres of solution—can be obtained by multiplying together the density and percentage weight determined as already described. It can, however, be ascertained directly by weighing a quantity of active substance in a measured flask and forming a solution of determinate volume. The latter method will suffice in cases where the object is simply to determine the specific rotation of some particular solution of the active substance. If, on the other hand, it be desired to know the variation of specific rotation consequent on changes in the proportion of inactive solvent present, and so to deduce the rotation-constant of the active substance itself, as described in §§ 24 and 25, then it is essential to know the percentage composition of the solutions. The measuring flask may indeed also serve for determining this, for which purpose all that is necessary is simply to take the weight also of the contained volume of solution. The method is somewhat simpler, requiring fewer weighings than the process of preparing an indeterminate volume of solution, and then taking the specific gravity. On the other hand, it has the disadvantage that the volume cannot be nearly so accurately known with graduated flasks as with the pycnometer, and as we have to deal with larger volumes of the solutions, it requires longer time to bring the whole to the normal temperature. Moreover, it is not so easy a matter to prepare solutions exactly of the percentage desired. For exact observations, the determination of specific gravity with the pycnometer is therefore far preferable.

Fig. 54.

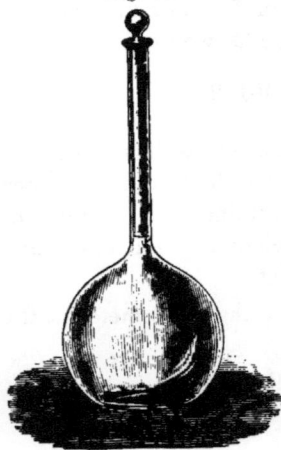

The graduated flask, Fig. 54, which may have a capacity of 20 to 50 cubic centimetres, according to the size of the polariscope-tube, should not be wider in the neck than 8 millimetres at the most, and the mark should be pretty far down, so as to avoid any want of uniformity in the solution. The solution having been prepared approximately in the flask, or in some other vessel from which it can be removed to the flask, a thermometer is inserted before finally adjusting to the mark, and the temperature brought to the standard (20° Cent.) by warming with the hand or some form of water-bath. When the temperature stands at 20° Cent. the thermometer should be removed, washed down with a little of the solvent employed, and the liquid then brought accurately to the mark.

From the weight of the known volume of solution so prepared we have at the same time approximately the specific weight, and by applying, as in § 69, the reduction to vacuum, we may obtain a more accurate value for the percentage composition of the solution.

§ 73. *Standardising the Flasks.*—The volume of the graduated terms of flasks must be determined accurately in the true cubic centimetre (the space, that is, occupied by 1 gramme of water weighed *in vacuo* at a temperature of 4° Cent.). If the measure is already marked, it should be filled almost to the mark with distilled water, a thermometer inserted, and the liquid warmed or cooled to the normal temperature at which the vessel is to be used (17·5° or 20° Cent.). Then withdrawing the thermometer, water should be added until the mark appears, to an eye looking horizontally, tangential to the concave surface of the liquid. The neck of the flask must be freed of all adhering drops, and the weight of water determined. The reduction to vacuum may then be made with sufficient accuracy, if we are using brass weights, by taking each gramme of water weighed in air as 1 milligramme too light.[1] For p grammes of water, the corrected weight P will thus be $P = p + 0·001 p$. If now the density Q of water at the temperature of the experiment $t°$, relative to water at 4° taken as unity, or, in other words, if the weight in grammes of 1 cubic centimetre of water at the temperature $t°$ is known, the volume in

---

[1] Provided, that is, the weight of water does not exceed about 100 grammes. For larger quantities, the value 0·00106 gramme or 1·06 milligrammes given in § 69 must be employed.

cubic centimetres V_t contained in the measure at the temperature $t°$ is given by the equation

$$V_t = \frac{P}{Q}.$$

The density Q of water at different temperatures is shown in the annexed table, prepared by Rosetti[1] from the results obtained by various observers (Kopp, Despretz, Hagen, Matthiessen, and Rosetti). The values are given from 0° to 50° Cent., so as to allow the capacities of graduated flasks or pycnometers to be calculated at any of these temperatures.

t.	Q.	t.	Q.
0°	0·99987	26°	0·99687
1°	0·99993	27°	0·99660
2°	0·99997	28°	0·99633
3°	0·99999	29°	0·99605
4°	1·00000	30°	0·99577
5°	0·99999	31°	0·99547
6°	0·99997	32°	0·99517
7°	0·99993	33°	0·99485
8°	0·99989	34°	0·99452
9°	0·99982	35°	0·99418
10°	0·99975	36°	0·99383
11°	0·99966	37°	0·99347
12°	0·99955	38°	0·99310
13°	0·99943	39°	0·99273
14°	0·99930	40°	0·99235
15°	0·99916	41°	0·99197
16°	0·99900	42°	0·99158
17°	0·99884	43°	0·99118
18°	0·99865	44°	0·99078
19°	0·99846	45°	0·99037
20°	0·99826	46°	0·98996
21°	0·99805	47°	0·98954
22°	0·99783	48°	0·98910
23°	0·99760	49°	0·98865
24°	0·99737	50°	0·98819
25°	0·99712		

[1] Rosetti: *Pogg. Ann., Erg.* Bd. **5**, 268.

Take the case of a flask holding 25·065 grammes of water at a temperature of 20° Cent., then the weight *in vacuo* $P = 25·065 + 0·025 = 25·09$ grammes, and the capacity at 20° Cent. will be $\dfrac{25·09}{0·99826} = 25·13$ cubic centimetres.

With a width of neck of 7 or 8 millimetres, there will be variations in the results of successive determinations, no matter how carefully performed, amounting to about 0·05 cubic centimetre. For example, three subsequent measurements of the above flask gave instead of 25·13 cubic centimetres, the values 25·16, 25·11, 25·17 respectively. Now a difference of 0·05 cubic centimetre has an appreciable effect on the determination of the amount of substance in 100 cubic centimetres. For instance, suppose the flask to hold 25 cubic centimetres, and to contain 5 grammes of active substance, so that the concentration $c = 20$ grammes, then a variation of 0·05 cubic centimetre in volume will represent a variation of 0·84 gramme in weight. The amount of error due to this source decreases the larger the flask, and the less the concentration, so that in a 50 cubic centimetre solution containing 5 grammes of active substance, or $c = 10$, the variation only amounts to 0·01 gramme. The degree to which this error in the determination of the concentration of solutions affects the value of specific rotation is stated in a subsequent section (§ 75).

To graduate a flask for a particular volume, the corresponding weight of water is weighed into it, and the level marked upon its neck by a line. The under-surface of the concavity of the fluid-meniscus is taken as the reference level. For example, to graduate a flask to hold exactly 50 cubic centimetres at a temperature of 20° Cent., there will be required—inasmuch as the previous table shows that 1 cubic centimetre of water at 20° Cent. weighs 0·99826 gramme— $50 \times 0·99826 = 49·913$ grammes of water to give the desired volume. Again, correcting for weighing in air, this amount will be reduced by one-thousandth, so that 49·863 grammes of water at 20° Cent. must be weighed into the previously tared flask. Since, however, the volume so determined is based on a single experiment only, it is necessary, if we wish to be strictly accurate, to fill the flask several times afterwards up to the mark, and take the mean of the several weights. The result so obtained almost invariably differs from a whole number, and since in this way the simplicity of even numbers in the calculations of concentration is lost, it is of no importance to bestow special care on accurate drawing of the mark.

When a measure is used at some temperature t' other than that at which it has been graduated t, allowance must be made for the cubic expansion of glass, the coefficient of which may be taken as 0.000025 for $1°$ Cent. In calculating the volume at t', the capacity at the temperature of graduation being represented by V_t, we have

$$V_{t'} = V_t \left[1 + 0.000025\,(t' - t)\right] \text{ where } t' > t.$$
$$V_{t'} = V_t \left[1 - 0.000025\,(t - t')\right] \text{ where } t' < t.$$

No account, however, need be taken of these variations in volume unless the difference of temperature is considerable and the measures of large size.

§ **74.** Mohr's[1] method of graduating measures used in titration is different. For example, in the case of a 100 cubic centimetre flask, 100 grammes of water at a temperature of $17.5°$ Cent. are weighed into the flask, and the resulting volume marked as 100 cubic centimetres, the reductions to volume at $4°$ Cent. and weight *in vacuo* being omitted.

The cubic centimetres thus obtained are somewhat larger than the true volumes, and, as the density of water at $17.5°$ Cent. $= 0.99875$, and a deduction of 0.00105 gramme has to be made for weighing in air, the ratio between the two will be $1 : 0.9977$. Hence, to reduce Mohr's cubic centimetres to their true value, they must be divided by 0.9977. If, therefore, we use Mohr's measures to determine the concentration-values of solutions, the values obtained will be greater, and so the specific rotations less than when true centimetre measures are employed, and this in the above-stated proportions. Accordingly, concentrations estimated in Mohr's cubic centimetres should be multiplied by 0.9977, and specific rotations calculated therefrom should be divided by 0.9977, to bring them to their proper values for true centimetres.

The definition of specific rotation presupposes the use of the true centimetre, and upon this supposition all scientific data thereto relating are based.

[1] Mohr: *Lehrbuch der Titrirmethode*, 4 Aufl. S. 37.

F. Influence of the several Observation-Errors on Specific Rotation Values.

§ 75. All the factors entering into the formula $[a] = \dfrac{10^t a}{L . d . p}$ or $\dfrac{10^t a}{L . c}$, are severally subject to observation-errors, the effects of which may be estimated as below :—

1. As regards the angle of rotation a, the experiments given in § 60 show that the values obtained by different observers with different instruments do not in general vary by more than the hundredth part of a degree, and that the mean error for a single observation may be taken as $\pm$ 0·025. In the calculations following a maximum error of 0·05° has been allowed.

2. The length L of the tubes can easily be determined by the method described in § 66, so that the variation in a length of 100 millimetres never exceeds 0·05 millimetre.

3. The specific gravity d of solutions can be found accurately to four places of decimals by means of the pycnometer (see §§ 70 and 71). The maximum error here, which may occur when the normal temperature has been taken more than 1° wrong, or, in particular cases, when the correction *in vacuo* has been omitted, may be taken as 0·001.

4. As regards the percentage composition p of solutions, neglect to reduce the weighings may involve an error of about 0·01. In cases, however, where filtering is necessary (see § 68), the percentage may be increased by evaporation to the amount of 0·005 in aqueous solutions, and 0·02 to 0·05 in alcoholic solutions for each 10 per cent. of active substance. In choosing a value as nearly as possible applicable to all liquids, 0·02 has been allowed in what follows as error under this heading.

5. In determining the concentration c (see § 73), an error of 0·04 gramme may occur in solutions containing 20 grammes of active substance. This amount has been taken as a mean.

In order approximately to estimate the actual influence of these errors individually on the value obtained for specific rotation, the following examples have been tabulated for substances having unequal rotatory powers in solutions of different degrees of concentration.

I. *Solutions of Oil of Turpentine in Alcohol.*

Error.	a.	L.	d.	p.	c.	$[a]$.	Diff.
			50 per cent. Solution.				
	15·49°	100	0·825	49·97	41·23	37·57°	—
$a = 0·05$	15·54°	100 .	0·825	49·97	—	37·70°	0·13°
$L = 0·05$	15·49°	100·05	0·825	49·97	—	37·55°	0·02°
$d = 0·001$	15·49°	100	0·826	49·97	—	37·53°	0·04°
$p = 0·02$	15·49°	100	0·825	49·99	—	37·56°	0·01°
$c = 0·04$	15·49°	100	—	—	41·27	37·53°	0·04°
			30 per cent. Solution.				
	9·23°	100	0·813	29·97	24·37	37·88°	—
$a = 0·05$	9·28°	100	0·813	29·97	—	38·09°	0·21°
$L = 0·05$	9·23°	100·05	0·813	29·97	—	37·86°	0·02°
$d = 0·001$	9·23°	100	0·814	29·97	—	37·83°	0·05°
$p = 0·02$	9·23°	100	0·813	29·99 .	—	37·85°	0·03°
$c = 0·04$	9·23°	100	—	—	24·41	37·81°	0·07°
			10 per cent. Solution.				
	3·09°	100	0·801	10·01	8·02	38·54°	—
$a = 0·05$	3·14°	100	0·801	10·01	—	39·16°	0·62°
$L = 0·05$	3·09°	100·05	0·801	10·01	—	38·52°	0·02°
$d = 0·001$	3·09°	100	0·802	10·01	—	38·49°	0·05°
$p = 0·02$	3·09°	100	0·801	10·03	—	38·46°	0·08°
$c = 0·04$	3·09°	100	—	—	8·06	38·34°	0·20°

II. *Solutions of Cane-Sugar in Water.*

Error.	a.	L.	d.	p.	c.	[a].	Diff.
			40 per cent. Solution.				
	31·17°	100	1·177	39·98	47·06	66·24°	—
a = 0·05	31·22°	100	1·177	39·98	—	66·35°	0·11°
L = 0·05	31·17°	100·05	1·177	39·98	—	66·21°	0·03°
d = 0·001	31·17°	100	1·178 ·	39·98	—	66·13°	0·06°
p = 0·02	31·17°	100	1·177	40·00	—	66·21°	0·03°
c = 0·04	31·17°	100	—	—	47·10	66·18°	0·06°
			17 per cent. Solution.				
	12·06°	100	1·068	16·99	18·15	66·46°	—
a = 0·05	12·11°	100	1·068	16·99	—	66·74°	0·28°
L = 0·05	12·06°	100·05	1·068	16·99	—	66·43°	0·03°
d = 0·001	12·06°	100	1·069	16·99	—	66·40°	0·06°
p = 0·02	12·06°	100	1·068	17·01	—	66·39°	0·07°
c = 0·04	12·06°	100	—	—	18·19	66·30°	0·16°
			5 per cent. Solution.				
	3·39°	100	1·018	5·00	5·09	66·60°	—
a = 0·05	3·44°	100	1·018	5·00	—	67·58°	0·98°
L = 0·05	3·39°	100·05	1·018	5·00	—	66·57°	0·03°
d = 0·001	3·39°	100	1·019	5·00	—	66·54°	0·06°
p = 0·02	3·39°	100	1·018	5·02	—	66·51°	0·26°
c = 0·04	3·39°	100	—	—	5·13	66·08°	0·52°

III. *Solutions of Nicotine in Alcohol.*

Error.	a.	L.	d.	p.	c.	[a].	Diff.

60 per cent. Solution.

Error.	a.	L.	d.	p.	c.	[a].	Diff.
	83·69°	100	0·920	59·93	55·14	151·79°	—
a = 0·05	83·74°	100	0·920	59·93	—	151·88°	0·09°
L = 0·05	83·69°	100·05	0·920	59·93	—	151·71°	0·08°
d = 0·001	83·69°	100	0·921	59·93	—	151·62°	0·17°
p = 0·02	83·69°	100	0·920	59·95	—	151·74°	0·05°
c = 0·04	83·69°	100	—	—	55·18	151·67°	0·12°

30 per cent. Solution.

Error.	a.	L.	d.	p.	c.	[a].	Diff.
	37·35°	100	0·855	30·03	25·68	145·47°	—
a = 0·05	37·40°	100	0·855	30·03	—	145·66°	0·19°
L = 0·05	37·35°	100·05	0·855	30·03	—	145·39°	0·08°
d = 0·001	37·35°	100	0·856	30·03	—	145·30°	0·17°
p = 0·02	37·35°	100	0·855	30·05	—	145·37°	0·10°
c = 0·04	37·35°	100	—	—	25·72	145·22°	0·25°

15 per cent. Solution.

Error.	a.	L.	d.	p.	c.	[a].	Diff.
	17·47°	100	0·825	14·96	12·34	141·55°	—
a = 0·05	17·52°	100	0·825	14·96	—	141·96°	0·41°
L = 0·05	17·47°	100·05	0·825	14·96	—	141·48°	0·07°
d = 0·001	17·47°	100	0·826	14·96	—	141·38°	0·17°
p = 0·02	17·47°	100	0·825	14·98	—	141·36°	0·19°
c = 0·04	17·47°	100	—	—	12·38	141·11°	0·44°

According to the foregoing examples, then, it is the angle of rotation above all that one needs to determine with the utmost accuracy, and the more so the less the angle is. The error in the measurement of tube length influences the result but slightly, that in the determination of specific gravity rather more, the effects of both being greater the higher the specific rotation. As regards the error in determining percentage composition and concentration of solutions, this of course exerts a greater effect the more dilute the solutions. In practice, however, the amount is not so great as shown in the foregoing calculations, because the amount of error in determining concentration and percentage composition is there assumed to be the same for all solutions, whereas it diminishes with increased dilution.

In general, indeed, the errors are seldom so large as they are assumed to be in the foregoing examples, and the smaller they can be made the less of course will the result obtained differ from the true value. Besides, errors made by any single observer may be partly positive and partly negative, and so become eliminated from the result. In any case it will be seen that careful working is requisite to get even the first place of decimals correct, and that in all researches of this kind it is necessary to know the degree of accuracy with which the several measurements have been made before any judgment can be formed on the value of the final result.

VI

PRACTICAL APPLICATIONS OF ROTATORY POWER.

A. Determination of Cane-Sugar.

Optical Saccharimetry.

§ 76. The determination of the percentage of sugar in aqueous solutions is based upon the following propositions established by Biot:—

1. The amount of deviation of the plane of polarization is proportional to the length of the liquid column.

2. The deviation is proportional to the concentration—that is, to the number of grammes of sugar in the unit-volume (100 or 1,000 cubic centimetres) of solution.

Accordingly, by determining once for all the angle of rotation given by a single saccharine solution of known concentration in a tube of a certain length, we are able by simple proportion to calculate the number of grammes of sugar in 100 cubic centimetres of any solution of unknown strength from its observed angle of rotation.

As before explained, §§ 23 and 37, Biot's second proposition is not strictly correct, inasmuch as the specific rotation of cane-sugar decreases somewhat with increase of concentration, so that a solution of double percentage does not give exactly double the angle of rotation, but rather less. The errors from this source are, however, small, and apart from very accurate experiments may be entirely neglected. In the following account of the saccharimeters, the proposition is at first taken as strictly true; corrections for varying rotation being always discussed separately.

It is evident that any of the forms of polariscope already described will serve as a saccharimeter, if the graduated arc be replaced by a scale on which the corresponding sugar percentages can be read off directly. Instruments, however, on a different optical principle have been constructed expressly for determining solutions of sugar. Of these so-called *saccharimeters*, now so extensively used in trade, the most important are the following :—

(a.) *The Soleil-Ventzke-Scheibler Saccharimeter.*

§ 77. This very ingenious instrument was originally devised in the year 1848, by the Paris optician Soleil,[1] and more recently improved by Soleil and Duboscq.[2] In Germany, various alterations were made in the instrument by Ventzke,[3] who introduced a different scale, and also by Scheibler,[4] who devised important improvements in the mechanical arrangements.

The optical principle of the instrument is based upon the following facts discovered by Biot :—1. That when a polarized ray is transmitted through several media possessing rotatory power in different directions, their separate activities may become either partially or wholly neutralized, according to the lengths of the media. 2. That the rotatory dispersion of cane-sugar is the same as that of quartz. White day or lamp-light is used with the instrument. The optical parts are as shown in Fig. 55, the course of the

Fig. 55.

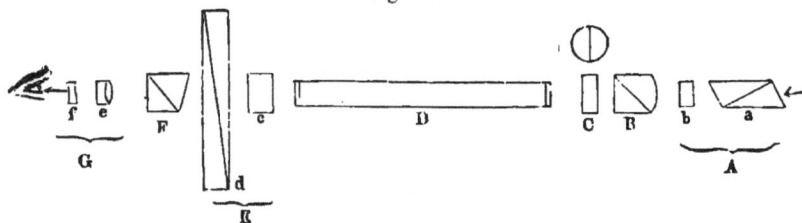

light being from right to left. Starting from the right-hand side, we have :—

A. The so-called *regulator* for restoring the sensitive tint, consisting of a rotating Nicol's prism *a*, and a plate of quartz

[1] Soleil: *Comptes Rend.*, 24, 973; 26, 162.
[2] Soleil and Duboscq: *Idem.* 31, 248.
[3] Ventzke: *Erdm., Journ. für prakt. Chem.* 25, 84; 28, 111.
[4] Scheibler: *Zeitsch. des Vereins für Rübenzuckerindustrie*, 1870, 609.

b (dextro-rotatory or lævo-rotatory), ground perpendicularly to the axis.

B. The *polarizer*, for which an achromatized calc-spar prism, a Sénarmont's prism, or double quartz prism can be used. It is arranged with its principal section vertical. The polarized extraordinary ray is allowed to proceed along the axis, but the sidewards-refracted ordinary ray is intercepted by a diaphragm. The face of the prism towards *A* is ground convex, so that light may emerge approximately parallel.

C. The *bi-quartz*, composed of two plates, the one of dextro-rotatory, the other of lævo-rotatory quartz, fitted accurately together. They may be either 3·75 millimetres or 7·5 millimetres thick, and are fixed in a brass case, so that the line of junction remains vertical. (A front view is shown above *C*.)

D. The *experimental tube*.

E. The so-called *rotation-compensator*. This consists of a plate of quartz *c*, which may be dextro-rotatory, in which case the other two plates, *d*, must be made of lævo-rotatory quartz. The latter are ground to a wedge shape, and are made to slide over one another, so that their combined thickness may be made either equal to that of *c*, or greater or less, the distance moved to effect any particular adjustment being shown by an attached scale.

The plate *c* may be made of left-handed quartz, but in that case the wedges must be right-handed.

F. The *analyzer*, which may consist of an achromatized calc-spar prism. Its principal section must be arranged parallel to that of the polarizer *B*, in case the thickness of the bi-quartz of the instrument, *C*, is 3·75 millimetres, and perpendicular thereto when the thickness is 7·5 millimetres.

G. A small Galilean telescope, consisting of objective *e* and eyepiece *f*. The latter is to be adjusted so that the line of junction of the plates of the bi-quartz *C* is sharply defined

§ 78. In order clearly to understand the action of the several parts, let us suppose first that we are merely dealing with the polarizer *B* and analyzer *F*, and that these are arranged with principal sections parallel, and the field at its maximum of illumination. Let the active bi-quartz *C* be now introduced between *B* and *F*; the white light coming from B will thus be rotated, and suffer decomposition into its component coloured rays. Now of the emergent rays, those whose plane of polarization is at right angles to that of the analyzer will not be transmitted; and should these be the yellow rays, the remainder will, in transmission, combine to a pale lilac mixed tint which, with the slightest alteration of the plane of polarization, passes either into pure red or pure blue. This intermediate colour has been already referred to in § 47 as the *sensitive* or *transition tint*. With the polarizer and analyzer undisturbed in their parallel position, the sensitive tint will appear when the bi-quartz C has the thickness requisite to rotate the yellow rays exactly 90° to the right or left. This requires a thickness of 3·75 millimetres, since, according to Biot, a thickness of 1 millimetre of quartz rotates mean yellow rays through an angle of 24°, whence we get the proportion, $24 : 1 = 90 : 3·75$. If, on the other hand, the corresponding planes of polarizer and analyzer were set at right angles to each other, a rotation through 180° would be required to eliminate the yellow rays, and the bi-quartz must then have a thickness of 7·5 millimetres. In one half of the bi-quartz the sensitive tint is produced by dextro-rotation, in the other by lœvo-rotation, and as the thickness of the two sides is equal, the tints produced will be the same.

Let the compensator E now be put in its place, the quartz wedges being so adjusted that their combined thickness is exactly equal to that of the quartz plate c. As the rotatory powers of c and d act in opposite directions, they neutralize each other, and the sensitive tint still occupies the field of vision. This position of the wedges corresponds with the zero-point of the scale. If, however, we shift the relative position of the wedges, the transmitted coloured rays will suffer rotation, and the analyzer will eliminate those of which the planes of polarization are perpendicular to its own. The other rays which pass on combine to produce a new chromatic mixture, which will, necessarily, be unlike for the rays transmitted by the respective halves of the bi-quartz C. Thus, if the compensator be so adjusted that its action is dextro-gyrate, the rays contributed by the dextro-rotatory half of the bi-quartz will undergo an increase, and those coming from the lœvo-rotatory half a decrease, in the amount of their rotation. Seen through the analyzer, the two halves will thus appear differently coloured—green and blue predominating in one, red and orange light in the other; and these tints will change places when the action of the compensator is lœvo-gyrate.

Lastly, having again set the compensator to zero and repro-
duced the sensitive tint, let a tube filled with an optically-active
solution be introduced. A splitting of the field of vision into two
different coloured halves will once more occur. By sliding the
quartz wedges, d, so as to produce rotation opposite to that of the
solution, a position may be found where the action of the latter is
annulled, and this will be indicated when the halves of the field of
vision again exhibit uniformly the sensitive tint. The rotatory power
of the substance introduced can thus be measured by the amount of
change in the combined thickness of the quartz wedges, d, as indicated
by the amount of adjustment required to bring into view the sensitive
tint.

For the proper action of the compensator, it is, however, requisite that the active solution should have the same dispersive power as
quartz, otherwise the effects of dispersion due to the solution will not
be exactly neutralized by the opposite dispersion of the quartz. As
before stated, § 18, this is the case with cane-sugar; but there are
many substances which do not fulfil this condition, as, for instance,
cholesterin, and in such cases, at least with strong rotation, the
compensator can no longer be adjusted so as to restore a perfectly
uniform colour to the field of vision. With Soleil's instrument,
exact data are therefore only attainable in the case of substances
whose rotatory dispersion does not materially differ from that of quartz.

The sensitive tint makes its appearance as above described when
ordinary white light is used. When, on the contrary, lamp-light is
employed, which contains the coloured rays in somewhat different
proportions, there appears instead not the blue-violet, but a red-
dish tint, and a similar alteration may occur when the active
solution is itself coloured. In this way arises the need for yet
another addition to the instrument, namely, the regulator A, by
which we can still produce the sensitive tint. It consists of a
rotating Nicol prism, a, and a quartz plate, b, both placed in front
of the polarizer B. When light which has been polarized by the
Nicol falls on the quartz plate, it undergoes rotatory dispersion, and
according to the position of the Nicol in reference to the polarizer, B,
so will certain rays suffer total extinction by the polarizer or be
transmitted by it with diminished intensity. In this way we are
able to weaken the red and yellow rays of lamp-light, and so admit
to the bi-quartz, C, a selected mixture of rays, which, in consequence
of the rotation there experienced, reproduces the sensitive tint. As

the polarizer, B, brings the transmitted rays into one plane of polarization, the two halves of the bi-quartz will appear coloured alike, the tint being variable at will by turning the Nicol a. A similar process is adopted in dealing with coloured solutions. The colour of the latter should, however, never be more than faint, otherwise too much light will be lost by absorption.

The two portions forming the regulator can be placed at the eye-end instead of the light-end if preferred, as is done in instruments of French make.

§ 79. The external form of Soleil's saccharimeter with Scheibler's improvements is shown in Fig. 56, which represents the instrument as supplied by Berlin opticians.[1]

On a brass stand rests, with horizontally rotating motion, a blackened metal trough, $h\,h$, in which the experimental tube R is

Fig. 56.

laid, the upper half of the trough serving as a cover, which can be shut down during observations so as to exclude extraneous light. The end of the trough next the light—in the Fig. the right-hand end—is connected with a brass tube enclosing at one end the bi-quartz D, at the other the polarizer C. Into this tube fits the rotatory tube, $A\,B$, containing the regulator with its Nicol at A, and the quartz plate at B. The rotation of the tube is effected by wheel and pinion movement, worked by means of a rod with the milled-head L. The ocular part of the instrument contains at G the quartz plate of the compensator, and at $E\,F$ the two quartz wedges. Each of the wedges is cemented to a similar wedge of glass, so as to form

[1] May be obtained of Schmidt and Haensch, Stallschreiberstrasse 4, Berlin; Dr. Steeg and Reuter, Homburg v. d. Höhe.

two plano-parallel plates, which are set in brass frames. One of
these plates is fixed, while the other has a horizontal sliding motion,
for which purpose it is provided at the bottom with a side to side
rack and pinion motion, worked by the milled-head M. The frame
on the upper edge of the movable plate carries the divided scale,
and that of the fixed plate a vernier. For reading the latter an
inclined mirror s throws the image of the scale along the axis of the
tube, which is fitted with a lens K. The piece, $H I$, contains the
analyzer and the telescopic lenses, the eye-glass being movable.
At H in the Fig. is shown a screw-head, which admits of the analyzer being adjusted to its proper position in relation to the fixed
polarizer. It is only to be used, however, either by the makers in
the first adjustment of the instrument, or when it has got out of
proper adjustment—when, that is, we cannot by any movement of
the quartz wedges bring a uniform tint into the two halves of the
field. When this is the case, either the polarizer or analyzer must
have been disturbed. To readjust the instrument, the two quartz
wedges along with the fixed plate are removed, and the screw H
turned until uniformity of tint in the field is attained. To fix the
zero-point, another screw, not shown in the Fig., is attached to the
quartz wedge frame which carries the scale, and allows of a certain
amount of the adjustment of the latter. Lastly, as regards the
length of the solution tubes, the instrument is generally arranged for
tubes 2 decimetres in length ; but saccharimeters capable of taking
4 and 6 decimetre tubes are made, and are of service in the analysis
of weak saccharine solutions.

§ 80. In using the saccharimeter, it is set up against a gas or
paraffin lamp furnished with a metal chimney having a side opening
(see Fig. 25, p. 102), care being taken to keep it a distance of at least
5 centimetres (2 inches) from the flame, to prevent the Nicol becoming heated. The manipulation is then as follows :—

1. A tube, either empty or filled with water, is laid in the
instrument, and the eye-piece of the telescope drawn out until the
vertical joining line bisecting the circular bi-quartz appears sharply
defined ; next taking the milled-head M (Fig. 56), the movable
quartz wedge must be adjusted, until over both halves of the field an
approximately uniform colour prevails.

2. The regulator is then moved by means of the milled-head L
of the side gearing, so as to bring into view that sensitive tint

which with the slightest displacement backwards or forwards of the movable quartz wedge produces the most distinct colour-differentiation in the two semi-circles. Thus it will be found that if we start with a uniform deep-red or deep-blue tint, the quartz wedge has to be moved much further before we can distinctly recognize a difference of colour in the halves, than if we had started from a uniform bright violet tint. For most eyes the best position of the regulator is that which gives a sensitive tint most nearly approaching to white. Here a very slight movement of the quartz wedge will cause one half to exhibit a faint greenish tinge, and the other a flesh-pink, which on further turning of the milled-head, *M*, passes into green and orange-red respectively. In arranging the regulator for this position we do not get exactly the ordinary sensitive tint—the reddish-purple, which denotes most perfect extinction of the yellow light. Some of the yellow rays are allowed to pass, thereby making possible the green and orange colours which appear. It is, however, in other respects immaterial which tint is chosen for purposes of observation.

3. Having once decided upon the sensitive tint, the quartz wedge is adjusted to give the greatest possible similarity of colour in the halves of the field and the corresponding reading on the scale noted. This should be several times repeated, the mean of the readings being taken as the zero-point of the instrument. Should this not coincide with the zero-point of the scale, the latter must be adjusted, by slightly moving the screw referred to in § 79 as being fixed on the brass frame of the wedge. The position of the zero-point must be verified from time to time. Moreover, it is found to vary considerably for different eyes.

4. If the tube be now filled with a solution of cane-sugar and laid in its place, colour-dissociation at once takes place, and may be made to vanish again by moving the quartz wedge in such a way that the 100 point on the scale is made to approach the zero-point of the fixed vernier. If the solution be perfectly colourless, the screw *L*, Fig. 56, which manipulates the regulator, need not be touched ; but if, on the contrary, as frequently happens, it has a yellowish tinge, we must disturb the screw a little to the right or left, until we obtain, as nearly as possible, the tint used in fixing the zero-point. By making several such adjustments of the quartz wedge to the position in which colour-uniformity is shown, we arrive at an accurate determination of the rotation. With a little practice, it will be found easy to get observa-

tious not varying from one another by more than 0·4 division of the scale at the most, provided that the instrument is well constructed. Where greater differences occur, as will be the case with coloured solutions, the number of observations must be increased. As a rule, five are enough to determine the mean value within ± 0·1 of a division.

This instrument is, of course, unsuited for the colour-blind.

§ 81. *The graduation of the saccharimeter scale*, according to Ventzkè, is made by laying in the instrument a tube 2 decimetres long, filled with an aqueous solution of pure sugar, having at the temperature of 17·5° Cent. a specific gravity of 1·1, and marking the point of the observed deviation as 100. A solution of the above density contains in 100 cubic centimetres exactly 26·048 grammes of sugar, so that it can be more easily prepared by weighing this quantity and adding the proper amount of water. The space between the 100 point and the zero-point is then divided into one hundred equal parts, and the graduation extended some way (30 or 40 divisions) on the other side of the zero-point. As cane-sugar is dextro-rotatory, the latter divisions of the scale will indicate lævo-rotation, and must be marked with the negative sign.

Assuming that deviation is exactly proportional to concentration, each division of Ventzke's scale will indicate a sugar value of 0·26048 gramme in 100 cubic centimetres of solution (or 2·6048 grammes per litre). Accordingly the process for estimating a saccharine solution is as follows :—

To determine *concentration* (the number of grammes of sugar in 100 cubic centimetres of solution), we fill a 2 decimetre tube with the solution, note the deviation on the scale, and multiply the number of degrees by 0·26048. The solution must not, of course, contain more than 26·048 grammes of sugar in 100 cubic centimetres, otherwise the indication on the scale would pass outside the 100 point.

To determine *per cent. composition* (the number of grammes of sugar in 100 grammes of solution), weigh out 26·048 grammes of the solution, dilute to the mark in a 100 cubic centimetre flask, with this latter solution fill a 2 decimetre tube, and note the degree of deviation on the scale. This will indicate directly the percentage of sugar in the original solution.

The analysis of solid saccharines may be effected similarly :—To estimate the value of a crude sugar, for instance, 26·048 grammes of substance are dissolved in water, diluted to 100 cubic centimetres, and observed in a 2 decimetre tube. The degree of deviation expresses directly the percentage of pure sugar in the substance.

If instead of 26·048 grammes we had used some other weight P of liquid, or solid saccharine matter, in preparing the 100 cubic centimetre solution, then, taking the degree of deviation observed in a 2 decimetre tube as a, the percentage composition will be

$$x = \frac{26·048 \times a}{P}.$$

Moreover, the percentage weight of a sugar solution can be calculated from the concentration as determined directly, provided we know the specific gravity—that is, the weight of 100 cubic centimetres.

When a tube 1 decimetre in length is used instead of a 2 decimetre tube, the scale-readings must, of course, be doubled; similarly, in using a 4 decimetre tube, they must be halved.

For the method of preparing sugar solutions for the saccharimeter, see § 92.

The original method as used by Ventzke was to prepare, by means of an areometer, a solution of the saccharine substance to be analyzed of a specific gravity 1·1, and to examine this in a 2 decimetre tube. The observed deviation was then taken as giving directly the weight per cent. of sugar in dry substance. This method, however, which was intended to dispense with all weighings, does not yield accurate results, as the salts contained in natural sugars always have a specific gravity different from that of the sugar itself. It has therefore been abandoned, but the awkward normal weight of 26·048 grammes continues in use. It would be much more convenient to substitute some simpler figure—as, for example, 20 grammes—in which case, however, a re-calculation of the tables prepared for Ventzke's instrument would be necessary.

§ 82. *Correction of Saccharimeter-Readings for slight Disproportionality between Rotation and Concentration.*—As shown by the researches of Schmitz and Tollens, already described § 37, the specific rotation of sugar is not constant for solutions of different concentrations, but increases inversely as the concentration. Thus, given that a solution containing 26·048 grammes of sugar in 100 cubic centimetres records 100 on the scale of the saccharimeter, the reading

50 on the same scale will not be recorded exactly by one containing 13·024 grammes, but by a solution containing a somewhat smaller sugar percentage. Hence the need, at least in the case of exact observations, of correcting the saccharimetric readings.

Schmitz[1] has calculated the required correction, on the basis of the following observations :—

No. of Solution.	Concentration c.	$[a]_D$ Observed.	$[a]_D$ Calculated.	Difference.
I.	27·6407	66·328°	66·308°	− 0·020
II.	18·1751	66·375°	66·388°	+ 0·013
III.	10·3994	66·460°	66·453°	− 0·007
IV.	5·0955·	66·495°	66·498°	+ 0·003

According to these experiments the increase of specific rotation with decrease of concentration may be expressed by the following interpolation-formula, from which the calculated values in the table were obtained :—

$$[a]_D = 66·541 − 0·0084153 \, c.$$

Putting $[a]_n$ as the specific rotation calculated from this formula, for a normal solution containing 26·048 grammes of sugar in 100 cubic centimetres of solution, and $[a]_c$ as that for some other solution of lower concentration c, then $\dfrac{[a]_c}{[a]_n}$ represents the proportion in which the concentration and per cent. composition will appear too high when estimated on the supposition that the rotation varies uniformly therewith. To obtain the true values the results must, therefore, be divided by the above fraction. This calculation is exemplified in detail in the table annexed.

Col. 1. N gives the scale-reading at every ten divisions.

,,   2. c, the number of grammes of sugar in 100 cubic centimetres of solution, assumed to correspond.

,,   3. $[a]_N$ and $[a]_c$, the specific rotation of sugar for the above-mentioned concentrations (26·048 and c), calculated from the interpolation-formula.

,,   4. $\dfrac{[a]_c}{[a]_N} = Q$, the ratio of the specific rotations.

[1] Schmitz : *Zeitsch. des Vereins für Rübenzuckerindustrie*, 1878, 63.

Col. 5. $\dfrac{c}{Q}$, the corrected number of grammes of sugar in 100 cubic centimetres of solution.

,, 6. $c - \dfrac{c}{Q}$, the difference between the assumed and corrected values of the concentration.

,, 7. $\dfrac{N}{Q}$ or $\dfrac{100 \cdot c}{26 \cdot 048 \, Q}$, the corrected scale-reading.

,, 8. $N - \dfrac{N}{Q}$, the difference between the corrected and uncorrected percentages.

1.	2.	3.	4.	5.	6.	7.	8.
N	c	$[\alpha]$	$\dfrac{[\alpha]_o}{[\alpha]_n} = Q$	$\dfrac{c}{Q}$	$c - \dfrac{c}{Q}$	$\dfrac{N}{Q}$	$N - \dfrac{N}{Q}$
100	26·048	$[\alpha]_n = 66·322°$	1·00000	26·048	0·000	100	0·000
90	23·443	$[\alpha]_o = 66·344°$	1·00035	23·435	0·008	89·969	0·031
80	20·838	,, 66·366°	1·00070	20·824	0·014	79·945	0·055
70	18·234	,, 66·388°	1·00101	18·216	0·018	69·930	0·070
60	15·629	,, 66·409°	1·00132	15·608	0·021	59·920	0·080
50	13·024	,, 66·431°	1·00165	13·003	0·021	49·917	0·083
40	10·419	,, 66·453°	1·00198	10·398	0·021	39·921	0·079
30	7·814	,, 66·475°	1·00232	7·796	0·018	29·930	0·070
20	5·210	,, 66·497°	1·00265	5·196	0·014	19·947	0·053
10	2·605	,, 66·519°	1·00297	2·597	0·008	9·970	0·030

As the 0 and 100 marks of the saccharimeter scale, indicating 0 gramme and 26·048 grammes concentration respectively, are given as fixed, the difference between the corrected and the uncorrected readings will be greater the further their distance from these points, and greatest midway between them. This is shown in cols. 6 and 8.

The fact that the formula above given for the calculation of specific rotation refers to ray D, whereas observations with the Soleil-Scheibler saccharimeter are taken with the transition tint, has no effect upon the results, the ratio $\dfrac{[\alpha]_o}{[\alpha]_n}$ being the same for all rays.

The differences between the corrected and uncorrected values are thus seen to be not inconsiderable for solutions of medium concentration. The corrections required have been calculated by Schmitz for each degree of the saccharimeter scale and embodied in the accompanying table :—

Degrees of Saccharimeter Scale.	Corrected per cent.	Grammes of Sugar in 100 cub. cent. Solution.		
		Uncorrected.	Corrected.	Difference.
1	1·00	0·261	0·260	0·001
2	1·99	0·521	0·519	0·002
3	2·99	0·781	0·779	0·002
4	3·99	1·042	1·039	0·003
5	4·98	1·302	1·298	0·004
6	5·98	1·563	1·558	0·005
7	6·98	1·823	1·817	0·006
8	7·98	2·084	2·078	0·006
9	8·97	2·344	2·337	0·007
10	9·97	2·605	2·597	0·008
11	10·97	2·865	2·857	0·008
12	11·97	3·126	3·117	0·009
13	12·96	3·386	3·376	0·010
14	13·96	3·647	3·637	0·010
15	14·96	3·907	3·896	0·011
16	15·96	4·168	4·156	0·012
17	16·95	4·428	4·416	0·012
18	17·95	4·689	4·676	0·013
19	18·95	4·949	4·936	0·013
20	19·95	5·210	5·196	0·014
21	20·95	5·470	5·456	0·014
22	21·94	5·731	5·716	0·015
23	22·94	5·991	5·976	0·015
24	23·94	6·252	6·236	0·016
25	24·94	6·512	6·496	0·016
26	25·94	6·773	6·756	0·017
27	26·94	7·033	7·016	0·017
28	27·93	7·293	7·276	0·017
29	28·93	7·554	7·536	0·018
30	29·93	7·814	7·796	0·018
31	30·93	8·075	8·056	0·019
32	31·93	8·335	8·316	0·019
33	32·93	8·596	8·577	0·019
34	33·93	8·856	8·837	0·019
35	34·92	9·117	9·097	0·020
36	35·92	9·377	9·357	0·020
37	36·92	9·638	9·618	0·020

Degrees of Saccharimeter Scale.	Corrected per cent.	Grammes of Sugar in 100 cub. cent. Solution.		
		Uncorrected.	Corrected.	Difference.
38	37·92	9·898	9·878	0·020
39	38·92	10·159	10·138	0·021
40	39·92	10·419	10·398	0·021
41	40·92	10·680	10·659	0·021
42	41·92	10·940	10·919	0·021
43	42·92	11·201	11·180	0·021
44	43·92	11·461	11·440	0·021
45	44·92	11·722	11·701	0·021
46	45·92	11·982	11·961	0·021
47	46·92	12·243	12·222	0·021
48	47·92	12·503	12·482	0·021
49	48·92	12·764	12·743	0·021
50	49·92	13·024	13·003	0·021
51	50·92	13·285	13·264	0·021
52	51·92	13·545	13·524	0·021
53	52·92	13·805	13·784	0·021
54	53·92	14·066	14·044	0·022
55	54·92	14·326	14·305	0·021
56	55·92	14·587	14·566	0·021
57	56·92	14·847	14·826	0·021
58	57·92	15·108	15·087	0·021
59	58·92	15·368	15·347	0·021
60	59·92	15·629	15·608	0·021
61	60·92	15·889	15·868	0·021
62	61·92	16·150	16·130	0·020
63	62·92	16·410	16·390	0·020
64	63·92	16·671	16·651	0·020
65	64·92	16·931	16·912	0·019
66	65·93	17·192	17·173	0·019
67	66·93	17·452	17·433	0·019
68	67·93	17·713	17·694	0·019
69	68·93	17·973	17·954	0·019
70	69·93	18·234	18·216	0·018
71	70·93	18·494	18·476	0·018
72	71·93	18·755	18·738	0·017
73	72·93	19·015	18·998	0·017
74	73·94	19·276	19·259	0·017

Degrees of Saccharimeter Scale.	Corrected per cent.	Grammes of Sugar in 100 cub. cent. Solution.		
		Uncorrected.	Corrected.	Difference.
75	74·94	19·536	19·519	0·017
76	75·94	19·797	19·781	0·016
77	76·94	20·057	20·042	0·015
78	77·94	20·317	20·302	0·015
79	78·94	20·578	20·564	0·014
80	79·95	20·838	20·824	0·014
81	80·95	21·099	21·085	0·014
82	81·95	21·359	21·346	0·013
83	82·95	21·620	21·608	0·012
84	83·95	21·880	21·868	0·012
85	84·96	22·141	22·130	0·011
86	85·96	22·401	22·391	0·010
87	86·96	22·662	22·652	0·010
88	87·96	22·922	22·912	0·010
89	88·97	23·183	23·174	0·009
90	89·97	23·443	23·435	0·008
91	90·97	23·704	23·696	0·008
92	91·98	23·964	23·957	0·007
93	92·98	24·225	24·219	0·006
94	93·98	24·485	24·480	0·005
95	94·98	24·746	24·742	0·004
96	95·98	25·006	25·002	0·004
97	96·99	25·267	25·265	0·002
98	97·99	25·527	25·525	0·002
99	98·99	25·788	25·787	0·001
100	100·00	26·048	26·048	0·000

As will be seen by comparing cols. 1 and 2, the corrected percentages at points between 17 and 84 on the scale, differ from the actual readings by amounts ranging from 0·05 to 0·08. Within these limits, therefore, and reckoning percentages to tenths only, it will suffice to deduct 0·1 per cent. from the actual readings. With rotations of less than 16 or more than 85 scale degrees, such as occur in the analysis of natural sugars and their refined products, the correction is superfluous.

§ 83. *To convert the degrees on Ventzke's scale into true rotation-degrees*, as is necessary when the specific rotation of a substance is to be determined with this instrument, the formula $[a]_D = 66\cdot541 - 0\cdot0084153\ c$, given in § 82, can be used. With concentration $c = 26\cdot048$, $[a]_D = 66\cdot322$, and determining the angle of rotation given by such solution, in a tube 2 decimetres long, from the equation $\dfrac{a \times 100}{2 \times 26\cdot048} = 66\cdot322$ we get $[a]_D = 34\cdot55°$. That is to say, a solution of 26·048 grammes of sugar in 100 cubic centimetres, which rotates mean yellow ray j to the amount of 100 divisions on the Ventzke scale, would record on instruments having angular graduation a rotation of the sodium ray D through an angle of 34·**55°**. Hence,

1° Ventzke's scale (ray j) = 0·3455° angular measurement (ray D).

And further assuming the dispersive power of the substance to be equal to that of quartz, in which, as we have seen (§ 18), the rotations for rays D and j are to one another as 1 to 1·1306, we obtain :—

1° Ventzke's scale (ray j) = 0·3906° angular measurement (ray j).

Another mode of arriving at these relations is afforded by the observed fact (see § 18) that a quartz plate 1 millimetre thick rotates ray D through 21·67° and ray j through 24·5° angular measure. In a subsequent table (§ 91) the angles of rotation of ray D for sugar solutions of various degrees of concentration, in 2 decimetre tubes are given, from which, by interpolating for the decimal figures, it appears that to give an angle of rotation of 21·67°, a solution must contain 16·302 grammes of sugar in 100 cubic centimetres. This concentration, it will be seen from the table already given, § 82, corresponds with a reading of 62·662 divisions on Ventzke's scale. It is evident, therefore, that by dividing the angular values 21·67° and 24·5° by that number we obtain the value of 1° Ventzke. Thus :—

1° Ventzke (ray j) = 0·3458 angular degrees (ray D),
1° Ventzke (ray j) = 0·3910 angular degrees (ray j),

values which agree almost exactly with those previously obtained.

§ 84. *Correction of Errors due to Imperfect Construction.*— In using a new instrument, it is necessary previously to test the correctness of the scale. When the zero-point of the instrument has been carefully fixed, and brought, by means of the adjustment screw, to coincide exactly with the zero-point of the graduation, the introduction of a 2 decimetre tube filled with a solution of 26·048 grammes of sugar in 100 cubic centimetres, should give a rotation of exactly 100 on the scale.

For this experiment it is necessary to use specially prepared pure sugar only. The best refined still contains from 0·1 to 0·2 per cent. of inorganic matter, and does not record more than 99·8 to 99·9. In sugar-candy, the frequent presence of invert-sugar, recognizable by its reduction of Fehling's copper solution, causes it to record a rotation likewise too low. To obtain a pure material, sugar-candy should be repeatedly crystallized from alcohol of about 85 per cent. Or it may be prepared by putting one part sugar-candy in half its weight of water and heating till it dissolves, filtering the hot solution into a porcelain dish, adding two parts absolute alcohol, and stirring frequently till it cools. The sugar crystallizes out as a fine powder, which should then be brought upon a filter and washed, first with dilute, then with strong alcohol, and dried at a temperature of about 60° Cent. The product thus obtained does not yield more than about 0·005 per cent. of ash, and has no action on Fehling's solution.

If now a solution of 26·048 grammes of sugar so purified does not give a rotation of exactly 100 divisions of the scale, the reading being either too high or too low, it will be necessary, before using the instrument, to determine the particular *normal sugar weight* belonging to it. Suppose, for instance, that the above solution has given the reading 100·3° on the scale, the sugar-percentage answering to a reading of 100 must be determined from the proportion

$$100\cdot3° : 26\cdot048 \text{ grammes} = 100° : x$$
$$\text{whence } x = 25\cdot970 \text{ grammes.}$$

So that in all applications of the instrument the normal weight must be taken as 25·970 instead of 26·048 grammes, and 1° of the scale must be understood to indicate 0·2597 gramme sugar in 100 cubic centimetres solution. In this way accurate results can be obtained with such an instrument, but it presents the disadvantage that all tables calculated for the number 26·048 are useless, and must be reconstructed.

If the 100 point has been found in its proper place, still the scale, which is supposed to be equally graduated throughout its length, must be tested at a few other points as well.

Proceeding, according to Schmitz's table, given in § 82, solutions are prepared containing respectively 19·519, 13·003, and 6·496 grammes of pure sugar, which should give rotations of 75°, 50°, and 25° respectively. If, however, important discrepancies should occur in the readings, then it becomes necessary to prepare a whole series of solutions of known saccharine strengths, note the degrees of rotation indicated by each, and then draw up a special correction-table for the instrument. This is best done by the graphic method.

Errors of this kind appear when the four faces of the two quartz wedges of the compensator have not been ground perfectly true, so that differences in the total thicknesses of the compensators, produced by the sliding of the wedges, are not perfectly proportional to the differences of reading on the scale. Scheibler [1] has given a method by which such errors, which are of frequent occurrence, can be eliminated, at least when we are dealing with rotations exceeding 80°.

This, the so-called method of *double observation*, most generally used in the analysis of natural sugars, is as follows :—A sample of 26·048 grammes is made into a 100 cubic centimetre solution, or more commonly 13·024 grammes are taken and a 50 cubic centimetre solution prepared, and the rotation observed in a 2 decimetre tube in the ordinary way. If the degrees indicated be, let us say, 94·2° (which should therefore be the sugar-percentage), this result will be correct provided the quartz wedges at the point corresponding to this reading are of the proper thickness. To test this, we must calculate the concentration required to give a rotation of 100° by the proportion

$$94·2° : 13·024 = 100° : x$$
whence $x = 13·826$.

A 50 cubic centimetre solution must then be prepared with 13·826 grammes of the sugar to be analyzed, and the rotation observed in a 2 decimetre tube. If it gives 100° on the scale, the first result, 94·2 per cent., is correct.

If, however, the second solution does not give exactly 100°, but some less number, as 99·6°, in that case the result of the first observation is incorrect. The correct sugar-percentage can, however, be easily ascertained, as it must stand in the proportion

$$13·826 : 99·6 = 13·024 : x$$
$$x = 93·8.$$

The true sugar-percentage is accordingly 93·8, and the number of degrees indicated on the scale in the first experiment (94·2) was 0·4 too high.

[1] Scheibler : *Zeitsch. des Vereins für Rübenzuckerindustrie*, 1870, 212 ; 1871, 318.

If, again, the second experiment gives, say, 100·2 instead of 100, the proportion will then stand :—

$$13·826 : 100·2 = 13·024 : x,$$

and the correct sugar-percentage will be

$$x = 94·4 \text{ per cent.}$$

In this way errors arising from imperfect construction of the quartz wedges can be eliminated, and as the second observation invariably lies close to the 100 point, the position of which has already been accurately fixed, it is found that the results given by different instruments correspond within $\pm$ 0·1 per cent., while by the ordinary method the differences between them may be much greater. As to the effect of the imperfect proportion between rotation and concentration upon this method, we have already said (§ 82) that when we are dealing with sugar-percentages recording more than 84°, such as we find in crude sugars, it may be neglected. When, however, the solutions indicate deviations ranging from 30° to 76°, this is no longer the case, and the method of double observation fails to afford correct results.

It will be seen from the foregoing remarks that the quartz-compensation principle in polariscope instruments involves considerable difficulties. Errors and corrections, like those just mentioned, do not occur in instruments with rotating Nicols (Wild's and Laurent's); moreover, these give generally more accurate results.[1] For the method of verifying the length of solution-tubes see § 66. For the influence of glass end-plates see § 64.

§ 85. *Influence of Temperature on Determinations by the Saccharimeter.*—The normal temperature at which the determination of the 100 point is made being 17·5° Cent. (63·5 F.), the experiments will ordinarily be made at some other temperature ; for the experimental tubes not being usually provided with water-jackets, but exposed to the air, are subject to any variations in the temperature of the apartment. It is true, indeed, as Tuchschmid's[2] researches have shown, that the specific rotatory power of sugar is not in itself affected by heat, but the directly observed deviation is influenced thereby. Thus, when the temperature rises, the length of the tube,

---

[1] Schmidt and Haensch, opticians, Berlin, have lately brought out a half-shade instrument, with quartz-wedge compensators and Ventzke's scale. This instrument admits of much more accurate adjustment than the usual bi-quartz colour instrument. The variations do not exceed 0·1 division and they can be used by colour-blind persons.

[2] Tuchschmid : *Journ. für prakt. Chem.*, New Ser. 2, 235.

on the one hand, is increased, while, on the other, the density of the contained solution is reduced by its increase of volume. The former tends to increase the deviation, but this tendency is overpowered by the larger decrease due to the latter. Mategczek,[1] taking as his basis the dilatation-coefficient of glass, § 66, along with Gerlach's [2] researches on the density of saccharine solutions at different temperatures, has calculated the variations arising from this source. He gives a table for Ventzke's scale, of which the following is an abridgment :—

Temperature.	Saccharimeter degrees recorded by a Solution normal at 17·5° Cent.	Grammes of Sugar in 100 cub. cent. of Solution, corresponding to 1 degree on the Scale.
10°	100·17	0·26004
11°	100·14	0·26010
12°	100·12	0·26016
13°	100·10	0·26022
14°	100·08	0·26028
15°	100·05	0·26034
16°	100·03	0·26039
17°	100·01	0·26045
17·5°	100·00	0·26048
18°	99·99	0·26051
19°	99·96	0·26057
20°	99·94	0·26064
21°	99·91	0·26071
22°	99·88	0·26078
23°	99·85	0·26086
24°	99·83	0·26093
25°	99·80	0·26100
26°	99·77	0·26108
27°	99·74	0·26116
28°	99·71	0·26124
29°	99·68	0·26132
30°	99·65	0·26139

[1] Mategczek : *Zeitsch. des Vereins für Rübenzuckerindustrie*, 1875, 877.
[2] Gerlach : *Idem.*, 1862, 283.

Hence, for example, if a thermometer plunged in the contents of the tube, after the rotation has been observed, indicates, say 20°, then, in calculating the sugar-percentage, the constant 0·26064 must be used instead of 0·26048.

(b.) *The Soleil-Duboscq Saccharimeter.*

§ 86. The original Soleil instrument, as used in France, and manufactured at the optical instrument works of J. Duboscq, Paris, agrees essentially in principle with that just described, except as regards the scale. This is so devised that the deviation produced by introducing a plate of dextro-rotatory quartz 1 millimetre thick is taken as a fixed point, and the space between this and the zero-point is divided into 100 equal parts. Now, according to Clerget,[1] the same amount of deviation is produced by a 2 decimetre column of a sugar solution containing 16·471 grammes in 100 cubic centimetres. This concentration was subsequently corrected by Duboscq[2] to 16·350 grammes, and the scales for this instrument are now so constructed that the 100 point is recorded by a solution containing 16·350 grammes of pure dry sugar-candy in 100 cubic centimetres. Consequently each degree of the scale represents 0·1635 gramme in 100 cubic centimetres.

In construction, the French-made instruments differ from the German (Ventzke-Scheibler) in having both quartz wedges movable. Besides this, the regulator for producing the transition-tint is placed in the eye-piece of the instrument, with its Nicol immediately behind the eye-glass of the telescope, and capable of partial rotation by means of a rim projecting through the telescope tube.

In using this form of instrument, 16·35 grammes of the substance to be examined are weighed, made into a 100 cubic centimetre solution, and the solution examined in a 2 decimetre tube. The degrees of rotation on the scale indicate directly the number of grammes of sugar in 100 cubic centimetres. In other respects, the method of observation is precisely the same as in using the Soleil-Ventzke-Scheibler saccharimeter. To avoid errors from imperfect

---

[1] Clerget: *Ann. Chim. Phys.*, [3] **26**, 175.

[2] According to the still later researches of Schmitz and Tollens, a still more correct value for this constant is 16·302 grammes. (From the equation for *c*, given on p. 180).

quartz-wedges, Scheibler's method of double observation for the 100 point may likewise be adopted.

In using Soleil's saccharimeter for the approximate determination of the specific rotation of other substances, the deviation of which has to be reckoned in angular measure, one has simply to bear in mind that, according to § 18, a quartz plate 1 millimetre thick rotates ray D through an angle of $21.67°$, and mean yellow light (ray j) through $24.5°$, or,

$$1° \text{ Soleil (ray } j) = 0.2167 \text{ angular degrees (ray } D),$$
$$1° \text{ Soleil (ray } j) = 0.245 \text{ angular degrees (ray } j).$$

§ **87.** Correction for imperfect proportionality between deviation and concentration will also require to be made in using the Soleil-Duboscq. The amount of correction necessary is given for every $10°$ in the table annexed:—

Degrees on Scale.	Corrected Reading.	Difference.	Grammes of Sugar in 100 cubic centimetres Solution.		
			Uncorrected.	Corrected.	Difference.
100	100.00	0.00	16.350	16.350	0.000
90	89.98	0.02	14.715	14.712	0.003
80	79.97	0.03	13.080	13.075	0.005
70	69.96	0 04	11.445	11.438	0.007
60	59.95	0.05	9.810	9.802	0.008
50	49.95	0.05	8.175	8.167	0.008
40	39.95	0.05	6.540	6.532	0.008
30	29.96	0.04	4.905	4.898	0.007
20	19.97	0.03	3.270	3.265	0.005
10	9.98	0.02	1.635	1.632	0.003

As to the influence of temperature upon determinations by this instrument, Mategczek [1] gives a table from which the following is an extract:—

---

[1] Mategczek: *Zeitsch. des Vereins für Rübenzuckerindustrie*, 1875, 891.

Tempera-ture.	Degrees observed.	Grammes of Sugar in 100 cub. cent. corresponding to 1 degree.
15	100·05	0·16341
16	100·03	0·16344
17	100·01	0·16348
17·5	100·00	0·16350
18	99·99	0·16352
19	99·96	0·16356
20	99·94	0·16360
21	99·92	0·16363
22	99·89	0·16367
23	99·87	0·16371
24	99·85	0·16375
25	99·82	0·16378

(c.) *Wild's Polariscope, with Saccharimetric Scale.*

§ 88. Wild's polariscope, already described (§ 49), can be fitted with a scale for use as a saccharimeter. As supplied from the works of Hermann and Pfister, of Berne, these instruments have a scale divided into 400 equal parts, and their construction is based upon the specific rotation of cane-sugar, $[a]_D = 66·417°$, as determined by Wild[1] in a solution containing 30·276 grammes of cane-sugar per 100 cubic centimetres. Then assuming the amounts of concentration and of rotation to be strictly proportional, he calculates the value of the angle of rotation a, which a solution containing 40 grammes of sugar in 100 cubic centimetres should give in a 2 decimetre tube. The equation $\dfrac{a \times 100}{2 \times 40} = 66·417$, gives $a = 53·134°$.

Accordingly, an angle of this amount, measured from one of the zero-points of the instrument, is divided into 400 equal parts, so that, tak-

---

[1] Wild: *Ueber ein neues Polaristrobometer u. eine neue Bestimmung der Drehungs-coutante des Zuckers.* Berne: 1865. Also, *Mélanges Phys. et Chim. Bull. de l'Acad. de St. Petersbourg,* 8, 33.

REESE LIBRA
OF THE
UNIVERSITY
OF
CALIFORNI.

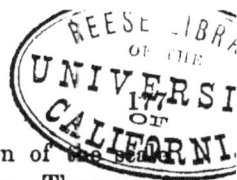

ing observations with a 2 decimetre tube, each division of the scale will represent 1 gramme of sugar in 1 litre of solution. The use of a sodium flame is here presupposed. According to this mode of graduating the scale, a solution containing 10 grammes of pure sugar in 100 cubic centimetres, observed in a 2 decimetre tube, should record 100 degrees, each divison of the scale indicating 1 per cent. of sugar. If a solution of 20 grammes of sugar in 100 cubic centimetres be used, the amount recorded should be 200 degrees, each division of the scale corresponding to one half per cent. of sugar, and so on for solutions containing up to 40 grammes of sugar.

Thus any weight of sugar may be chosen as normal weight, but a solution of 20 grammes in 100 cubic centimetres, or 10 grammes in 50 cubic centimetres, will be found most convenient. The observed rotation must then be divided by 2, and as with a little practice the scale can be read to one-fifth of a division, we can bring out values to 0·1 per cent. For example, if a solution of 20 grammes of a crude sugar in a 2 decimetre tube recorded 184·6 degrees, the percentage weight of sugar would be 92·3.

In dealing with substances of low sugar-percentage, such as beet-root juice, it is more convenient to weigh out as much as 60 or 80 grammes, and dilute to 100 cubic centimetres. The degrees recorded must then, of course, be divided by 6 or 8 to get the correct sugar-percentage. The greater the weight of substance taken for solution, the greater will be the accuracy of the determination.

In other respects, the mode of observation is similar to that described in § 51. The lamp, figured in § 46, with a bead of salt or soda, can be employed as the source of light.

To convert the readings on the saccharimetric scale into angular measure, we have only to remember that as an angle of 53·134° was divided into 400 equal parts, 1° of the scale = 0·1328 angular degrees.

§ 89. The effect of inconstancy of specific rotation on the saccharimetric scale is shown in the annexed table of corrections, calculated by Schmitz,[1] and based on the assumption that a solution containing 20 grammes of pure sugar records exactly 200 degrees.

[1] Schmitz: *Zeitsch. des Vereins für Rübenzuckerindustrie*, 1878, 48.

N

Divisions of Scale.	Grammes of Sugar in 100 cub. cents. of Solution (Calculated).	Grammes of Sugar in 100 cub. cents. of Solution (Corrected).	Percentage of Sugar by weighing out 20 grammes of Substance.	
			Uncorrected.	Corrected.
200	20	20·000	100	100
190	19	18·997	95	94·99
180	18	17·995	90	89·98
170	17	16·993	85	84·97
160	16	15·992	80	79·96
150	15	14·990	75	74·95
140	14	13·989	70	69·95
130	13	12·988	65	64·94
120	12	11·988	60	59·94
110	11	10·987	55	54·94
100	10	9·987	50	49·94
90	9	8·987	45	44·94
80	8	7·988	40	39·94
70	7	6·988	35	34·94
60	6	5·989	30	29·95
50	5	4·991	25	24·95
40	4	3·992	20	19·96
30	3	2·993	15	14·97
20	2	1·995	10	9·98
10	1	0·998	5	4·99

It will be seen from the two last columns that no correction is needed for percentages under 20 or over 80, the error in such cases not affecting the results to the amount of a tenth per cent. For percentages between 25 and 75, the indications directly observed are from 0·05 to 0·06 too high.

(d.) *Saccharimeter with Angular Graduation on Mitscherlich's, Wild's or Laurent's Principle.*

§ 90. Any one of the forms of polariscope described in §§ 45, 49 and 57 may be used for determining the concentration c of a given sugar-solution by observing the angle of rotation a for a column l decimetres long, and substituting values in the equation :—

$$c = \frac{100\,a}{l \times [a]}.$$

Disregarding the variation of specific rotation $[a]$ with the concentration of the solutions, a mean value may be assigned to it, which will serve for most of the sugar-solutions met with in practice. For example, by adopting, in analyses of natural sugars and their refined products, a normal solution of 15 grammes in 100 cubic centimetres, we shall obtain percentages ranging between 80 and 100, and the mean concentration $c = 14$ grammes. But according to the observations of Schmitz and Tollens recorded in § 91, $c = 14$ should yield an angle, which by the above equation corresponds to a specific rotation $[a]_D = 66\cdot50$. Introducing this value for $[a]$ in the equation, we get the annexed formula for determining the number of grammes of sugar in 100 cubic centimetres solution :—

$$c = 1\cdot504\,\frac{a}{l},$$

in which a is the angle of rotation observed with the sodium light, and l the length of tube employed. With a 2 decimetre tube

$$c = 0\cdot752\,a.$$

In many cases this may be simplified to $0\cdot75$. Thus a solution of 15 grammes of pure sugar should give a rotation of $20°$ exactly. When, therefore, 15 grammes of any saccharine (natural sugar) are made into a 100 cubic centimetre solution, and the angle of rotation observed in a 2 decimetre tube, the sugar percentages may be obtained by simply multiplying by 5.

In weighing out any other number of grammes, P, of substance the percentage can be calculated from the observed deviation a from the proportion

$$P : 0\cdot752\,a = 100 : x.$$

§ 91. Here, again, in exact determinations the imperfect proportionality between the angle of rotation and the concentration must be taken into account. Schmitz has prepared a table based upon the following observations,[1] partly his own and partly those of Tollens,[2] at a temperature of $20°$ Cent.[3]

[1] Schmitz : *Zeitsch. des Vereins für Rübenzuckerindustrie*, 1878, 53 and 58.

[2] Tollens : *Ber. der deutsch. chem. Gesellsch.* 10, 1403.

[3] The angles of rotation were all observed at $20°$ Cent. ; the concentrations by Schmitz at $20°$ Cent., those by Tollens at $17\cdot5°$ Cent. The latter trifling difference does not affect the results.

Number of Solution.	Grammes of Sugar in 100 cub. cents. of Solution. c.	Grammes of Sugar in 100 grammes Solution. p.	Observed angle of Rotation with a 2-decimetre Tube. a.	Observer.
1	2·014	2·000	2·69°	Schmitz
2	4·460	4·384	5·93°	Tollens
3	5·096	5·000	6·78°	Schmitz
4	8·195	7·945	10·92°	Tollens
5	10·399	10·004	13·82°	Schmitz
6	15·010	14·200	19·94°	Tollens
7	18·175	16·999	24·13°	Schmitz
8	27·641	25·010	36·67°	Schmitz
9	31·151	27·803	41·26°	Tollens
10	40·175	34·833	53·29°	Tollens
11	47·120	39·988	62·35°	Schmitz
12	54·132	44·914	71·65°	Tollens

Hence the following interpolation-formulæ for determining from the observed angle of rotation a the concentration c and percentage p of solutions have been derived by the method of least squares:—

$$c = 0.75063\, a + 0.0000766\, a^2,$$
$$p = 0.74730\, a - 0.001723\, a^2.$$

From these equations the annexed table has been prepared:[1]—

Observed Rotation with 2-decimetre Tube. a.	Grammes of Sugar in 100 cub. cents. Solution. c.	Difference for 0·1° of Rotation	Observed Rotation with 2-decimetre Tube. a.	Grammes of Sugar in 100 grammes Solution. p.	Difference for 0·1° of Rotation.
1°	0·751		1°	0·745	0·074
2°	1·501	0·075	2°	1·488	0·074
3°	2·253		3°	2·226	0·073
4°	3·004		4°	2·961	0·073

[1] The value p is for solutions of pure sugar. In beet-root solutions where other substances are present, the sugar-percentages must be determined from the values for c, and the observed specific gravity of the solutions.

Observed Rotation with 2-deci-metre Tube. *a.*	Grammes of Sugar in 100 cub. cents. Solution. *c.*	Difference for 0·1° of Rotation.	Observed Rotation with 2-deci-metre Tube. *a.*	Grammes of Sugar in 100 grammes Solution. *p*.	Differ-ence for 0·1° of Rota-tion.
5°	3·755		5°	3·693	0·073
6°	4·507		6°	4·422	0·073
7°	5·259		7°	5·147	0·072
8°	6·010		8°	5·868	0·072
9°	6·762		9°	6·586	0·072
10°	7·514		10°	7·301	0·071
11°	8·266		11°	8·011	0·071
12°	9·019		12°	8·719	0·071
13°	9·771		13°	9·424	0·070
14°	10·524	0·075	14°	10·124	0·070
15°	11·277		15°	10·821	0·070
16°	12·030		16°	11·516	0·069
17°	12·783		17°	12·206	0·069
18°	13·536		18°	12·893	0·068
19°	14·290		19°	13·576	0·068
20°	15·044		20°	14·257	0·068
21°	15·797		21°	14·933	0·067
22°	16·551		22°	15·606	0·067
23°	17·306		23°	16·277	0·067
24°	18·059		24°	16·943	0·066
25°	18·814		25°	17·605	0·066
26°	19·568		26°	18·265	0·066
27°	20·323		27°	18·921	0·065
28°	21·078		28°	19·573	0·065
29°	21·833		29°	20·223	0·065
30°	22·588		30°	20·868	0·064
31°	23·343	0·076	31°	21·510	0·064
32°	24·098		32°	22·149	0·064
33°	24·853		33°	22·784	0·063
34°	25·611		34°	23·416	0·063
35°	26·366		35°	24·044	0·068
36°	27·122		36°	24·670	0·068
37°	27·878		37°	25·291	0·062
38°	28·635		38°	25·909	0·062
39°	29·392		39°	26·523	0·061

Observed Rotation with 2-decimetre Tube. a.	Grammes of Sugar in 100 cub. cents. Solution. c.	Difference for 0·1° of Rotation.	Observed Rotation with 2-decimetre Tube. a.	Grammes of Sugar in 100 grammes Solution. p.	Difference for 0·1° of Rotation.
40°	30·148		40°	27·134	0·061
41°	30·905		41°	27·743	0·061
42°	31·662		42°	28·347	0·060
43°	32·420		43°	28·948	0·060
44°	33·176		44°	29·515	0·060
45°	33·933	0·076	45°	30·139	0·059
46°	· 34·691		46°	30·729	0·059
47°	35·449		47°	31·317	0·059
48°	36·207		48°	31·900	0·058
49°	36·966		49°	32·481	0·058
50°	37·724		50°	33·057	0·058

For example, suppose an angle of 16·4° has been observed, we see that 16° = 12·030 and 0·4° = 4 × 0·075, hence the solution contains in 100 cubic centimetres 12·330 grammes of sugar. (The constant 0·752 given in § 90 would show a concentration of 12·333, and its approximate value 0·75, a concentration of 12·300.) Again, 100 parts by weight of the same solution, since 16° = 11·516 and 0·4° = 4 × 0·069, contain 11·792 parts by weight of sugar.

As regards the *influence of temperature*, the researches of Tuchschmid and the calculations of Mategczek (§ 85) show that when the concentration amounts to about 25 grammes of sugar in 100 cubic centimetres, and observations are taken between 15° and 25° Cent. in glass tubes 2 decimetres long, a rise of 1° Cent. causes a decrease of 0·011° angular measure. Thus, when the temperature of observation differs from 20° Cent., the foregoing value may be used to correct the angle of rotation observed. For example, suppose the temperature of the solution were 17° Cent., the amount 3 × 0·011 must be subtracted, or if the temperature were 23° Cent., the same amount must be added to the angle observed. For sugar-solutions of less concentration, the amount of this correction is less, and when the temperature is not far removed from 20° Cent. it may be neglected altogether.

(e.) *Preparation of Solutions for the Saccharimeter.*

§ 92. In saccharimetry, measuring flasks of 50 and 100 cubic centimetres are employed which are usually provided with an additional mark, indicating capacities of 55 and 110 cubic centimetres respectively. These marks are fixed by weighing into the flasks water at some determinate mean temperature, usually $17\frac{1}{2}°$ Cent. ($63\frac{1}{2}°$ Fahr., 14° Reaum.). As will be seen from the table given in § 73, page 146, the following weights of water must be introduced in order to fix the levels of the several marks :—

For the 50 cubic centimetre mark, 49·938 grammes.

„	55	„	„	54·932	„
„	100	„	„	99·875	„
„	110	„	„	109·863	„

The correction for weight *in vacuo* is here disregarded.[1]

The solutions necessary for the different forms of saccharimeter, taking observations in each case with 200 millimetre tubes, are as given below : 50 cubic centimetres will generally be found sufficient.

Saccharimeter.	Weight of Substance required for Preparation of :	
	100 cubic centimetre Solution.	50 cubic centimetre Solution.
Soleil-Ventzke-Scheibler	26·048 grammes	13·024 grammes
Soleil-Duboscq	16·35 „	8·175 „
Wild, with Saccharimetric scale . . .	20 „	10 „
Mitscherlich, Wild, and Laurent, with angular graduation.	15 „	7·5 „

For weighing the samples, Scheibler[2] recommends basins with a lip made of German silver, a material not readily wetted by aqueous liquids.

---

[1] Reduced to vacuo, the weights required, according to § 69, are for the 50 cubic centimetre mark, 49·888 grammes, and for the 100 cubic centimetre mark, 99·775 grammes of water at a temperature of $17\frac{1}{2}°$ Cent.

[2] Scheibler: *Zeitsch. des Vereins für Rübenzuckerindustrie*, 1870, 614.

The weighed substance should be dissolved directly in the basin, the contents poured into the flask, and the basin itself then carefully washed with a stream of water into the flask, care being taken that the latter does not get more than three-quarters filled.

§ 93. *Decoloration of Solutions.*—If the solution so prepared, as is usually the case with natural sugars, beet-juice, and the like, is more or less coloured and turbid, the addition of some clarifying substance becomes necessary. The substance most commonly employed for the purpose is *basic acetate of lead*, in quantities of one or more cubic centimetres, according to the impurity of the sugar and the concentration of the solution. This usually throws down a heavy precipitate, by which various non-saccharine substances, as malic acid, aspartic acid, etc., are removed as lead salts, carrying down with them the particles which produce the turbidity. If the precipitate formed is small, it is convenient to add besides a few drops of solution of alum, so as to produce a precipitate of sulphate of lead. Too great excess of the basic acetate must be avoided, otherwise the filtered liquid will, in contact with the air, again become cloudy by the formation of carbonate of lead. This turbidity may, however, be dispelled by the addition of a drop of acetic acid. The basic acetate of lead is prepared by putting a finely powdered mixture of 3 parts ordinary acetate of lead and 1 part litharge along with 10 parts water in a closed flask, and allowing the mixture to remain until, aided by frequent shaking, only a small white residue remains undissolved. The filtered liquid should then have a specific gravity of from 1·23 to 1·24.

Aluminium hydrate, as recommended by Scheibler,[1] can also be employed as a decolorant. This can be prepared by precipitating a solution of sulphate of aluminium or of alum by means of ammonia, and washing the precipitate by decantation until the wash-water no longer gives a blue colour to red litmus paper. The voluminous pulpy mass so obtained is to be preserved in a closed flask, from which, by means of a pipette with wide stem, quantities may be removed as required. For the clarification of 13·024 grammes of a sugar dissolved in a 50 cubic centimetre flask, from 3 to 5 cubic centimetres are generally required. The alumina is especially adapted for the removal of turbidity; but less so for the removal of colouring matters; and there-

[1] Scheibler: *Zeitsch. des Vereins für Rübenzuckerindustrie*, 1870, 223.

fore we may add besides, if necessary, some basic acetate of lead and a little alum-solution.

After introducing the clarifier, the contents of the flask are made to mix by a gentle motion, and then left undisturbed for five to ten minutes. The surface of the liquid is then brought to the mark by adding water from the wash-bottle. If foam forms on the surface of the solution and prevents the exact adjustment of the level, it can be removed by touching it with a drop of ether or by merely pouring upon it a small quantity of ether-vapour from the bottle (Scheibler). The flask is then closed with the thumb and shaken vigorously for some time, after which the mixture is to be filtered. For this purpose round filters 13 to 14 centimetres (about 5½ inches) diameter, which will easily take 50 cubic centimetres, should be used. Of course, they must not be wetted before use, and the funnel and receiver below must be perfectly dry. Evaporation must be prevented during filtration by covering the funnel and receiver with glass plates. Very often the first few drops of the filtrate are turbid, and should be received in another vessel. The clear solution should at once be put into the polariscope-tube.

The mode of using flasks with double marks (as 50 and 55 cubic centimetres) is as follows :—The sugar-solution is made up to the 50 cubic centimetre mark, the acetate of lead, etc., added to the upper mark and the flask then shaken, etc., as before. The solution is then too dilute by one-tenth of its volume, and the angle of rotation observed in a 200 millimetre tube must be increased by one-tenth to get the correct value; or the rotation may be directly observed in a tube 220 millimetres long—a tube of this size being sometimes supplied with the instrument.

On the other hand, in using flasks with single marks only, an error occurs whenever the addition of basic acetate of lead produces a precipitate, from the diminution thereby caused in the volume of liquid contained by the flask, when the mixture stands at the 50 cubic centimetre or 100 cubic centimetre mark. The liquid will then be too concentrated, and its rotatory power too high. The error from this source will obviously vary in magnitude, with the amount of precipitate furnished by different saccharine products. Scheibler[1] examined by a variety of methods the volume of the precipitate obtained in the decoloration of 100 cubic centimetres of beet-root liquors with 10 cubic centimetres of basic acetate of lead. His experiments showed a mean value

---

[1] Scheibler: *Zeitsch. des Vereins für Rübenzuckerindustrie*, 1875, 1054.

of 1·3 cubic centimetres. Hence sugar-percentages, estimated from the observed rotation, appear 0·15 per cent. too high. According to Nebel and Sostmann,[1] the error in beet-juices averages 0·17 ; in diffusion-juices 0·27 per cent. Pellet[2] obtained the following error-values :— For beet-juices 0·15 to 0·2 per cent. ; cane-sugar juices 0·1 per cent. ; thick syrups, 0·25 per cent. ; sugars from the second and third crops, 0·25 per cent.; molasses, 0·63 per cent.

Accordingly, the results obtained by direct observation must be reduced by the above amounts to get the true sugar-percentages.

§ 94. Sometimes, indeed, as in the case of molasses and dark-coloured bye-products, the colour is so intense that even the basic acetate of lead fails to sufficiently decolorize the solution. In such cases an attempt should be made to observe the rotation by employing a 100 millimetre tube, or by diluting the solution to double its original volume, the observed rotation being, of course, doubled. Where this is found impracticable the solutions should be cleared with animal charcoal after preliminary treatment with acetate of lead. For this purpose, 30 to 40 cubic centimetres of the filtrate are placed in a flask, with 3 to 6 grammes of powdered and strongly dried bone-charcoal, and either shaken vigorously for some time or allowed to stand for twelve to twenty-four hours. In most cases, we shall then obtain, after filtration, a perfectly clear liquid.

The charcoal has, however, the disadvantage of abstracting not only colouring matter, but some sugar as well, so that the rotation results will be considerably too small. It will be necessary, therefore, in order to apply the proper correction, to make a preliminary determination of the absorptive power of the particular charcoal used, by a few experiments with sugar-solutions of known strength. Thus, Scheibler[3] found that, with 50 cubic centimetre solutions, containing 13·024 grammes of various natural sugars, cleared with basic acetate, after standing over 5·5 grammes of desiccated bone-charcoal for from twelve to twenty-four hours, the observed rotation gave sugar-percentages averaging 0·4 to 0·5 too low. He also showed that the amount of sugar absorbed is proportional to the quantity of charcoal used.

[1] Nebel and Sostmann : Zeitsch. des Vereins für Rübenzuckerindustrie, 1876, 724.
[2] Pellet : Idem., 1876, 730.
[3] Scheibler : Idem., 1870, 218.

(f.) *Determination of Cane-Sugar in the Presence of other Active Substances.*

§ 95. Another source of inaccuracy met with in the saccharimetric analyses of beet-liquors, inferior natural sugars, and molasses, is the occurrence along with cane-sugar of a whole series of other substances, capable of affecting in different ways the plane of polarization of light. Of such substances the following have been found, viz. :—Malic acid (−), asparagin and aspartic acid (both + in acid solutions and − in alkaline), glutamic acid (+), invert-sugar (−), beet-gum (−), dextran (+), the two last-named possessing very high rotatory powers. In molasses these substances may be present in such quantity as to render the determination of the sugar in the highest degree inaccurate, and even with beet-liquors some uncertainty is thereby involved in the results. It is true that by clearing with basic acetate of lead such substances are precipitated in part; but a method of certainly removing them entirely or of optically neutralizing them is still a desideratum. Eisfeldt and Follenius[1] attempt this, by heating with a solution of copper sulphate and caustic soda, so as partly to precipitate and partly decompose them by oxidation. Sickel[2] has proposed a method, which consists in adding to 13·024 grammes of beet-juice 1 cubic centimetre of basic acetate of lead, and diluting to 50 cubic centimetres with absolute alcohol. In this way the asparagin, aspartic acid, malic acid, gum, and dextran are precipitated, and the rotatory power of the invert-sugar almost completely annulled by the presence of the alcohol. This method appears to be serviceable, but requires further confirmation.

§ 96. When invert-sugar alone accompanies the cane-sugar, the effect of which is to reduce the rotation, the correct percentage of the cane-sugar present can be determined by employing the so-called *inversion method* of Clerget.[3] When a saccharimeter with Soleil scale is used, the method is as follows :—The usual normal solution, containing 16·35 grammes of the sugar is prepared in the ordinary way, cleared, if necessary, with basic acetate of lead, and the rotation determined. Then 50 cubic centimetres of the solution

[1] Eisfeldt and Follenius: *Zeitsch. des Vereins für Rübenzuckerindustrie*, 1877, 728 and 794.
[2] Sickel: *Idem.*, 1877, 779 and 800.
[3] Clerget: *Ann. Chim. Phys.* [3], **26**, 175.

are heated with 5 cubic centimetres of concentrated hydrochloric acid for ten minutes, on a water-bath at a temperature of about 68° Cent. (154° Fahr.), whereby the whole of the cane-sugar is transformed into invert-sugar. After cooling, the rotation (in this case left-handed) is observed in a 220 millimetre tube, the temperature of the solution being also noted by introducing a thermometer.

The calculation of the percentage of cane-sugar from the above two observations is performed in the following manner :—

According to Clerget's experiments, a solution containing 16·35 grammes of pure sugar in 100 cubic centimetres—which indicates a rotation of + 100° with a Soleil saccharimeter—will, after inversion, and observed at a temperature of 0° Cent., indicate a rotation of − 44° on the scale, the total change of rotation indicated being 144°.

Moreover, it is found that the amount of lævo-rotation of a solution of invert-sugar varies very markedly with the temperature of observation. The above solution, for instance, would undergo, for every rise of 1° Cent., a reduction of 0·5 division on the Soleil scale, so that at a temperature $t°$, its amount would be $144 - \frac{1}{2} t$. Putting S for the sum of the opposite saccharimetric readings (the readings before and after inversion), that is, for the total decrease of rotation, and t for the temperature at which the inverted solution is observed, the required percentage of cane-sugar R may be found from the proportion

$$(144 - \tfrac{1}{2} t) : 100 = S : R,$$

whence :—

$$R = \frac{100 \, S}{144 - \frac{1}{2} t}.$$

If, for example, the amounts of rotation were :—

By direct observation . . . . 94·1° right
After inversion at temperature 20° Cent.   37·2° left
then $S = 131\cdot3°$

which gives

$$R = \frac{100 \times 131\cdot3}{144 - 10} = 98\cdot0 \text{ per cent. cane-sugar,}$$

instead of the incorrect value, 94·1 per cent., directly obtained from the original solution.

Tuchschmid[1] has studied minutely the change of rotation produced by the inversion of sugar-solutions, as well as the influence of

[1] Tuchschmid : *Journ. für prakt. Chem.* [2], **2**, 235.

temperature, and gives the subjoined formula for the calculation of cane-sugar percentages :—

$$R = \frac{100\ S}{144\cdot16 - 0\cdot506\ t}.$$

If, instead of Soleil's saccharimeter, one with angular graduation be employed, Tuchschmid's formula becomes

$$R = \frac{21\cdot719\ S}{31\cdot31 - 0\cdot11\ t}.$$

To determine at the same time the percentage of invert-sugar originally present in the substance, the following method can be adopted :—As we have already said, a solution of 16·35 grammes of cane-sugar in 100 cubic centimetres gives, by inversion, a liquid, which at a given temperature, t°, rotates towards the left through $(44 - \frac{1}{2}\ t)$ divisions of the scale. Now, since 171 parts of cane-sugar, when treated with acid, yield 180 parts invert-sugar, the above rotation corresponds to a percentage of 17·21 grammes invert-sugar in 100 cubic centimetres. Putting A for the result of direct observation, R for the sugar-percentage found after inversion, and J for the proportion of invert-sugar to be determined, we have the proportion

$$44 - \tfrac{1}{2}\ t : 17\cdot21 = R - A : J,$$

whence,

$$J = \frac{17\cdot21\ (R - A)}{44 - \tfrac{1}{2}\ t}.$$

Taking, as in the former example, $A = 94\cdot1$, $R = 98\cdot0$, and assuming the rotation to have been observed, both before and after inversion, at 20° Cent., we get the result :—

$$J = \frac{17\cdot21\ (98\cdot0 - 94\cdot1)}{44 - 10} = 2\cdot0 \text{ per cent.}$$

Employing the more exact constants determined by Tuchschmid, we have :—

1. For Soleil's saccharimeter :

$$J = \frac{17\cdot21\ (R - A)}{44\cdot16 - 0\cdot506\ t},$$

2. For saccharimeters with angular graduation :

$$J = \frac{17\cdot21\ (R - A)}{9\cdot59 - 0\cdot11\ t}.$$

This method will, of course, cease to furnish correct results when other optically-active substances are present in addition to invert-sugar. As this generally is the case, the method is not of much practical utility.

Nevertheless, the inversion process is convenient for ascertaining whether a sugar is or is not contaminated with other active substances. It will be known to be free of such substances when the amounts of rotation before and after inversion correspond, whilst if they do not correspond we may reckon on the presence of impurities. Scheibler[1] in this way detected the presence of dextrin in natural sugars. As the direction of the rotatory power is unaffected when that substance is treated with acids, the lævo-rotation indicated by his invert-solution, and, therefore, also the calculated percentage of saccharose, was found too low.

B. Determination of Glucose.

GRAPE SUGAR—DIABETIC SUGAR.

§ 97. The specific rotation of dextro-rotatory glucose has been determined by various observers, the earlier of whom assumed that it was independent of the strength of the solutions. We have already seen, however (§ 38), that, according to the careful and minute investigations of Tollens,[2] the specific rotation rises with increase of concentration. Nevertheless, in dilute solutions not containing more than about 14 grammes of anhydrous grape-sugar in 100 cubic centimetres, the differences are of small account, and in such cases $[a]_D = 53.0^\circ$ may be taken as constant. Putting this value in the equation $[a] = \dfrac{100\,a}{l.\,c}$, we get the subjoined formula for determining the concentration c, from the angle of rotation a, observed with a tube l decimetres in length:—

$$c = 1.8868 \frac{a}{l}.$$

We are supposing here that the angle of rotation is taken with a Mitscherlich, Wild, or Laurent polariscope, and with sodium flame. The temperature of the solution must not differ much from 20° Cent. Then, if a 2 decimetre tube be employed,

$$c = 0.9434\,a.$$

Hence, 1° rotation represents 0.9434 gramme of anhydrous grape-sugar in 100 cubic centimetres of solution. This constant will serve

[1] Scheibler: Zeitsch. des Vereins für Rübenzuckerindustrie, 1871, 322.
[2] Tollens: Ber. der deutsch. chem. Gesellsch. 1876, 1631.

for all degrees of rotation up to 15°, which is sufficient for most purposes.

§ 98. With higher amounts of concentration (c greater than 14), the specific rotation of glucose undergoes a rather appreciable increase, as will be seen from the table given in § 38. In consequence of this variation, it was necessary to construct interpolation-formulæ for calculating the degrees of concentration corresponding to various angles of rotation. For this purpose the following observations of Tollens have been taken as a basis:—

No. of Solution.	Grammes of Glucose in 100 cub. cents. Solution. $= c$	Grammes of Glucose in 100 grammes Solution. $= p$	Observed Angle of Rotation, a, for $l = 2$ deci-metres.
1	9·634	9·292	10·20°
2	20·039	18·621	21·88°
3	35·898	31·614	38·46°

These figures yield the formulæ[1]

$$c = 0·94727\,a - 0·0004233\,a^2$$
$$p = 0·94096\,a - 0·0031989\,a^2$$

By means of these, the table given on page 192 has been prepared, showing the number of grammes of anhydrous glucose in (1) 100 grammes, and (2) 100 cubic centimetres, of solution, indicated by various angles of rotation observed for ray D, and in a tube 2 decimetres long. For angles under 10°, the constants given in § 97 have been employed.

[1] The extent to which these formulæ agree with other observations made by Tollens, may be gathered from the annexed table:—

a	c Calculated.	c Employed.	p Calculated.	p Employed.
8·43°	7·96 grm.	7·91 grm.	7·71 grm.	7·68 grm.
11·09°	10·45 ,,	10·46 ,,	10·04 ,,	10·06 ,,
11·72°	11·01 ,,	11·08 ,,	10·59 ,,	10·63 ,,
14·47°	13·62 ,,	13·63 ,,	12·95 ,,	12·95 ,,

Angle of Rotation for a Column 2 decimetres long and Ray *D*.	Grammes of Anhydrous Glucose in 100 grammes Solution.	Amount for 0·1° Rotation.	Grammes of Anhydrous Glucose in 100 cub..cents. Solution.	Amount for 0·1° of Rotation.
1°	0·93		0·94	
2°	1·86	0·093	1·89	
3°	2·79	0·093	2·83	0·095
4°	3·71	0·092	3·77	
5°	4·62	0·091	4·72	
6°	5·52	0·090	5·66	
7°	6·42	0·090	6·60	
8°	7·32	0·090	7·55	
9°	8·21	0·089	8·49	
10°	9·09	0·088	9·43	0·094
11°	9·96	0·087	10·37	
12°	10·83	0·087	11·31	
13°	11·69	0·086	12·24	
14°	12·55	0·086	13·18	
15°	13·40	0·085	14·11	
16°	14·24	0·084	15·05	
17°	15·07	0·083	15·98	
18°	15·90	0·083	16·91	
19°	16·72	0·082	17·85	
20°	17·54	0·082	18·78	
21°	18·35	0·081	19·71	0·093
22°	19·15	0·080	20·64	
23°	19·95	0·080	21·56	
24°	20·74	0·079	22·49	
25°	21·53	0·079	23·42	
26°	22·30	0·077	24·34	
27°	23·07	0·077	25·27	
28°	23·84	0·077	26·19	
29°	24·60	0·076	27·12	
30°	25·35	0·075	28·04	
31°	26·10	0·075	28·96	
32°	26·84	0·074	29·88	0·092
33°	27·57	0·073	30·80	
34°	28·30	0·073	31·72	
35°	29·02	0·072	32·64	

To express the result in terms of *glucose-hydrate*, $C_6 H_{12} O_3 + H_2 O$, since the molecular weights, $C_6 H_{12} O_6$ and $C_6 H_{12} O_6 + H_2 O$, are as 180 : 198 or 1 : 1·1, the values found for the anhydride must be increased by one-tenth.

§ 99. Determinations of glucose may also be made with a saccharimeter of quartz-wedge compensation form, as its dispersive power, according to Hoppe-Seyler's[1] measurements, agrees approximately with that of quartz. In dealing with dilute solutions not containing more than about 10 grammes in 100 cubic centimetres, the specific rotations of cane- and grape-sugars may be assumed to stand in the constant proportion of 66·5 : 53·0. Hence, (1) with *Ventzke's scale*, in which the 100 point indicates 26·048 grammes of cane-sugar in 100 cubic centimetres, the same amount of rotation will, in a solution of grape-sugar, indicate a concentration $\frac{66·5}{53·0} \cdot 26·048$ = 32·683. This gives 0·3268 gramme of anhydrous glucose for each division of the scale when the rotation is observed in a 2 decimetre tube. Or, (2) with a *Soleil-Duboscq scale*, 1 division of the scale will correspond with $\frac{66·5}{53·0} \cdot 0·1635 = 0·2051$ gramme of anhydrous glucose in 100 cubic centimetres.

Thus in the examination of grape-sugars a 100 cubic centimetre solution should be prepared, containing

for Ventzke's saccharimeter, 32·68 grammes,
for Soleil's   ,,   20·51   ,,

and observed in a 2 decimetre tube. The number of degrees recorded indicate directly the percentage of anhydrous glucose in the weighed samples.

Polariscopes are also constructed on Soleil's principle with scales expressly graduated for grape-sugar.[2] In these so-called *diabetometers* the index reading gives the number of grammes of glucose in 100 cubic centimetres solution, observed in a tube 1 decimetre long. Where a 2 decimetre tube is used the reading must, of course, be

---

[1] Hoppe-Seyler (*Fresenius', Zeitsch. für analyt. Chem.* 1866, 412) found :—

	$[\alpha]_c$	$[\alpha]_D$	$[\alpha]_E$	$[\alpha]_F$
For glucose (G)	42·45	53·45	67·9	81·3
Rotation for 1 millimetre quartz (Q) is	17·22	21·67	27·16	32·69
Whence ratio $\frac{G}{Q}$ is	2·46	2·47	2·17	2·49

[2] May be obtained of Schmidt and Haensch, Berlin.

divided by 2. Usually these instruments have the graduation continued on the other side of the zero-point, so that the scale then serves for determinations of albumen as well, which has a lævo-rotatory power equal in amount to the dextro-rotatory power of glucose.

§ 100. *Determination of Grape-sugar in Diabetic Urine.*—The urine should, if possible, be examined in its natural state, or if the colour interferes it may be diluted to twice its volume, or observations made with a tube of 1 decimetre length. Turbid urine must of course be filtered. If it be too dark in colour we may take 100 cubic centimetres, add to it 10 cubic centimetres basic acetate of lead, filter, and examine the filtrate, or we may attempt to decolorize it with animal charcoal.   In either case, the urine may lose some of its grape-sugar ; this has been proved to occur when the basic acetate of lead[1] is employed, and is probable in the case of charcoal (compare § 94).

The presence of albumens, exercising their lævo-rotatory power, may interfere by causing the sugar to appear too low.   They must, therefore, be previously removed.   For this purpose, 100 cubic centimetres of the urine are heated to boiling in a basin, and very dilute acetic acid added till the urine exhibits an acid reaction, and the albumen separates as a flocculent precipitate.   It is then filtered, the filter washed, and the filtrate again made up to 100 cubic centimetres ; or, having acidified with acetic acid a determinate volume of the urine, it is mixed with an equal volume of a concentrated solution of sodium sulphate.   On boiling the mixture, the albumen separates completely, and can be removed by filtering.

Bile-acids, which are dextro-rotatory, are never present in such urine in sufficient quantity to cause error in the above processes.

In *ordinary urines,* and in general when the proportion of grape-sugar in a urine is less than about 0·2 gramme in 100 cubic centimetres, it can no longer be accurately determined by observing the rotation of the urine itself directly.   In such cases, we must take from 1 to 2 litres for an analysis.   To this we must add, first, some solution of ordinary acetate of lead, and then, after filtering, some *basic* acetate together with a little ammonia.   The second precipitate will contain the whole of the grape-sugar.   It must be filtered off, mixed with alcohol, and excess of lead removed by sulphuretted hydrogen.   The solution separated from the lead sulphide by filtration

[1] See Brücke: *Sitzungber. der Wiener Akad.* **39**, 10.

must be decolorized with animal charcoal and concentrated by evaporation to a small determinate volume, which can then be placed in the polariscope. By this method, any bile-acids present in the urine pass over into the alcoholic solution finally obtained, and exercise their dextro-rotatory action. Their presence may be detected by evaporating a portion of the solution, dissolving the residue in a little water and mixing with some yeast. In two or three days all the sugar present will be decomposed, so that if, after filtration, the liquor still exhibits dextro-rotatory. power this must be attributed to the presence of bile-acids.

§ 101. The polariscope has also been employed by Neubauer to detect grape-sugar in "chaptalized" wine. The potato-sugars of commerce invariably contain from 10 to 20 per cent. of imperfectly known substances (*amylin* of Béchamp), characterized by high dextro-rotatory power, and great resistance to fermentation. When the must has been *chaptalized*, these substances pass over into the wine.

Pure natural wines of moderate age do not contain such substances, and, therefore, when submitted to polariscopic examination in tubes 2 or 2·2 decimetres in length, either exhibit no rotatory power, or at most a rotation to the right through an angle of 0·1 to 0·4 degree. Choice wines from very highly saccharine must, on the other hand, containing lœvulose still unfermented, may appear more or less lœvo-rotatory. This is the case with wine "chaptalized" with cane-sugar.

The following is the mode of examination recommended by Neubauer :—

Fifty cubic centimetres of the wine (whether red or white) are placed in a flask with 5 cubic centimetres basic acetate of lead ; some animal charcoal which has been purified by extraction with hydrochloric acid added, and the whole shaken up for a few minutes, and then filtered. The colourless solution is then introduced into the polariscope in a 2 or 2·2 decimetre tube. If it appears dextro-rotatory to the extent of 1° or more, it may safely be concluded that the wine has been chaptalized with potato-sugar.

If, however, the result appears doubtful, 100 to 200 cubic centimetres of the wine may be concentrated by evaporation to 25 (or 50)

[1] Neubauer: *Fresenius', Zeitsch. für analyt. Chem.* 1876, 188; 1877, 201 ; 1878, 321.

cubic centimetres, treated with basic acetate of lead and animal charcoal as before, and the rotation again examined. A rotation of from 1° to 4° at the least will now be obtained if the wine has been chaptalized. When 400 to 500 cubic centimetres of such wines are reduced by evaporation to 50 cubic centimetres, amounts of dextro-rotation of from 5° to 8° are not unfrequently obtained.

If the result of the first examination shows a dextro-rotation of not more than 0·4° to 0·6°, a further investigation may be made by the following method, which is based on the fact that the unfermentable matters accompanying the potato-sugar are, for the most part, soluble in alcohol, and can be precipitated therefrom by the addition of ether —250 to 350 cubic centimetres of the wine are first concentrated till the salts crystallize out. This liquid is decanted, decolorized with animal charcoal, diluted to 50 cubic centimetres, and finally filtered. Almost all pure natural wines treated in this way will exhibit a feeble dextro-rotatory power, which in tubes of 2 or 2·2 decimetres may amount to as much as about 2°. Wines that have been chaptalized, on the other hand, yield deviations of from 4° to 11°.

After this preliminary examination the 50 cubic centimetre solution must be reduced on a water-bath to a syrupy consistency, and alcohol of 90 per cent. added with constant stirring so long as any deposit forms. The mixture is then allowed to stand for several hours, until the liquid is perfectly clear, when it is poured off from the generally tough gelatinous residue. If, however, the precipitate formed is flocculent it must be filtered. The precipitate A, and the alcoholic solution B, so obtained are then treated separately as follows:—

The precipitate A is dissolved in cold water, decolorized with animal charcoal, and filtered. The solution must then be diluted to a volume corresponding with the capacity of the polariscope-tube and placed in the polariscope. In all pure wines, the bulk of the dextro-rotatory substances will be found in this solution, which may therefore give an angle of rotation of from 0·5° to 1·8°.

The alcoholic solution B is evaporated on the water-bath, till about one-fourth of the alcohol originally added remains. This is then placed in a small flask, and after cooling is mixed with from four to six times its volume of ether and vigorously shaken. If after standing the ether is found to have separated from the more or less thick watery liquid beneath, it can be removed by decanting, or by the help of a separating funnel. The watery solution is then diluted somewhat with water, warmed to expel any ether still remaining, and decolorized with charcoal. The filtrate, which now contains the unfermentable substances in the original potato-sugar, is then examined in a 2 or 2·2 decimetre tube. If the wine has been chaptalized this filtrate will exhibit dextro-rotation to the extent of from 3° to 11° or more. In pure natural wines of average quality, on the contrary, the filtrate will, in most cases, appear inactive, or may rotate at most from 0·2 to 0·5° to the right. For this optical examination of wines any sensitive polariscope, such as Wild's or Laurent's, may be used. Special polariscopes of simple form (called optical wine-testers) are manufactured for this purpose at the Optical Institute of Dr. Steeg and Reuter, Homburg v. d. Höhe. These instruments are in construction essentially similar to that described in § 43, Fig. 20.

C. Determination of Milk-Sugar.

§ 102. Milk-sugar, $C_{12} H_{22} O_{11} + H_2 O$, exhibits, in freshly prepared cold solutions, the property known as *bi-rotation* (see § 27). The following numbers apply to solutions reduced by heating to constant rotation.

Hesse[1] examined four aqueous solutions in a Wild's polariscope with a 2 decimetre tube, and found :—

for $c = 2$	$a_D = 2·144°$	$[a]_D = + 53·60°$
,, $c = 3$	3·19°	53·16°
,, $c = 5$	5·29°	52·90°
,, $c = 12$	12·64°	52·67°

Thus the specific rotation decreases with increased concentration ; but for solutions of the above strengths $[a]_D = 53°$ may be taken as the mean value. Moreover, since for each decrease of c by 1, a shows a constant increase of 1·05°, the concentration of such solutions may be obtained from the subjoined table, in which :—

a is the angle of rotation observed in a 2 decimetre tube with sodium light,

c the corresponding amount in grammes of milk-sugar ($C_{12} H_{22} O_{11} + H_2 O$) in 100 cubic centimetres of solution.

a.	c.	a.	c.
1°	0·92	7°	6·63
2°	1·87	8°	7·58
3°	2·82	9°	8·54
4°	3·77	10°	9·49
5°	4·73	11°	10·44
6°	5·68	12°	11·39

[1] Hesse : *Liebig's Ann.* 176, 93.

In observations made with a Ventzke's saccharimeter, and taking for the specific rotation of milk-sugar the constant value of 53°, whereby it is made to agree with that of anhydrous glucose, the data afforded by § 99 show that each division of the scale will represent 0·3268 gramme of milk-sugar in 100 cubic centimetres solution, assuming the rotation to have been observed in a 2 decimetre tube. With the French scale the corresponding value is 0·205 gramme.

To Determine the Milk-sugar in Milk.—For this purpose the fat and lævo-rotatory casein must first be removed. Fifty cubic centimetres of milk are placed in a porcelain basin along with 25 cubic centimetres of a moderately strong solution of ordinary acetate of lead, heated to the point of incipient boiling, and afterwards allowed to become perfectly cold. The mixture, together with the coagulum, is then poured into a 100 cubic centimetre flask, and water added to bring it up to the mark. After shaking and filtering, the rotation is observed in a 2 decimetre tube, and the result so obtained doubled on account of the solution having been diluted to half its original strength. When the milk exhibits a strong acid reaction, it should first be neutralized with a few drops of soda solution. · As the volume of precipitate is considerable, the result will be somewhat too high (see § 93).

D. Determination of Cinchona Alkaloids.

§ 103. The specific rotation of the cinchona alkaloids and of their most important salts has been studied in detail by Hesse.[1] The values so determined serve both as a means of testing the purity of other samples, and in determining the composition of mixtures. Oudemans[2] has also made a number of observations on the same subject.

The rotation constants which have been determined with the greatest accuracy are those of quinine, cinchonidine, conchinine (quinidine), and cinchonine. In each of these four alkaloids the specific rotation varies considerably with the nature of the solvent, and decreases, moreover, with increase of concentration. Hesse has investigated the rotation of solutions containing, according to their respective solvent powers, from 1 to 10 grammes of substance in 100 cubic centimetres of solution, and has found that within these limits the variations are represented by the formula $[a] = A - Bc$. As solvent, alcohol of 97 per cent. by volume was employed for the pure

[1] Hesse: *Liebig's Ann.* 176, 203; 182, 128.   [2] Oudemans: *Idem.*, 182, 33.

alkaloids, and for their salts either water, dilute hydrochloric or sulphuric acid of known strength, the latter being added in such quantity that the solutions contained for 1 molecule alkaloid not more than 3 molecules H Cl or $H_2 S O_4$. This was the proportion of acid used also in the solution of the free alkaloids. In calculating the number of cubic centimetres of standard acid to be added to a given weight of alkaloid, Hesse took 316 for the molecular weight of all four bases, being the mean of 308 ($C_{20} H_{24} N_2 O_4$, cinchonine and cinchonidine), and 324 ($C_{20} H_{24} N_2 O_2$, quinine and quinidine). The error arising from the slight difference from the true molecular weight is trifling.

As the rotatory power of solutions containing alkaloids decreases more or less with a rise of temperature, the solutions must be kept at a constant temperature. Hesse took 15° Cent. as a standard.

The following tabular arrangement shows the constants obtained by Hesse with preparations of the highest possible degree of purity.

The numbers have reference—

1. To compounds of the alkaloids having the chemical formulæ respectively assigned to them (water of crystallization included).

2. To the alkaloid contained in these compounds—the latter numbers being calculated from the former.[1]

(As in previous cases c stands for the number of grammes of active substance in 100 cubic centimetres of solution; and for subsequent reference the formulæ are numbered.)

Quinine (lævo-rotatory).

Quinine hydrate, $C_{20} H_{24} N_2 O_2 + 3 H_2 O$.

Solution in alcohol 97 per cent. by vol. $c = 1$ to 10.

(1) $\qquad [a]_D = - (145 \cdot 2 - 0 \cdot 657\, c)$.

Quinine hydrochloride, $C_{20} H_{24} N_2 O_2 . H Cl + 2 H_2 O$.

Solution in water. $c = 1$ to 3.

---

[1] If $[a]_v$ be the specific rotation of a compound and $[a]_a$ that of the active group (*e.g.*, alkaloid) contained in it, then putting M and m as the respective molecular weights:

$$[a]_a = [a]_v \cdot \frac{M}{m}.$$

The equation which expresses the value of constant for compounds

$$[a]_v = A \pm B c$$

must be transformed for active groups into

$$[a]_a = \left(A \pm B \frac{M}{m} \cdot c' \right) \frac{M}{m} = A \frac{M}{m} \pm B \left(\frac{M}{m} \right)^2 \cdot c,$$

where c' is the amount of essential active substance (alkaloid) in c parts by weight of the compound. (Hesse: *Liebig's Ann.* 182, 131.)

(2)   Compound $[a]_D = -(144 \cdot 98 - 3 \cdot 15\,c)$.

(3)   Alkaloid $[a]_D = -(167 \cdot 41 - 4 \cdot 71\,c)$.

Solutions in hydrochloric acid :

     1 mol. hydrochloride + 2 mols. H Cl, or 1 mol. quinine hydrate + 3 mols. H Cl + water to 100 cub. cent. $c = 1$ to 7.

(4)   Hydrochloride $[a]_D = -(229 \cdot 46 - 2 \cdot 21\,c)$.

(5)   Alkaloid $[a]_D = -(280 \cdot 78 - 3 \cdot 31\,c)$.

Quinine sulphate (neutral), $2\,(C_{20}\,H_{24}\,N_2\,O_2)\,.\,H_2\,S\,O_4 + 8\,H_2\,O$.

(6)   Solution in alcohol of 80 per cent. by vol. $c = 2$. $[a]_D = -162 \cdot 95$.

(7)   Solution in alcohol of 60 per cent. by vol. $c = 2$. $[a]_D = -166 \cdot 36$.

Solution in hydrochloric acid :

     1 mol. sulphate + 4 mols. H Cl + water.

(8)   Anhydrous salt. $c = 2$. $[a]_D = -239 \cdot 2$.

Quinine sulphate (mono-acid), $(C_{20}\,H_{24}\,N_2\,O_2)\,.\,H_2\,S\,O_4 + 7\,H_2\,O$.

Solution in water. $c = 1$ to 6.

(9)   Salt $[a]_D = -(164 \cdot 85 - 0 \cdot 31\,c)$.

(10)   Alkaloid $[a]_D = -(278 \cdot 71 - 0 \cdot 89\,c)$.

Quinine disulphate (di-acid), $(C_{20}\,H_{24}\,N_2\,O_2)\,.\,2\,H_2\,S\,O_4 + 4\,H_2\,O$.

Solution in water. $c = 2$ to 10.

(11)   Salt $[a]_D = -(155 \cdot 69 - 1 \cdot 14\,c)$.

(12)   Alkaloid $[a]_D = -(284 \cdot 48 - 3 \cdot 79\,c)$.

For the rotation of this salt with 7 mols. $H_2 O$, see Chapter VII.

Solutions in sulphuric acid :

     I. 1 mol. quinine hydrate + 3 mols. $H_2\,S\,O_4$ + water to 100 cub. cent.

(13) $c = 1$ to 5.   Hydrate $[a]_D = -246 \cdot 63 - 3 \cdot 08\,c)$.

(14)   Alkaloid $[a]_D = -287 \cdot 72 - 4 \cdot 19\,c)$.

     II. 1 mol. sulphate + 2 mols. $H_2\,S\,O_4$ + water to 100 cub. cent.

(15) $c = 1$ to 10.   Sulphate $[a]_D = -(171 \cdot 68 - 0 \cdot 78\,c)$.

(16)   Alkaloid $[a]_D = -(290 \cdot 36 - 2 \cdot 23\,c)$.

     III. 1 mol. disulphate + 1 mol. $H_2\,S\,O_4$ + water to 100 cub. cent.

(17) $c = 2$ to 6.   Disulphate $[a]_D = -(153 \cdot 87 - 0 \cdot 92\,c)$.

(18)   Alkaloid $[a]_D = -(281 \cdot 15 - 3 \cdot 11\,c)$.

Cinchonidine (lævo-rotatory).

Cinchonidine, $C_{20}H_{24}N_2O$.
Solution in alcohol of 97 per cent. by vol. $c = 1$ to 5.

(19) $\qquad [a]_D = - (107\cdot48 - 0\cdot297\ c)$.

Cinchonidine hydrochloride, $C_{20}H_{24}N_2O . HCl + H_2O$.
Solution in water. $c = 1$ to 3.

(20) $\qquad$ Salt $\quad [a]_D = - (105\cdot34 - 0\cdot76\ c)$.

(21) $\qquad$ Alkaloid $[a]_D = - (123\cdot98 - 1\cdot05\ c)$.

Solutions in hydrochloric acid : 1 mol. hydrochloride + 2 mols.
HCl or 1 mol. alkaloid + 3 mols. HCl + water to 100
cub. cent. $c = 1$ to 10.

(22) $\qquad$ Salt $\quad [a]_D = - (154\cdot07 - 1\cdot39\ c)$.

(23) $\qquad$ Alkaloid $[a]_D = - (181\cdot32 - 1\cdot925\ c)$.

Cinchonidine sulphate (neutral), $2\ (C_{20}H_{24}N_2O) . H_2SO_4 + 6H_2O$.

(24) Solution in water. $c = 1\cdot06$. Salt $\quad [a]_D = - 106\cdot77$.
$\qquad\qquad\qquad\qquad\qquad\qquad$ Alkaloid $[a]_D = - 142\cdot31$.

Cinchonidine sulphate (mono-acid), $(C_{20}H_{24}N_2O) . H_2SO_4 + 5H_2O$.

(25) Solution in water. $c = 2$. Salt $\quad [a]_D = - 110\cdot5$.
$\qquad\qquad\qquad\qquad\qquad\qquad$ Alkaloid $[a]_D = - 177\cdot95$.

Cinchonidine disulphate (di-acid), $(C_{20}H_{24}N_2O) . 2H_2SO_4 + 2H_2O$.
Solution in water. $c = 1$ to 7.

(26) Salt $\quad [a]_D = - (105\cdot96 - 1\cdot0267\ c + 0\cdot03376\ c^2 - 0\cdot00104\ c^3)$.

(27) Alkaloid $[a]_D = - (185\cdot77 - 3\cdot1557\ c + 0\cdot18158\ c^2 - 0\cdot00981\ c^3)$.

Quinidine or Conchinine (dextro-rotatory).

Quinidine hydrate, $C_{20}H_{24}N_2O_2 + 2\frac{1}{2}H_2O$.
Solution in alcohol of 97 per cent. by vol. $c = 1$ to 3.

(28) $\qquad$ Hydrate $\quad [a]_D = + (236\cdot77 - 3\cdot01\ c)$.

(29) $\qquad$ Anhydride $[a]_D = + (269\cdot57 - 3\cdot90\ c)$.

Quinidine hydrochloride, $C_{20}H_{24}N_2O_2 . HCl + H_2O$.
Solution in water. $c = 1$ to 2.

(30) $\qquad$ Salt $\quad [a]_D = + (205\cdot83 - 4\cdot93\ c)$.

(31) $\qquad$ Alkaloid $\quad [a]_D = + (240\cdot45 - 6\cdot60\ c)$.

Solutions in hydrochloric acid : 1 mol. hydrochloride + 2 mols.
HCl, or 1 mol. alkaloid + 3 mols. HCl + water to 100 cub.
cent. $c = 1$ to 5.

(32)   Salt $[a]_D = + (292 \cdot 56 - 3 \cdot 09\ c)$.
(33)   Alkaloid $[a]_D = + (338 \cdot 37 - 4 \cdot 52\ c)$.

Diquinidine sulphate, $2\ C_{20} H_{24} N_2 O_2 . H_2 S O_4 + 2\ H_2 O$.

(34) Solution in water. $c = 1$. Salt. $[a]_D = + 179 \cdot 54$.
                   Alkaloid $[a]_D = + 215 \cdot 55$.
   Solution in hydrochloric acid: 1 mol. salt + 4 mols. H Cl
     + water.
(35) Anhydrous salt. $c = 2$. $[a]_D = + 286 \cdot 4$.
                   Alkaloid $[a]_D = 329 \cdot 8$.
   Solution in sulphuric acid: 1 mol. salt + 5 mols. H$_2$ S O$_4$
     + water.
(36) Anhydrous salt. $c = 2$. $[a]_D = + 281$.
                   Alkaloid $[a]_D = + 323$.

Quinidine sulphate, $C_{20} H_{24} N_2 O_2 S O_4 + 4 H_2 O$.
   Solution in water. $c = 2$ to 8.
(37)   Salt $[a]_D = + (212 \cdot 0 - 0 \cdot 8\ c)$.
(38)   Alkaloid $[a]_D = + (323 \cdot 23 - 1 \cdot 86\ c)$.
   Solution in sulphuric acid : 1 mol. disulphate + 2 mols. H$_2$ S O$_4$
     + water to 100 cub. cent. $c = 1$ to 10.
(39)   Sulphate $[a]_D = + (215 \cdot 49 - 1 \cdot 41\ c)$.
(40)   Alkaloid $[a]_D = + (328 \cdot 55 - 3 \cdot 27\ c)$.

Cinchonine (dextro-rotatory).

Cinchonine, $C_{20} H_{24} N_2 O$.
   Solution in alcohol of 97 per cent. by vol. $c = 0 \cdot 5$. $c = 1$. Mean.
(41) $[a]_D = + 226 \cdot 36 \quad 225 \cdot 96 \quad 226 \cdot 13$.

Cinchonine hydrochloride, $C_{20} H_{24} N_2 O . H Cl + 2 H_2 O$.
   Solution in water. $c = 0 \cdot 5$ to 3.
(42)   Salt $[a]_D = + (165 \cdot 50 - 2 \cdot 425\ c)$.
(43)   Alkaloid $[a]_D = + (204 \cdot 46 - 3 \cdot 7\ c)$.
   Solutions in hydrochloric acid : 1 mol. hydrochloride + 2 mols.
     H Cl, or 1 mol. alkaloid + 3 mols. H Cl + water to 100
     cub. cent. $c = 1$ to 7.
(44)   Hydrochloride $[a] = + (214 \cdot 0 - 1 \cdot 72\ c)$.
(45)   Alkaloid $[a]_D = + (264 \cdot 37 - 2 \cdot 625\ c)$.

Dicinchonine sulphate, $2\ C_{20} H_{24} N_2 O . H_2 S O_4 + 2 H_2 O$.
   Solution in water. $c = 1$ to 2.
(46)   Salt $[a]_D = + (170 \cdot 3 - 0 \cdot 855\ c)$.

(47)   Alkaloid $[a]_D = + (206\cdot79 - 1\cdot26\ c)$.

Solution in sulphuric acid : 2 mols. dicinchonine sulphate $+ 5$ mols. $H_2SO_4 +$ water to 100 cub. cent. or 3 mols. sulphuric acid to 1 mol. alkaloid. $c = 0\cdot5$ to 6.

(48)   Sulphate $[a]_D = + (219\cdot10 - 1\cdot85\ c)$.

(49)   Alkaloids $[a]_D = + (266\cdot07 - 2\cdot69\ c)$.

For the rotatory powers of the other cinchona bases, see Chapter VII.

§ 104. Oudemans[1] has determined the specific rotation $[a]_D$ of anhydrous quinine and cinchonidine (both lævo-rotatory) in solution in absolute alcohol for different degrees of concentration (c grammes of substance in 100 cubic centimetres), and at different temperatures t, with the following results :—

Quinine.

$t.$	$c = 1.$	$c = 2.$	$c = 3.$	$c = 4.$	$c = 5.$	$c = 6.$
0°	171·4°	169·6°	167·9°	166·1°	164·2°	162·4°
5°	170·5°	168·7°	167·0°	165·2°	163·4°	161·6°
10°	169·6°	167·8°	166·1°	164·4°	162·7°	160·9°
15°	168·9°	167·1°	165·4°	163·7°	162·1°	160·4°
20°	168·2°	166·6°	164·8°	163·2°	161·6°	159·8°

Cinchonidine.

$t.$	$c = 1\cdot5.$	$c = 2.$	$c = 2\cdot5.$	$c = 3.$	$c = 3\cdot5.$	$c = 4.$
15°	110·0°	109·6°	109·2°	108·8°	108·4°	108·0°
20°	109·0°	108·6°	108·2°	107·8°	107·4°	107·0°

By employing a dilute alcohol as the solvent, the specific rotation of the cinchona alkaloids decreases with the amount of the dilution.

[1] Oudemans: *Liebig's Ann.* 182, 46. The results obtained by Oudemans do not admit of strict comparison with those given by Hesse, as the former employed absolute, and the latter 97 per cent., alcohol.

Oudemans gives the following observations on this point made at a temperature of 17° Cent.

Quinine.		Quinidine.		Cinchonidine.	
Per cent. of Alcohol by Weight.	$[a]_D$ for $c = 1.62$.	Per cent. of Alcohol by Weight.	$[a]_D$ for $c = 1.62$.	Per cent. of Alcohol by Weight.	$[a]_D$ for $c = 1.54$.
100·0	− 167·5°	100·0	+ 255·4°	100·0	− 109·6°
94·9	− 169·7°	95·3	+ 257·6°	90·5	− 115·0°
93·5	− 170·4°	90·5	+ 259·0°	80·2	− 117·8°
90·5	− 171·9°	85·0	+ 259·4°	70·8	− 120·4°
83·3	− 174·3°	80·0	+ 259·3°	69·0	− 121·1°
73·9	− 176·1°	75·0	+ 259·4°	—	—
65·1	− 176·5°	—	—	—	—

The following values, also given by Oudemans, for the specific rotation of the various salts, apply to solutions containing each 0·308 to 0·324 gramme (molecular weight of $C_{20} H_{24} N_2 O$ and $C_{20} H_{24} N_2 O_2$) of the particular alkaloid in 20 cubic centimetres of solution, or 1·54 to 1·62 grammes per 100 cubic centimetres.   Temperature, 17° Cent.

Salt.	Solvent.	Specific Rotation $[a]_D$	
		Anhydrous Salt.	Alkaloid.
Quinine Neutral Sulphate $2 (C_{20} H_{24} N_2 O_2) . H_2 SO_4 + 7\frac{1}{2} H_2 O$	Absolute Alcohol	− 157·4°	− 214·9°
Acid Sulphate $C_{20} H_{24} N_2 O_2 . H_2 SO_4 + 7 H_2 O$	Water	− 213·7°	− 278·1°
„ „ „	Absolute Alcohol	− 134·5°	− 227·6°
Neutral Hydrochloride $C_{20} H_{24} N_2 O_2 . H Cl + 2 H_2 O$	Water	− 133·7°	− 163·6°
„ „ „	Absolute Alcohol	− 138·0°	− 169·0°
Neutral Oxalate $2 (C_{20} H_{24} N_2 O_2) . C_2 H_2 O_4 + 3 H_2 O$	Absolute Alcohol	− 131·4°	− 160·5°

Salt.	Solvent.	Specific Rotation $[\alpha]_D$.	
		Anhydrous Salt.	Alkaloid.
Cinchonidine			
Neutral Sulphate			
$2 (C_{20}H_{24}N_2O)H_2SO_4 + 6H_2O$	Absolute Alcohol	$- 118 \cdot 7°$	$- 157 \cdot 5°$
,, ,, ,,	Alcohol 89 per cent. by Weight	$- 128 \cdot 7°$	$- 171 \cdot 8°$
,, ,, ,,	Alcohol 80 per cent. by Weight	$- 131 \cdot 2°$	$- 175 \cdot 1°$
Neutral Nitrate			
$C_{20} H_{24} N_2 O . H N O_3 + H_2 O$	Water	$- 99 \cdot 9°$	$- 126 \cdot 3°$
,, ,, ,,	Absolute Alcohol	$- 103 \cdot 2°$	$- 130 \cdot 4°$
,, ,, ,,	Alcohol 89 per cent. by Weight	$- 119 \cdot 0°$	$- 150 \cdot 4°$
,, ,, ,,	Alcohol 80 per cent. by Weight	$- 127 \cdot 0°$	$- 160 \cdot 4°$
Neutral Hydrochloride			
$C_{20} H_{24} N_2 O . H Cl + 2 H_2 O$	Water	$- 104 \cdot 6°$	$- 129 \cdot 2°$
,, ,, ,,	Absolute Alcohol	$- 99 \cdot 9°$	$- 123 \cdot 5°$
,, ,, ,,	Alcohol 89 per cent. by Weight	$- 119 \cdot 6°$	$- 147 \cdot 7°$
,, ,, ,.	Alcohol 80 per cent. by Weight	$- 128 \cdot 7°$	$- 159 \cdot 0°$
Conchinine (Quinidine)			
Neutral Sulphate			
$2(C_{20}H_{24}N_2O_2).H_2SO_4 + 2H_2O$	Absolute Alcohol	$+ 211 \cdot 5°$	$+ 255 \cdot 2°$
Neutral Nitrate			
$C_{20} H_{24} N_2 O_2 . H N O_3$	Absolute Alcohol	$+ 199 \cdot 3°$	$+ 232 \cdot 6°$
Neutral Hydrochloride			
$C_{20} H_{24} N_2 O_2 . H Cl + 2 H_2 O$	Water	$+ 190 \cdot 8°$	$+ 233 \cdot 6°$
,, ,, ,,	Absolute Alcohol	$+ 199 \cdot 4°$	$+ 244 \cdot 1°$
,, ,, ,,	Alcohol 90·5 per cent. by Weight	$+ 213 \cdot 0°$	$+ 260 \cdot 7°$

§ 105. The constants given by Hesse and Oudemans serve for testing the purity of commercial samples. For this purpose it is necessary, of course, that the conditions in respect of solvent and concentration should be the same as in the determination of these normal values, and the temperature should not differ very much. Moreover, it is necessary, in dealing with compounds containing water of crystallization, to determine directly the amount of the latter, in order either to establish the identity of the same with the formula assigned, or, if it disagrees, to be able to find the percentage of anhydrous substance, viz., of alkaloid present.

To give some idea of the amount of variation that may be expected, a few determinations of specific rotations obtained by Hesse and Oudemans respectively, for the same substances are given below. This is only practicable in the experiments made on aqueous solutions, as the other solutions employed by these observers are not comparable. Employing the constants given in §§ 103, 104, the specific rotation for the pure alkaloid, in a solution of 1·6 grammes in 100 cubic centimetres, appears for the undermentioned salts as follows :—

	Hesse.	Oudemans.	Difference.
Quinine sulphate	− 277°	− 278°	1°
Quinine hydrochloride . . .	− 160°	− 164°	4°
Quinidine hydrochloride .	+ 230°	+ 234°	4°
Quinidine hydrochloride .	− 122°	− 129°	7°

It thus appears, that with samples equally pure, different observers may obtain values differing by seven degrees, or even more.

The differences, however, which occur when impurities are present, are much larger in amount when another alkaloid is present, the specific rotations of the four cinchona bases differing very considerably from one another. This is shown in a number of experiments by Hesse, of which the results are as follows :—A solution of a specimen of neutral diquinine sulphate, which a separate analysis showed to contain 15 per cent. of water of crystallization was prepared with hydrochloric acid having a concentration $c =$ 2 grammes anhydrous base. To prepare 50 cubic centimetres of this solution, since 85 parts of anhydrous base were equivalent to 100 parts of the hydrated substances, $\frac{1 \times 100}{85} = 1·176$ grammes, was required. This amount was put in a 50 cubic centimetre flask, together with so much standard hydrochloric acid to give for 1 molecule $(C_{20} H_{24} N_2 O_2)_2 . H_2 S O_4$ four molecules H Cl, and the resulting solution diluted up to the mark with water. Observed in a Wild's polariscope with a 2 decimetre tube, the solution gave an angle of rotation $a_D = − 9·58°$, whence the specific rotation of the anhydrous substance $= \dfrac{9·58 \times 100}{2 \times 2} = − 239·5°.$

Now the figures in § 103 (8), show that perfectly pure sulphate of quinine in such a solution gives $[a]_D = − 239·2°$ (as a mean of three observations which gave respectively 239·1° ; 239·1° ;

239·3°) and hence the preparation under examination must have been free from admixture.

Three other samples of commercial sulphate of quinine, examined under like conditions, gave the following specific rotations respectively :—

$$- 236·6°, - 235·5°, + 109·5°.$$

Of these the two first agree so nearly with the normal value that no admixture of foreign substance, at least in any appreciable quantity, could have been present; the third, on the contrary, shows such a marked rotation to the right, that it must contain a notable amount of quinidine and cinchonine.

In this way Hesse showed that a so-called quinidine disulphate of English houses is an essentially different product from pure quinidine (conchinine) sulphate as prepared by the firm of F. Jobst, of Stuttgart.

These preparations exhibited the following differences :—

Quinidine (conchinine) sulphate, with 4 mols. H Cl $c = 2$ (anhydrous salt) . . . . } $[\alpha]_D = + 286°.$

English quinidine sulphate, with 4 mols. H Cl, $c = 2$ (anhydrous salt) . . . . } $[\alpha]_D = + 264°.$

Quinidine (conchinine) sulphate, with 4 mols. $H_2 SO_4$, $c = 2$ (anhydrous salt) . . . . } $[\alpha]_D = + 280°$ to $282°.$

English quinidine sulphate, with 4 mols. $H_2 SO_4$, $c = 2$ (anhydrous salt) . . . . } $[\alpha]_D = + 247·5°.$

The direct optical analysis of alkaloids in cinchona-bark extracts is, as Hesse remarks, beset with difficulty from the solutions containing a yellow colouring matter, which cannot be removed alone, and which impedes accurate observation of rotation. Hence the polariscope cannot serve for the direct valuation of cinchona-barks, although it provides a useful check on the results obtained by other modes of analysis.

§ 106. *Optical Analysis of Mixtures of Cinchona Alkaloids.*—The quantitative composition of a mixture of two alkaloids may be deduced from its specific rotation with the aid of the values given in § 103.

To test this method Hesse[1] determined the specific rotations of a number of mixtures of known composition, to ascertain whether the former could be deduced from the rotatory powers of the constituents. This was found to be sufficiently exact, and there-

[1] Hesse: *Liebig's Ann.* 182, 146.

fore the association of alkaloids in solution does not materially interfere with their individual optical properties. Hence a quantitative analysis by this method is practicable.[1]

---

[1] The mode of calculating the specific rotation of a mixture from those of its components will be best understood by taking an example. Let us suppose a mixture of 2 grammes of quinine hydrochloride with 1 gramme of cinchonine hydrochloride dissolved in water to 100 cubic centimetres. The specific rotation of the separate salts is given by the formulæ in § 103 :—

For quinine hydrochloride   (Form. 2) $[a]_D = - (144\cdot98 - 3\cdot15 \ c)$,
And for cinchonine hydrochloride (Form. 42) $[a]_D = + (165\cdot50 - 2\cdot425 \ c)$.

Whence for quinine hydrochloride, when $c = 2$ $[a]_D = - 138\cdot68$,
and for cinchonine hydrochloride when $c = 1$ $[a]_D = - 163\cdot07$.

Introducing these values for $[a]$ in the equation $a = \dfrac{[a] \ l \, . \, c}{100}$ we obtain the angles of rotation a, which solutions of these strengths ought to give, when observed in a tube whose length $l = 1$ decimetre :

For quinine hydrochloride when $c = 2$, angle $a = - 2\cdot774°$
,, cinchonine hydrochloride when a, $c = 1$,   ,, $= + 1\cdot631°$
For the mixture   when $c = 3$, angle $a = - 1\cdot143°$.

And hence the specific rotation of the mixture should be

$$[a]_D = - \frac{1\cdot143 \times 100}{1 \times 3} = - 38\cdot1°.$$

Actual experiment gave for this solution

$$[a]_D = - 37\cdot4°,$$

thus closely approximating to the calculated value.

In dealing, however, with mixtures of unknown composition, we are unable to arrive at the true specific rotations corresponding to the amounts of the separate substances present ; and we are forced to take for c in each case the total weight of mixture employed. But even so we obtain numbers closely agreeing with the results of actual experiment. Thus, suppose in the example given above we had taken c in each case as 3, and introduced into the equation the values for specific rotation corresponding thereto, viz :—

For quinine hydrochloride . $c = 3$. $[a]_D = - 135\cdot53$,
,, cinchonine hydrochloride $c = 3$. $[a]_D = + 158\cdot23$,

we should have obtained from—

Quinine hydrochloride, $c = 2$. $a = - 2\cdot711°$
Cinchonine   ,, $c = 1$. $a = + 1\cdot582°$
Mixture $c = 3$. $a = - 1\cdot129°$,

as the angle of rotation for the mixture in a 1 decimetre tube, whence the specific rotation—

$$[a]_D = - \frac{1\cdot129 \times 10}{1 \times 3} = - 37\cdot6°.$$

This latter mode of calculation, although not strictly accurate, is found to yield fairly good results, so long as the 100 cubic centimetre mixture does not contain more than a few grammes of substance. In such cases the true values of c for the components do not differ greatly from the value of c for the whole mixture, and so the departure from the true specific rotations is small.

The mode in which the percentage composition of a mixture of two alkaloids is deduced from the observation of its specific rotation, will be seen from the following examples given by Hesse :—

I. Four grammes of a mixture of quinine hydrate and cinchonidine were dissolved in alcohol of 97 per cent. by volume, to form 100 cubic centimetres of solution. Examined in a Wild's polariscope with a 2 decimetre tube, the solution gave an angle of rotation $a_D = -9.95°$, and thus the specific rotation of the mixture was

$$[a]_D = -\frac{9.95 \times 100}{2 \times 4} = -124.37°.$$

The specific rotation of the individual alkaloids is, according to § 103, when $c = 4$,

Quinine hydrate, according to Form. (1) : $[a]_D = -142.57°.$
Cinchonidine, according to Form. (19) : $[a]_D = -106.29°.$

Putting x for the required percentage of quinine hydrate, whereby that of cinchonidine $= 100 - x$, we get the equation

$$-142.57\,x - 106.29\,(100 - x) = 124.37 \times 100,$$

whence

$$x = \frac{100\,(124.37 - 106.29)}{142.57 - 106.29} = 49.8.$$

According to this, the mixture consists of

Quinine hydrate   49·8 parts.
Cinchonidine   50·2 ,,
                 ───────────────
                 100·0 parts.

The real composition of the mixture actually consisted of equal parts of the alkaloids, with which the results of optical analysis coincide almost exactly.

II. A mixture of 3 parts quinine sulphate and 1 part quinidine sulphate gave in an aqueous solution containing 4 grammes in 100 cubic centimetres, the specific rotation $[a]_D = -71.87°$:—

For quinine sulphate, according to Form. (9), when $c = 4$:
   $[a]_D = -163.61°.$
For quinidine sulphate, according to Form. (37), when $c = 4$:
   $[a]_D = +208.80°.$
Indicating by x the percentage of quinine sulphate we have :
   $-163.61\,x + 208.80\,(100 - x) = 71.87 \times 100;$
whence $x = 75.4$ per cent.

P

Thus we have:—

	By optical analysis.	Actual composition.
Quinine sulphate	75·4	75·0
Quinidine sulphate	24·6	25·0
	100·0	100·0

III. A mixture of 1·714 grammes quinine and 1·756 grammes quinidine was dissolved in so much standard hydrochloric acid to give for 1 molecule of the two alkaloids ($C_{20}H_{24}N_2O_2 = 324$), 3 molecules hydrochloric acid, and diluted with water to 100 cubic centimetres. This solution, of which the concentration $c = 3·470$, gave the specific rotation $[a]_D = +27·92°$.

In hydrochloric acid solution, when $c = 3·47$, the specific rotation is—

For quinine, according to Form. (5): $[a]_D = -269·29°$.
    „ quinidine,   „   „   (33): $[a]_D = +323·73°$.

Putting x for the proportion of quinidine, we have:—

$$323·73\,x - 269·29\,(100 - x) = 27·92 \times 100.$$

This gives x the values 50·1, the results appearing as hereunder:—

	By optical analysis.	Actual composition.
Quinine	50·1	50·6
Quinidine	49·9	49·4
	100·0	100·0

IV. 1·878 grammes of a mixture of quinidine (0·878 gramme) and cinchonine (1 gramme) made into a 100 cubic centimetre solution, containing 3 molecules hydrochloric acid for 1 molecule alkaloid (316), showed a specific rotation $[a]^D = +291·80°$.

With concentration $c = 1·878$, we have—

For quinidine, according to Form. (33): $[a]_D = +330·44°$.
    „ cinchonine   „   „   (45): $[a]_D = +259·44°$.

Putting x for the percentage of quinidine, we have—

$$330·44\,x + 259·44\,(100 - x) = 291·80 \times 100,$$

from which we have the following :—

	By optical analysis.	Actual composition.
Quinidine	45·6	46·8
Cinchonine	54·4	53·2
	100·0	100·0

To analyze a mixture of three cinchona alkaloids in a similar manner, the specific rotation must be determined by means of two different solvents. Thus, Hesse[1] has investigated weighed mixtures of cinchonidine, quinidine, and cinchonine, in the following solutions :—

I. 0·5 gramme of the mixture dissolved in alcohol of 97 per cent. by volume, into a 25 cubic centimetre solution, so that the concentration $c = 2$ gave in a 2 decimetre tube an angle of rotation $a = + 2·78°$, whence $[a]_D = + 69·5°$.

II. 0·5 gramme of the same mixture dissolved in hydrochloric acid to a 25 cubic centimetre solution, containing 3 molecules H Cl to 1 molecule alkaloid, gave in a 2 decimetre tube $a = + 2·82°$, whence $[a]_D = + 64·09°$.

Now the specific rotations of the three alkaloids, with a concentration $c = 2$, are as follows :—

	Solution in alcohol.	Solution in hydrochloric acid.
Cinchonidine $[a]_D =$	$- 106·89$ Form. (19)	$- 177·47$ Form. (23)
Quinidine $[a]_D =$	$+ 261·77$,, (29)	$+ 329·94$,, (33)
Cinchonine $[a]_D =$	$+ 226·13$,, (41)	$+ 259·12$,, (45)

Putting x, y, z, for the several required proportions, cinchonidine $= x$, quinidine $= y$, and cinchonine $= z$, we get the equations :

$$x + y + z = 100$$
$$- 106·89\,x + 261·77\,y + 226·13\,z = 100 \times 69·50$$
$$- 177·47\,x + 329·94\,y + 259·12\,z = 100 \times 64·09$$

From the solution of which we obtain :—

	By optical analysis.	Actual composition.
Cinchonidine	51·5	51·3
Quinidine	42·9	38·3
Cinchonine	5·6	10·4
	100·0	100·0

While the values for cinchonidine agree, there are considerable

---

[1] Hesse: *Liebig's Ann.* 182, 152.

differences between those for each of the other two alkaloids. These differences disappear when in the analysis of solution I. the angle of rotation 2·80° is substituted for 2·78°, and in that of solution II. $a = 2·80°$ instead of 2·82°. Errors of observation thus exert considerable influence over the results, and, consequently, the optical analysis of a mixture of three alkaloids is attended with some uncertainty.

§ 107. Other experiments for determining by the polariscope the composition of various mixtures of cinchona bases have been made by Oudemans,[1] jun., and with fairly satisfactory results.

The following special method for determining the quinine in a mixture of quinine and cinchonidine, has been given by Oudemans:—

If to any solution (e. g., bark-extract) containing the above alkaloids, as sulphate or chloride, we add a solution of neutral tartrate of potash or Rochelle salt, the tartrates, being insoluble in water, are precipitated:—

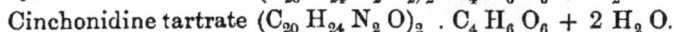

Quinine tartrate $\quad (C_{20} H_{24} N_2 O_2)_2 . C_4 H_6 O_6 + H_2 O.$

Cinchonidine tartrate $(C_{20} H_{24} N_2 O)_2 . C_4 H_6 O_6 + 2 H_2 O.$

Each of these compounds is readily soluble in dilute hydrochloric acid, and Oudemans took the rotations in three solutions of different concentrations. Each of the solutions contained for every 0·4 gramme of the tartrates, 3 cubic centimetres normal hydrochloric acid (containing 36·4 grammes H Cl per litre), and was made up with water to 20 cubic centimetres. At a temperature of 17° Cent. the following rotation-constants were observed:—

	Quinine tartrate.	Cinchonidine tartrate.
0·4 grm. tartrate + 3 cub. cent. normal hydrochloric acid + water to 20 cub. cent. solution	$[a]_D = -215·8.$	$[a]_D = -131·3.$
0·8 grm. tartrate + 6 cub. cent. normal hydrochloric acid + water to 20 cub. cent. solution	$[a]_D = -211·5.$	$[a]_D = -129·6.$
1·2 grm. tartrate + 9 cub. cent. normal hydrochloric acid + water to 20 cub. cent. solution	$[a]_D = -207·8.$	$[a]_D = -128·1.$

[1] Oudemans: *Liebig's Ann*. **182**, 63, 65.

By means of these figures the percentage composition may be determined of a mixture of the two tartrates, obtained as evaporated extract from a solution. For this purpose we dissolve 0·4, 0·8, or 1·2 gramme of the dried extract, in 3, 6, or 9 cubic centimetres normal hydrochloric acid, dilute the solution to 20 cubic centimetres with water, and observe the specific rotation at 17° Cent. Denoting this by M, and putting x for the required percentage of quinine tartrate, we can calculate its value from one of the following formulæ:—

$$\text{Concentration } 0\cdot4 \quad x = \frac{100\,(M - 131\cdot3)}{215\cdot8 - 131\cdot3},$$

$$\text{Concentration } 0\cdot8 \quad x = \frac{100\,(M - 129\cdot6)}{211\cdot5 - 129\cdot6},$$

$$\text{Concentration } 1\cdot2 \quad x = \frac{100\,(M - 128\cdot1)}{207\cdot8 - 128\cdot1},$$

A number of experiments by Oudemans showed, however, that the proportion of quinine calculated as above is, in most cases, too high. This is seen in the following results:—

Concentration.	Observed Specific Rotation.	Tartrate of Quinine calculated from Specific Rotation.	True percentage of Tartrate of Quinine.	Difference.
0·4	152·5°	25·1 per cent.	25·0 per cent.	+ 0·1
	165·7°	40·7 ,,	39·7 ,,	+ 1·0
	183·5°	61·8 ,,	60·0 ,,	+ 1·8
	195·3°	75·7 ,,	75·2 ,,	+ 0·5
	203·3°	85·2 ,,	85·2 ,,	0
0·8	150·2°	25·0 per cent.	25·3 per cent.	− 0·3
	170·8°	50·3 ,,	50·0 ,,	+ 0·3
	183·7°	66·0 ,,	64·3 ,,	+ 1·7
	195·3°	80·2 ,,	79·6 ,,	+ 0·6
	195·1°	80·0 ,,	80·2 ,,	− 0·2
1·2	147·5°	24·3 per cent.	24·9 per cent.	− 0·6
	171·3°	54·1 ,,	54·6 ,,	− 0·5
	189·1°	76·4 ,,	75·4 ,,	+ 1·0

VII

ROTATION CONSTANTS OF ACTIVE SUBSTANCES.

— ·◦· —

§ 108. The following collection of rotation values embraces only such substances as have been properly examined—that is to say, those respecting which information has been recorded on the following points :—[1]

1. Name of ray used (generally D or j).

2. Nature of the solvent.

3. Concentration or percentage composition of solutions.

4. Temperature at which the angle of rotation has been observed. (In some cases this has been omitted through want of data.)

5. In the case of interpolation formulæ, the limits within which they are correct.

Accompanied with such data, specific rotation values, as marked in § 40, afford, when the given experimental conditions are strictly observed, striking characters for the substances, useful in determining the nature of the substances as well as in examining the purity of particular samples. Moreover, they may serve other purposes, as follows :—

(a) For obtaining at least an approximate estimation of the percentage composition of solutions of active substance, in cases where the specific rotation of the substance does not alter very much with the concentration of its solutions. Thus given, the angle of rotation a of any solution in a tube l decimetre long, the number of

---

[1] The list has no pretension to absolute completeness. Other substances which have been adequately examined may have escaped notice.

grammes c of active substance per 100 cubic centimetres of solution may be found from the equation

$$c = \frac{a \times 100}{[a] \times l} \cdot$$

Similarly, the proportion of án active solid substance in admixture with inactive substance can be found by dissolving a known weight, p, to a given volume, v cubic centimetres; then the percentage of solid substance present is represented by :—

$$x = \frac{a \times v \times 100}{[a] \times l \times p} \cdot$$

Of course, it is here assumed that the inactive ingredients present in solution exert no influence on the specific rotation of the active substance.

(b) In certain cases also, for the quantitative analysis of mixtures of two active substances, as in the case of the cinchona alkaloids, §§ 106, 107.

The values cannot, however, be applied in any way for comparison of the rotatory powers of different active substances with each other, as they refer only to solutions of a given composition, and do not represent the actual specific rotation of the pure substance itself.

All the specific rotations hereafter given, refer to the compounds of the exact composition denoted by the chemical formulæ annexed; so that, where the substance contains water of crystallization, the values refer to the hydrated substance. If the specific rotation of the anhydrous substance is required, it can be calculated from the formula

$$[a] \text{ anhydrous} = \frac{M}{m} [a] \text{ hydrated,}$$

in which M represents the molecular weight of the hydrated substance, and m that of the anhydrous substance.

As before :—

$[a]_D$, $[a]_j$ denote respectively the specific rotation for ray D and for the transition tint j.

a denotes angle of rotation for a layer 1 decimetre thick.

c ,, (concentration) the number of grammes of active substance in 100 cubic centimetres of solution.

p ,, (percentage composition) the number of grammes of active substance in 100 grammes of solution.

d ,, specific gravity of the solution.

t ,, temperature (Cent.) at which the rotation has been observed.

In referring to memoirs the following abbreviations have been adopted :—

L. A.   stands for *Liebig's Annalen der Chemie u. Pharmacie.*
J. P. C.   „   *Journal für praktische Chemie.*
D. C. G.   „   *Berichte der deutschen chemischen Gesellschaft.*
R. Z. J.   ,,   *Zeitschrift des Vereins für Rübenzuckerindustrie des deutschen Reichs.*
A. C. P.   ,,   *Annales de chimie et de physique.*
C. R.   „   *Comptes rendus.*
J. B.   ,,   *Jahresbericht der Chemie.* (Giessen.)

§ 109. Sugars, $C_{12} H_{22} O_{11}$.

Cane-Sugar, $C_{12} H_{22} O_{11}$.   Dextro-rotatory.   Aqueous solutions.   Tollens (D. C. G. 1877, 1403) gives the annexed formulæ :—

1. Specific gravity of solutions at $17·5°$ Cent. referred to that of water of $4°$ Cent. Rotation observed at $20°$ Cent.

$$p = 4 \text{ to } 18. \ [\alpha]_D = 66·810 - 0·015553\,p - 0·000052462\,p^2.$$
$$p = 18 \text{ to } 69. \ [\alpha]_D = 66·386 + 0·015035\,p - 0·0003986\,p^2.$$
$$q = 82 \text{ to } 96. \ [\alpha]_D = 64·730 + 0·026045\,q - 0·000052462\,q^2.$$
$$q = 31 \text{ to } 82. \ [\alpha]_D = 63·904 + 0·064686\,q - 0·0003986\,q^2.$$

2. Specific gravity of solutions at $17·5°$ Cent. referred to that of water at $17·5°$ Cent. Rotation at $20°$ Cent.

$$p = 5 \text{ to } 18. \ [\alpha]_D = 66·727 - 0·015534\,p - 0·000052396\,p^2.$$
$$p = 18 \text{ to } 69. \ [\alpha]_D = 66·303 - 0·015016\,p - 0·0003981\,p^2.$$

Schmitz (D. C. G. 1877, 1414. R. Z. J. 1878, 48), from the experiments referred to in § 37 gives the formulæ :—

1. Specific gravity of solutions at $20°$ Cent. referred to that of water at $4°$ Cent. Rotation observed at $20°$ Cent.

$$q = 35 \text{ to } 98. \ [\alpha]_D = 64·156 + 0·051596\,q - 0·00028052\,q^2.$$

2. Specific gravity of solutions at $20°$ Cent. referred to that of water at $17·5°$ Cent. Rotation at $20°$ Cent.

$$c = 10 \text{ to } 86. \ [\alpha]_D = 66·453 - 0·0012362\,c - 0·00011704\,c^2.$$
$$c = 3 \text{ to } 28. \ [\alpha]_D = 66·639 - 0·020820\,c + 0·00034603\,c^2.$$
$$c = 3 \text{ to } 28. \ [\alpha]_D = 66·541 - 0·0084153\,c.$$

Hesse (L. A. 176, 97) found for

$$c = 0 \text{ to } 10. \ [\alpha]_D = 68·65 - 0·828\,c + 0·115415\,c^2 - 0·0051167\,c^3.$$

Previously, the following values of $[a]_D$ had been given, without taking into account the alterability of the rotation[1] :—

Observers.	c	$[a]$	
Arndtsen . . .	77·394	67·02	A. C. P. (3) 54, 403
,, . . .	47·276	67·33	A. C. P. (3) 54, 403
,, . . .	33·891	66·86	A. C. P. (3) 54, 403.
Stefan	33·762	66·37	Wiener Ak. 52, II. 486.
Wild	30·276	66·42	Polaristrob. Bern, 1865.
Weiss	30·090	65·98	Wiener Ak. 69, III. 162.
Tuchschmid . .	27·441	66·48	J. P. C. (2) 2, 235.
Stefan	21·608	66·75	Ibid.
Calderon . . .	19·971	67·08	C. R. 83, 393.
Krecke . . .	16·470	67·02	Dissert. Utrecht. 1867.
Girard and Luynes .	16·350	67·31	C. R. 80, 1354.
Weiss	14·570	66·04	Ibid.
Stefan	10·375	66·12	Ibid.
Calderon . . .	9·986	67·12	Ibid.
Oudemans . . .	5.877	66·90	Pogg. Ann. 148, 350.

REESE LIBR. UNIVERSITY OF THE CALIFORNIA

The *specific rotation for different rays* has been determined, according to Broch's method, by Arndtsen (A. C. P. (3) 54, 403) and by Stefan (*Sitz.-Ber. d. Wiener Akad.* 52, II. 486).

Arndtsen, employing solutions with 30 to 60 per cent. of sugar, obtained the following mean values:—

Lines	C	D	E	b	F	e
$[a]$	53·41	67·07	85·41	88·56	101·38	126·33.

Stefan, with solutions wherein $p = 10$ to 30 per cent., obtained as mean values :—

Lines	A	a	B	C	D
$[a]$	38·47	43·32	47·56	52·70	66·41.
Lines	E	b	F	G	H
$[a]$	84·56	87·88	101·18	131·96	157·06.

The latter values can be derived from the equation $[a] = \dfrac{2538}{\lambda^2} - 5\cdot58$, taking the wave-length λ in ten-thousandths of a millimetre.

[1] As the specific rotation observed by Girard and Luynes, and also by Calderon with sugar solutions, in which $c = 10$ to 20, viz., ($[a]_D = 67\cdot1$ to $67\cdot3$) differs considerably from 66·5, the result obtained by Tollens and Schmitz, Tollens crystallized afresh some pure sugar, and examined the rotation in several crops (D. C. G., 1878, 1800). He obtained with $c = 10$ the value $[a]_D = 66\cdot48$. (Specific gravity of the solutions at 17·5° Cent. with reference to that of water at 4° Cent.) The value 67 is certainly too high.

For the *transition tint* the following values have been obtained :—

$$p = 25. \quad [a]_j = 72\cdot92$$
$$p = 50. \quad [a]_j = 72\cdot61$$
$$p = 65. \quad [a]_j = 70\cdot59$$
Biot, *Mém. de l'Acad.* **13**, 118).

$$c = 10 \text{ to } 20. \quad [a]^j = 73\cdot20. \quad \text{Calderon (C. R. } \textbf{83}, 393).$$

Adopting Montgolfier's ratio, $a_D : a_j = 1 : 1\cdot129$, in the case of sugar (§ 18, p. 43) we get, if $[a]_D = 66\cdot5$, $[a]_j = 75\cdot08$. With Weiss' co-efficient $1\cdot034$ (§ 18), $[a]_j = 68\cdot76$.

Stefan's dispersion-formula gives for $\lambda_j = 5\cdot5$, $[a]_j$ $78\cdot32$.

In *alcoholic solutions*, cane-sugar exhibits higher rotatory power than in aqueous solutions (R. Z. J. 1877, 803). Hesse (L. A. **176**, 97) found that in solutions containing 50 per cent. of alcohol, with $c = 5$, and $t = 15°$, $[a]_D = 66\cdot70$. No further estimations were made.

Sulphuric acid appears to increase the rotation. Hesse (L. A. **176**, 97) employing the proportion : 1 mol. sugar ($c = 6$) to 1 mol. H_2SO_4 + water to 100 cubic centimetres with $t = 15°$, obtained $[a]_D = 66\cdot67$.

Alkalies diminish the specific rotation of cane-sugar. 1 mol. sugar + 1 mol. $Na_2O . c = 5 . t = 15°$. $[a]_D = 66\cdot00$. Hesse (L. A. **176**, 97).

According to Sostmann (R. Z. J. 1866, 272) the saccharimetric estimations of sugar in presence of alkalies, come out too low by the following amounts :—

Alkali.	In solution of the concentration.		
Per 100 cubic centimetres solution	$c = 5$	$c = 10$	$c = 20$ to 25
1 gramme potash (K_2O)	0·426	0·650	0·915
1 ,, soda (Na_2O)	0·450	0·907	1·217

Pellet (R. Z. J. 1877, 1036) gives the following values :—

	$c = 5\cdot4$	$c = 17\cdot3$
1 gramme caustic potash (K O H)	0·170	0·500
1 ,, caustic soda (Na O H)	0·140	0·450
1 ,, ammonia	0·073	0·085

Lime causes a remarkable reduction of rotatory power. Müntz (R. Z. J. 1876, 737) records the following observations :—

Pure Sugar. $c = 10$.		$[a]$ = 67·0
With 0·409 gramme	= ¼ mol. lime	,, = 64·9
,, 0·818 ,,	= ½ ,,	,, = 61·3
,, 1·637 ,,	= 1 ,,	,, = 56·9
,, 3·274 ,,	= 2 ,,	,, = 51·8

Saccharimetric observations by the following observers have shown that the addition of 1 part of lime destroys the rotatory power of :—

0·64 part sugar according to Jodin (C. R. **58**, 613).
0·79 ,, ,, ,, Dubrunfaut.
1·12 ,, ,, ,, Bodenbender (R. Z. J. 1865, 167).
1·22 ,, ,, ,, Stammer (Dingler, P. J. **156**, 40).
1·25 ,, ,, ,, Michaelis (Dingler, P. J. **124**, 358).
0·90 ,, ,, in solution with $c = 5·4$ } Pellet
1·00 ,, ,, ,, ,, $c = 17·3$ } (R. Z. J. 1877, 1036).

The action of the lime is removed by neutralization with acetic acid.
Baryta and *strontia* similarly diminish rotatory power :—

1 part baryta destroys the rotation of 0·426 parts sugar, according to Bodenbender.
1 ,, ,, ,, ,, in 0·190 ,, ,, when $c = 5·4$ } Pellet.
1 ,, ,, ,, ,, in 0·430 ,, ,, when $c = 17·3$ }
1 ,, strontia ,, ,, in 0·597 ,, ,, according to Bodenbender.

Salts of the alkalies exert a similar action. Müntz (R. Z. J. 1876, 735) investigated the effects of the following salts on the specific rotation, which was taken for the pure sugar as $[a]_D = 67·0$, and obtained the results subjoined :—

Additions of		Concentration of Solutions.		
		$c = 5.$	$c = 10.$	$c = 20.$
Common Salt {	5 grm.	$[a]_D = 66·1$	$[a] = 66·2$	$[a]_D = 66·3$
	10 ,,	65·3	65·3	65·6
	20 ,,	63·8	63·7	61·0
	25 ,,	—	62·8	—
Anhydrous Soda. {	2·5 grm.	—	65·2	—
	5 ,,	—	63·8	63·8
	10 ,,	—	62·1	62·6
	15 ,,	—	60·4	59·8
	20 ,,	—	58·5	58·1
Anhydrous Borax. {	0·5 grm.	—	65·9	—
	1 ,,	64·7	65·0	—
	2 ,,	62·7	63·5	—
	3 ,,	62·1	62·5	64·2
	4 ,,	—	61·6	—
	5 ,,	60·8	61·1	63·0
	7 ,,	—	—	62·2

The presence of 1 gramme of each of the following salts was found to reduce the concentration of sugar solutions, as determined by the saccharimeter, by the respective amounts as below :—

Carbonate of potash	0·143	with c	=	10		Sostmann
„	„	0·185	„	c =	20 to 25	(R. Z. J. 1866, 272).
„	„	0·044	„	c =	5·4	Pellet
„	„	0·065	„	c =	17·3	(R. Z. J. 1877, 1036).
Carbonate of soda	0·093	„	c =	10		Sostmann.
„	„	0·254	„	c =	20 to 25	
„	„	0·040	„	c =	5·4	Pellet.
„	„	0·132	„	c =	17·3	
Carbonate of ammonia	0·040	„	c =	5·4		Pellet.
„	0·067	„	c =	17·3		
Phosphate of ammonia (crystallized)	0·016	„	c =	5·4		Pellet.
	0·036	„	c =	17·3		

Acetate of lead added in the proportion of 25 grammes to 100 cubic centimetres of sugar-solution produces no change in rotation. Müntz (R. Z. J. 1876, 737).

Milk-sugar, $C_{12}H_{22}O_{11} + H_2O$. Dextro-rotatory.

Freshly prepared aqueous solutions exhibit bi-rotation (§ 27). Hesse obtained for constant rotation by heating the solutions (L. A. **176**, 100) :—

$$c = 2 \text{ to } 12. \quad t = 15°. \quad [a]_D = 54 - 0·557\,c + 0·05475\,c^2 - 0·001774\,c^3.$$

Alkalies reduce the specific rotation considerably.

Acetyl-derivatives of milk-sugar. Schützenberger (L. A. **160**, 91).

   (1) $C_{12}H_{18}(C_2H_3O)_4O_{11}$ in water. $c = 7·46.$ $[a]_D = 50·1.$
   (2) $C_{12}H_{14}(C_2H_3O)_6O_{11}$ in Alcohol. $c = 2·18.$ $[a]_D = 32.$
       „   „ $c = 9·68.$ $[a]_D = 31.$

Micose. Trehalose, $C_{12}H_{22}O_{11} + 2H_2O$. Dextro-rotatory.

Aqueous solutions, after standing twenty-four hours, exhibit no change of rotatory power :—

Mycose. $c = 10·03.$ $[a]_j = 173·2.$ Mitscherlich (L.A. 106·17).
Trehalose. $c = 8·4$ to 14·8. $t = 15°.$ $[a]_j = 199$ (anhydrous 220). Berthelot (A. C. P. [3] **55**, 276).

Melitose, $C_{12}H_{22}O_{11} + 3H_2O$. Dextro-rotatory.

Water. $c = 17·27.$ $t = 25°.$ $[a]_j = 88$ (anhydrous 102). Berthelot (A. C. P. [3] **46**, 69).

Melizitose, $C_{12}H_{22}O_{11} + H_2O$. Dextro-rotatory.

Water. $c = 18·6$ anhydrous substance. $t = 20°.$ $[a]_j = 94·1$ (anhydrous), 89·4 hydrated. Berthelot (A. C. P. [3] **55**, 284).
Water. c not given. $[a]_D$ 88·8 (hydrated). $[a]_j = 94·8$ (anhydrous). Villiers (A. C. P. [5] **12**, 434).

§ 110. Sugars, $C_6 H_{12} O_6$.

Glucose. Dextrose, $C_6 H_{12} O_6 + H_2 O$. Dextro-rotatory.

Fresh solutions, prepared in the cold, exhibit bi-rotation (§ 27). The following numbers refer to the reduced constant of rotation.

The most accurate observations are those of Tollens (§ 38), whence the following formulæ are derived :—

For $C_6 H_{12} O_6 + H_2 O$. Aqueous solution. $t = 20°$.

$p = $ 8 to 91. $[a]_D = 47\cdot925 + 0\cdot015534 \ p + 0\cdot0003883 \ p^2.$

$q = $ 9 to 92. $[a]_D = 53\cdot362 - 0\cdot093194 \ q + 0\cdot0003883 \ q^2.$

For $C_6 H_{12} O_6$.

$p = $ 7 to 83. $[a]_D = 52\cdot718 + 0\cdot017087 \ p + 0\cdot0004271 \ p^2.$

$q = $ 17 to 93. $[a]_D = 58\cdot698 - 0\cdot10251 \ q + 0\cdot0004271 \ q^2.$

Hoppe-Seyler (*Med. chem. Untersuchungen* I., 163) determined the specific rotation of diabetic sugar by Broch's method, employing a solution in which $c = 36\cdot277$ anhydrous substance, and found :—

For rays	C	D	E	b	F
$[a] = $	42·45	53·45	67·9	71·8	81·3 (?)

In more recent investigations, Hoppe-Seyler (*Fresenius, Zeitsch. für analyt. Chem.* **14**, 305), employing solutions of diabetic sugar with $c = 14$ to 29, obtained a mean value for $[a]_D = 56\cdot4$ (anhydrous substance).

Hesse (L. A. **176**, 102) has investigated the specific rotation of $[a]_D$ of various glucoses, and found, at $t = 15°$, the following :—

Glucose-hydrate. c	Honey-sugar.	Grape-sugar.	Starch-sugar.	Salicin-sugar.
1	49·77	—	50·00	50·00
3	47·33	47·87	48·03	48·48
6	46·58	—	46·79	47·96
12	46·34	—	46·83	47·66
Anhydride	—	51·78	51·67	51·80
	—	for $c = 2\cdot8$	for $c = 3$	for $c = 2\cdot5$

These sugars are therefore identical with one another. Along with them must also be included amygdalin-sugar, which, with $c = 2$, gave $[a]_D = 49\cdot25$ (hydrate).   On the other hand, phlorhizin-sugar (hydrate) showed a lower rotation, viz., with $c = 3$, $[a]_D\, 40\cdot9$, and with $c = 6$, $[a]_D = 40\cdot08$.   Another sample gave, with $c = 8$, $[a]_D = 39\cdot9$; and with $c = 10\cdot52$, $[a]_D = 39\cdot7$.   Hesse (L. A. **192**, 174).

For the *transition tint* $[a]_j$, the following rotation values are recorded :—52, Bondonneau.   52·5, Clerget Listing.   53·2, Dubrunfaut.   55·1, Pasteur.   56, Berthelot.   57, Schmidt.   57·4, Béchamp. 57·7, Jodin.

Lime reduces the rotatory power.   A solution which contained, in 100 cubic centimetres, 0·98 gramme of lime for 6·9 grammes of grape-sugar, gave $[a]_j = 33\cdot3$.   Jodin (C. R. **58**, 613).

Fruit-sugar.   Lævulose, $C_6H_{12}O_6$.   **Lævo-rotatory.**

The observations under this head are very incomplete, as the effects of concentration have not been investigated.   This substance exhibits a marked decrease of rotation with increase of temperature. For lævulose obtained from invert-sugar by the lime process, Dubrunfaut (C. R. **42**, 901) found the following values (c not recorded) :—

t	14°	52°	90°
$[a]_j =$	− 106	− 79·5	− 53.

When the temperature exceeds 90°, a chemical change begins in the solutions.

According to Neubauer (D. C. G. 1877, 829), at a temperature of 14°, $[a]_D = -100$ (c not given).

Jodin (C. R. **58**, 613) gives the following values (t not given, 14° ?) :—

Aqueous solution	$c = 12\cdot8$.	$[a]_j = -104$.
Alcoholic ,,	$c = 12\cdot8$.	$[a]_j = -92$.

Lime causes a considerable reduction of the rotatory power.   A solution with $c = 5$, giving $[a]_j = -106$, on the addition of 0·64 lime gave $[a]_j = -63$.—Jodin.

Invert-sugar, $C_6H_{12}O_6$.   **Lævo-rotatory.**

The rotatory power decreases rapidly with increase of temperature.   Dubrunfaut (C. R. **42**, 901) found for a solution, the strength of which is not stated :—

t	14°	52°	90°
$[a]_j =$	− 26·65	13·33	0

Tuchschmid's observations (J. P. C. [2] **2**, 235) show that an aqueous solution of invert-sugar with $c = 17\cdot21$ has at 0° Cent. a specific rotation $[a]_D = -27\cdot9$, and that this value decreases with increase of temperature according to the formula $[a]_D^t = -(27\cdot9 - 0\cdot32\ t)$. According to which, at $t = 87\cdot2°$, rotation will be 0.

Alcohol, according to Jodin (C. R. **58**, 613), causes an important reduction in the lævo-rotation of invert-sugar, which can, moreover, by the application of heat, be converted into dextro-rotation. Lime also causes a decrease.

According to Maumené (C. R. **80**, 1139) different specimens of invert-sugar exhibit similar properties only when in their preparation the proportions of water and acid, the temperature and duration of the action, and the mode of neutralization employed have been strictly identical.

Galactose, $C_6H_{12}O_6$. Dextro-rotatory.

Exhibits bi-rotation. According to Fudakowski (Hoppe-Seyler's *Med. chem. Untersuchungen* I., 164), it is a mixture of two different sugars of unequal rotatory powers.

Sorbin, $C_6H_{12}O_6$. Lævo-rotatory.

Water. $c = 23\cdot9$. $[a]_j = -46\cdot9$. Berthelot. Pelouze (A. C. P. [3] **35**, 222).

§ 111. Mannite Group.

Mannite, $C_6H_{14}O_6$. Pasteur (C. R. **77**, 1192) and Bouchardat (C. R. **80**, 120; A. C. P. [5] **6**, 100) have shown that in aqueous solutions ($c = 15$) with a tube-length of 3 to 4 metres, this substance gives a left-handed rotation of $0\cdot1°$ to $0\cdot3°$, whence $[a]_j = -0\cdot03$. Vignon (A. C. P. [5] **2**, 433), also Müntz and Aubin (A. C. P. [5] **10**, 533) consider mannite as inactive.

The addition of various substances to aqueous solutions of mannite renders them optically active, dextro-rotatory in the case of boracic acid, borax, and borate of lime, and more feebly so with chloride and sulphate of sodium; lævo-rotatory in the case of caustic potash, caustic soda, potassium carbonate, potassium, and hydrogen arsenate, lime, baryta, and magnesia. After saturation with acetic acid the solution either remains feebly lævo-rotatory or shows slight dextro-rotation. Sulphuric acid or acetic acid added even in large proportion to mannite solutions produces no activity (Bouchardat, Vignon, Müntz, and Aubin, *loc. cit.*).

Certain derivatives of mannite are active, as, for example, nitro-mannite, diacetyl- and hexacetyl-mannite, which are dextro-rotatory, and mannite dichlorhydrin which is lævo-rotatory.

Mannitan is a variable mixture of dextro-rotatory and lævo-rotatory isomers, the proportions varying with the mode of production. (Bouchardat).

The mannite obtained by the action of nascent hydrogen on inactive glucose, invert-sugar, dextrose, lævulose (from invert-sugar or inulin) is also inactive, as well as that from manna. But each of them becomes active on addition of borax or conversion into nitro-mannite (Müntz and Aubin).

Nitro-mannite, $C_6 H_8 (O . NO_2)_6$. Dextro-rotatory.

Ether. $p = 4.2.$ $[\alpha]_j = 70.2.$ Krecke (*Arch. Néerl.* VII, 1872).
Alcohol. $p = 2.$ $[\alpha]_j = 63.7.$,,   ,,   ,,   ,,
Alcohol. $c = 7.5.$ $[\alpha]_D = 40.$ Krusemann (D. C. G. 1876, 1468).

Treated with ammonium sulphide, nitro-mannite passes into inactive (?) mannite, which by the action of nitric acid becomes once more active. Loir (*Bull. soc. chim.* 1861, 113). Krecke, *loc. cit.*

Dulcite is inactive. Biot. Jaquelain (J. B. 1850, 536). The acetyl-derivatives of dulcite and dulcitan rotate feebly to the right. Bouchardat (A. C. P. [4], 27 ; **68**, 145).

Isodulcite, $C_6 H_{12} O_5 + H_2 O$. Dextro-rotatory.

Water. $c = 10.2.$ $[\alpha]_j = 7.6.$ Hlasiwetz and Pfaundler (L. A. **127**, 362).

Quercite, $C_6 H_{12} O_5$. Dextro-rotatory.
Water. $c = 1$ to $10.$ $t = 16°.$ $[\alpha]_D = 24.3.$ Temperature without influence.
Prunier (*Bull. soc. chim.* **28**, 555, and C. R. **85**, 808).

§ 112.   Carbo-hydrates, $(C_6 H_{10} O_5)_n$.

The rotation data given by different observers for substances of this class differ so widely from each other, that they cannot be used as characteristic marks for the substances.

Cellulose, dissolved in cadmium, or zincoxide-ammonia (obtained by treating cellulose dissolved in cupric-oxide ammonia, with cadmium or zinc until a colourless solution is obtained) is inactive. Krecke (*Arch. Néerl.* VI. 1871). Collodion, according to Krecke, is inactive. According to Schützenberger (L. A. **160**, 77) it is dextro-rotatory.

Starch boiled for a few hours in water, solution of potash or of chloride of zinc, gives dextro-rotatory liquids. For $p = 2.22$

to 3·88. $[a]_j = + 211°$. Starch solution heated with dilute sulphuric acid at 100°, gives first $[a]_j = 216$, but the rotatory power quickly decreases with the formation of dextrin and dextrose. Béchamp (A. C. P. [3] **48**, 458).

Glycogen. Dextro-rotatory. Various modifications. $[a]_j = +140$ to 211. Tichanowitsch. Stscherbakoff (J. B. 1870, 848).

Inulin. Lævo-rotatory. The data for dahlia- and elecampane-inulin, range from $[a]_j = -26$ to 72. The *inuloid* of Popp (L. A. **156**, 190), $C_6 H_{10} O_5 + H_2 O$, with $c = 2$, gives $[a]_j = -30·5$.

The rotatory powers of the acetic ethers of cellulose, starch, glycogen, and inulin have been determined by Schützenberger (L. A. **160**, 74).

Dextrin. Dextro-rotatory. The data range from $[a]_j = 139$ to 213.

Gum Arabic Acid. The varieties of gum arabic met with in commerce, are partly dextro-rotatory, and partly lævo-rotatory. Scheibler (D. C. G. 1873, 618) found, on examination of five samples in aqueous solution with $c = 5$, the specific rotations for $[a]_j = + 37·3 + 46·1; -28·8; -29·2; -30·0$ respectively. On heating with dilute sulphuric acid, all these solutions become dextro-rotatory by the formation of gum-sugar (arabinose), $C_6 H_{12} O_6$.

The gum of beet-root is, in general, dextro-rotatory, but at certain seasons and in individual plants it is found to be lævo-rotatory. Scheibler, *loc. cit.*

Dextro-rotatory gum is further found in the stag-truffel, and lævo-rotatory gum in couch-grass. Ludwig (J. B. 1869, 791; 1872, 803).

§ 113. Glucosides.

Salicin, $C_{13} H_{18} O_7$. Lævo-rotatory.

Water. $c = 1$ to 3. $t = 15°$. $[a]_D = -(65·17 - 0·63 c)$. Hesse (L. A. **176**, 116).
Water. $p = 2·78$. $[a]_j = -73·4$. Biot and Pasteur (C. R. **34**, 607).

Populin, $C_{20} H_{22} O_8$. Lævo-rotatory.

Water. $p = 1$. $[a]_j = -53$. Biot and Pasteur, *loc. cit.*

Phlorhizin, $C_{21} H_{30} O_{11} + 2 H_2 O$. Lævo-rotatory.

Alcohol of 97 per cent. (by vol.) $c = 1$ to 5. $t = 22·5$. $[a]_D = -(49·40 + 2·41 c)$.
Hesse (L. A. **176**, 117).

Alcohol. $p = 4·6$. $[a]_D = -52$
Wood Spirit. $p = 3·9$. $[a]_D = -52$ } Oudemans (L. A. **166**, 69).

Q

§ 114. Derivatives of the Sugars.

Besides the previously-mentioned nitric and acetic ethers, active amyl-alcohol and active lactic acid belong to this group.

Active amyl-alcohol. Lævo-rotatory.

Commercial fermentation. Amyl-alcohol gave, in a layer 1 decimetre deep, an angle of rotation $[a]_D = -1 \cdot 97$ (Le Bel); $[a]_D = -2 \cdot 76$ (-40 of Ventzke's scale for a length of 5 decimetres. (Ley.)

For the active amyl-alcohol obtained from the commercial product by separation as completely as possible from admixture with the inactive alcohol by Pasteur's or Le Bel's method, we have the following data :—

(*a*) Specimens obtained by Pasteur's method of fractional crystallization of amyl-sulphate of barium.

	Angle of Rotation for a Layer 1 decimetre in thickness.	Specific Gravity.	Specific Rotation $[a]_D$.	Boiling Point.	Observers.
1	$a_j = -4°$	—	—	127° to 128°	Pasteur (C.R. 41, 296).
2	$a_j = -3 \cdot 4°$	—	—	128°	Pedler (L.A. 147, 245).
3	$a_D = -3 \cdot 18$ (46 divisions of Ventzke[1] for 5 decimetres)	0·808 at 15°.	$-3 \cdot 9$	128°	Ley (D. C. G. 1873, 1365).
4	$a_D = -1 \cdot 4$ to $1 \cdot 6$ (20 to 23 divisions of Ventzke for 5 decimetres)	0·812 at 19°.	$-1 \cdot 8$	127·5°	Erlenmeyer and Hell (L. A. 160, 283).
5	$a_D = -0 \cdot 92$ (8·5 divisions of Soleil for 2 decimetres)	0·825 at 0°	$-1 \cdot 1$	130°	Pierre and Puchot (C. R. 76, 1332).

(*b*) Amyl-alcohol, purified by Le Bel's method, by repeated treatment with hydrochloric acid gas, whereby the inactive portion is first converted into amyl-chloride and can then be removed by distillation. Boiling-point, 127°.

[1] 1° Ventzke [ray j] = 0·3457 angular degrees for ray D.
    1° Soleil [ray j] = 0·2167   ,,   ,,   ,,   ,,

$a_D = -4\cdot53$ to $4\cdot63$ for 1 decimetre. (Specific gravity not given; but taking it as $= 0\cdot81$, we get $[a]_D = -5\cdot6$ to $5\cdot7$.) Le Bel (*Bull. soc. chim.* [2] 21, 542). .

By repeated distillation with caustic potash, and still more quickly with metallic sodium, active amyl-alcohol is rendered inactive. Le Bel (*Bull. soc. chim.* [2] 25, 545). Lævo-amyl-alcohol by repeated distillation with caustic soda becomes dextro-rotatory. $a_D = +2°$ for 1 decimetre according to Beignes Bakhoven (*Pogg. Supp.* Bd. 6, 329); but this is denied by Le Bel (*Bull. soc. chim.* [2] 25, 199) and Balbiano (D. C. G. 1876, 1692).

According to Pierre and Puchot aqueous amyl-alcohol gives a stronger rotation than the anhydrous alcohol. Ley (D. C. G. 1873, 1370) is, however, unable to confirm this.

DERIVATIVES OF ACTIVE AMYL-ALCOHOLS.

Active valerianic acid. Dextro-rotatory.

(*a*) From lævo-amyl-alcohol.

	Angle of Rotation for a Layer 1 decimetre in thickness.	Specific Gravity.	Specific Rotation $[a]_D$.	Boiling Point.	Observers.
1	$a_j = 8\cdot60$	—	—	170	Pedler (L. A. 147, 246).
2	$a^D = 2\cdot7$ to $2\cdot76$ (39 to 40 divisions of Ventzke's scale for 5 decimetres) ..	—	—	168—171	} Erlenmeyer and Hell (L. A. 160, 284, 293).
3	$a_D = 3\cdot37$ (48·7° Ventzke for 5 decimetres) ..	0·933 at 19·5°	$+3\cdot6$	173	
4	$a_D = 4\cdot24$ (61·2° Ventzke 5 decimetres ..	0·917 at 15°	$+4\cdot6$	173	} Ley (D. C. G. 1873, 1368).
5	$a_D = 3\cdot12$ (45° Ventzke for 5 decimetres ..	0·917 at 15°	$+3\cdot4$	174—175	
6	$a_D = 0\cdot54$ (5° Soleil for 2 decimetres)	0·947 at 0°	$+0\cdot6$	178	Pierre and Puchot (C. R. 76, 1332).

(b) From Leucin.

a_D = 1·18 for 1 decim. (17 divs. Ventzke for 5 decims.) Erlenmeyer and Hell
(L. A. **160**, 286).

Valerianate of amyl.   Dextro-rotatory.

a_D = 7·6 for 1 decim. (44 divs. Ventzke for 2 decims.) d = 0·869 at 15°. $[a]_D$ = 8·7.
Boiling-point 186°. Ley (D. C. G. 1873, 1369).

a_D = 4·3 for 1 decim. (40 divs. Soleil for 2 decims.) d = 0·874 at 0°. $[a]_D$ = 4·9.
Boiling-point 190°.   Pierre and Puchot (C. R. **76**, 1332).

a_D = 2·3 for 1 decim. (33 to 34 divs. Ventzke for 5 decims.) Erlenmeyer and Hell.
(L. A. **160**, 289).

The following derivatives from lævo-amyl-alcohol, giving an
angle of rotation a_D = 4·63 for a length of 1 decimetre, have been
prepared and examined by Le Bel (*Bull. soc. chim.* [2] **21**, 542) :—

Amyl-chloride (boiling-point 97° to 99°).   Dextro-rotatory.

a_D = 1·10 for 1 decim. d = 0·886 at 15°. $[a]_D$ = 1·24.

Amyl-bromide (boiling-point 117° to 120°).   Dextro-rotatory.

a_D = 4·60 for 1 decim. d = 1·225 at 15°. $[a]_D$ = 3·75.

Amyl-iodide (boiling-point 144° to 145°).   Dextro-rotatory.

a_D = 8·22 to 8·33 for 1 decim. d = 1·54 at 15°. $[a]_D$ = 5·34 to 5·41.

Methyl-amyl and amylene from active amyl-alcohol are inactive.
(Le Bel.)

The following other derivatives of amyl-alcohol and valerianic
acid have been examined by Pierre and Puchot (C. R. **76**, 1332):—

	Boiling-point.	Divs. *Soleil* for 2 decims.	a_D for 1 decim.	Sp. gr. at 0°.	$[a]_D$
Amyl aldehyde	92·5°	+ 6	= 0·65°	0·8209	+ 0·8
Butyrate of amyl	170·3°	8·5	= 0·92°	0·8769	+ 1·05
Valerianate of butyl	173·4°	3	= 0·325°	0·8884	+ 0·4
,, propyl	157°	9	= 0·975°	0·8862	+ 1·1
,, ethyl	135·5°	12·5	= 1·35°	0·8860	+ 1·5
,, methyl	117·5°	8·5	= 0·92°	0·9005	+ 1·0

The stability of the active amylic grouping is shown by the
researches of Wurtz (L. A. **105**, 295), in which amyl-iodide was converted into the cyanide (a_{red} = 1·59° for a length of 1 decimetre),
this into capronic acid (a_{red} = 1·22°). and this, again, by electrolysis,
into di-amyl (a_{red} = 3·20°).

Para-lactic acid. Active sarco-lactic acid. Dextro-rotatory. The
investigations of Wislicenus (L. A. **167**, 302) have shown that an

exact determination of the specific rotation of para-lactic acid is impossible as this substance, even at ordinary temperatures, passes gradually into the ether-anhydride, $C_6H_{10}O_5$, and into lactide, $C_3H_4O_2$, both of which are strongly lævo-rotatory.

Freshly prepared, it shows in aqueous solutions, in which the degrees of concentration were determined by titration (reckoned as $C_3H_6O_3$), the following specific rotation :—

$$c \ = \ 7\cdot38 \qquad 25\cdot57 \qquad 39\cdot94 \text{ grammes.}$$
$$[a]_D = + \ 2\cdot78° \qquad 1\cdot64° \qquad 2\cdot63°$$

By standing, the rotation of the solutions gradually increases.

If they be then diluted with water a sudden decrease of rotation takes place, which, however, gradually increases again, but without returning to its original amount. For example, two specimens of equal concentration gave the following results :—

Original solution	$c = 21\cdot24$.	$[a]_D = + 1\cdot41$	$[a]_D = + 1\cdot85$
After keeping 1 month	,,	,, 2·07	2·26
,, 2 months	,,	,, 2·21	2·45
,, 3 ,,	,,	,, 2·66	2·64
Diluted with water	$c = 15\cdot75$.	,, 1·99	2·13
,, after 1 month	,,	,, 2·29	2·41

The gradual increase here observed is due to the gradual conversion of the lævo-rotatory molecules of the anhydride, $C_6H_{10}O_5$, originally present in the solution, into the dextro-rotatory molecules of the acid $C_3H_6O_3$, which, moreover, is proved by the fact that, if we neutralize the liquid with alkali, it gradually recovers its acidity.

The decrease of rotatory power on the adition of water, is explained by Wislicenus to be caused by the formation of the hydrate, $C_3H_6O_3 + H_2O$, to which he assigns a rotatory power inferior to that of the acid itself.

It requires but 2 or 3 per cent. of the anhydride to be present to give the solution of lactic acid distinct lævo-rotation. But in preparations that have been long kept the amount of rotation becomes considerable. Thus, a solution which had been kept *in vacuo* over sulphuric acid for a year and nine months, and whose composition approximated very closely to 84 per cent. $C_6H_{10}O_5 + 16$ per cent. $C_3H_4O_3$, gave in alcoholic solution with $c = 19\cdot54$, the specific rotation $[a]_D = -85\cdot9°$.

Zinc para-lactate, $Zn(C_3H_5O_3)_2 + 2H_2O$. Lævo-rotatory.

The specific rotation in aqueous solution appears to increase with decrease of concentration.

Wislicenus (L. A. 167, 332) obtained the following results:—

Solution.	Hydrated Salt.		Anhydrous Salt.	
1.	$c = 16.05.$	$[\alpha]_D = - 6.36$	$c = 13.98.$	$[\alpha]_D = - 7.30$
2. ,,	11.01. ,,	6.36	,, 9.60. ,,	7.29
3. ,,	7.47. ,,	6.83	,, 6.51. ,,	7.83
4. ,,	6.13.' ,,	7.41	,, 5.36. ,,	8.49
5. ,,	5.45. ,,	7.34	,, 4.75. ,,	8.43
6. ,,	5.26. ,,	7.60	,, 4.58. ,,	8.73

Solutions 1 and 2 were supersaturated.

Para-lactate of lime, Ca $(C_3 H_5 O_3)_2 + 9 H_2 O$. Lævo-rotatory.

Hydrated salt.	$c = 7.23.$	$[\alpha]_D = - 3.87$	} Wislicenus.
Anhydrous ,,	$c = 5.35.$	$[\alpha]_D = - 5.25$	}

§ 115. Vegetable Acids.

Dextro-tartaric acid, $C_4 H_6 O_6$.

See § 19, p. 47, Arndtsen. The formulæ apply for $q = .50$ to 95 per cent.

See also § 26, p. 60, Krecke (*For Influence of Temperature*).

Water	$c = 5$ to 15.	$t = 15.$	$[\alpha]_D = + (14.90 - 0.14 c)$ } Hesse
,,	,, ,,	$t = 22.5.$	$[\alpha]_D = + (15.22 - 0.14 c)$ } (L. A. 176, 129).
,,	$c = 0.5$ to 15.	$t = .20.$	$[\alpha]_D = + (15.06 - 0.131 c)$ Landolt (D. C. G. 1873, 1073).

The specific rotation of tartaric acid is considerably diminished by the presence of other acids. Biot (*Mém. de l'Acad.* 16, 229).

Dextro-tartrates. Solutions in water.

The following determinations (Landolt) are for the anhydrous salts. Temperature always at 20°.

K . H	$C_4 H_4 O_6$	$c = 0.615$	$[\alpha]_D = + 22.61$
Na . H	,,	4.409	23.95
N H$_4$. H	,,	1.712	25.65
Li . H	,,	7.998	27.43
K$_2$	,,	11.697	28.48
Na$_2$	,,	9.946	30.85
(N H$_4$)$_2$	,,	9.433	34.26
Li$_2$	,,	8.305	35.84
K . Na	,,	10.771	29.67
K . N H$_4$	,,	10.515	31.11
Na . N H$_4$	,,	9.690	32.65
Mg	,,	8.818	35.86
K . Bo	,,	2.744	51.48
,, ,,	,,	5.488	58.35
Na . Bo	,,	2.538	55.02
,, ,,	,,	5.075	63.48
,, ,,	,,	10.151	71.47
K . As O	,,	0.563	21.13
Na . As O	,,	3.358	20.64
K . Sb O	,,	7.982	142.76
K . C$_2$ H$_5$	,,	11.079	29.91
Ba$_{\frac{1}{2}}$. C$_2$ H$_5$	,,	12.586	25.68

For sodium tartrate, $Na_2 C_4 H_2 O_6 + 2 H_2 O$, Hesse (L. A. 176, 122) found :—Water, $c = 5$ to 15, $t = 15$. $[a]_D = 27\cdot85 - 0\cdot17\ c$. Krecke (*Arch. Néerl.* VII., 1872) has determined the specific rotation of a few tartrates. At temperatures between $0°$ and $100°$, and for various rays, with concentration $c = 20$, and the temperature $= 25°$, he obtained the following results :—

	$[a]_c$	$[a]_D$	$[a]_E$	$[a]_b$	$[a]_F$
$2 (K_2 . C_4 H_4 O_6) + H_2 O$	22·04	26·84	32·95	34·96	39·92
$Na_2 . C_4 H_4 O_6 + 2 H_2 O$	20·52	25·79	31·67	32·70	38·49
$(N H_4)_2 . C_4 H_4 O_6$	31·08	37·09	43·05	45·27	53·76
$K . Na . C_4 H_4 O_6 + 4 H_2 O$	18·52	22·42	26·49	27·67	32·08
$2 (K . SbO . C_4 H_4 O_6) + H_2 O\ (c = 5)$	111·82	138·66	180·39	187·39	218·74

Increase of temperature had little or no effect on the three first-mentioned salts ; Rochelle salt showed increase of specific rotation, but tartar emetic decrease.

Tartrate of ethyl, $(C_2 H_5)_2 . C_4 H_4 O_6$. See § 33, p. 77.

Lævo-tartaric acid.

Pasteur (A. C. P. [3] 24, 442 ; 28, 56 ; 38, 437) gives the following data, which in part apply to Biot's red ray (see § 18, p. 43.

Free acid, $p = 35\cdot7$. $t = 17$. $[a]_r = - 8\cdot43$. (Dextro-rotatory tartaric acid under like conditions gives $[a]_r = + 8\cdot53$. Biot.)

Lævo-tartrate of ammonia, $[a]_r = - 29\cdot3$. (The dextro-rotatory salt gives $[a]_r = + 29\cdot0$.)

Lævo-tartrate of sodium and ammonium, $Na . N H_4 . C_4 H_4 O_6 + 4 H_2 O$. $[a]_j = - 26\cdot0$. (The dextro-rotatory salt gives $[a]_j = + 26\cdot0$.)

Lævo-rotatory tartar emetic, $[a]_j = - 156\cdot2$. (Dextro-rotatory salt gives $[a]_j = + 156\cdot2$.)

Lævo-tartrate of lime dissolved in hydrochloric acid (20 grammes tartrate for 63 cubic centimetres of acid, of specific gravity 1·08) gives a decidedly dextro-rotatory solution.

In presence of boracic acid, lævo-tartaric acid behaves exactly as does the dextro-rotatory acid. Biot (A. C. P. [3] 28, 99).

Malic acid, $C_4 H_6 O_5$. Lævo-rotatory.

From mountain-ash berries. A slight lævo-rotation increased by dilution with water as well as by increase of temperature.

Water. $p = 16\cdot6$. $[a]_D = - 1\cdot04$.   Ritthausen (*Journ. für prakt. Chem.* [2] 5, 354).

„ $p = 32\cdot907$. $t = 10°$. $[a]_j = - 5\cdot00$. Pasteur (A. C. P. [3] 31, 81).

The specific rotation is increased considerably by the addition of boracic acid. On the other hand, mineral as well as organic acids diminish it, and may even convert it into dextro-rotation.

MALATES.

Neutral malate of soda, $Na_2 . C_4 H_4 O_5$.

Water. $c = 15.976$. $t = 22$. $[a]_D = - 7.64$. Landolt.

Acid malate of ammonia, $N H_4 . C_4 H_5 O_5$.

Water. $c = 6.5$. $t = 20$. $[a]_D = - 6.65$ Landolt.
Water. $c = 23.025$. $t = 20$. $[a]_j = - 7.22$
Nitric acid of 21° Baumé. $p = 26.803$. $t = 20$. $[a]_j = + 5.60$
   „   after addition of ammonia, lævo-rotatory.

Malate of lime, $Ca . C_4 H_4 O_5 + 3 H_2 O$.

Hydrochloric acid of 9·5° Baumé. $p = 12.474$. $t = 22$. $[a]_j = + 10.88$
Ammonia (per cent. ?) $p = 23.506$. $t = 17$. $[a]_j = + 4.34$

Antimon-ammonium malate, $N H_4 . Sb O . C_4 H_4 O_5$.
Water . . . . $p = 6.845$. $t = 17$. $[a]_j = + 115.47$

Pasteur

(A. C. P.

[3] 31, 85).

Amides of malic and tartaric acids. Pasteur, (A. C. P. [3] **38**, 466).

Glutanic acid, $C_5 H_8 O_5$. Lævo-rotatory.

Water. $p = 18.81$. $[a]_D = - 1.98$. Ritthausen (*Journ. für prakt. Chem.* [2] **5**, 354).

Glutamic acid, $C_5 H_9 N O_4$. Dextro-rotatory.

Dilute nitric acid. $p = 5.45$. $t = 18$. $[a]_D = 34.7$. Ritthausen (*Journ. für prakt. Chem.* [1] **107**, 238).

Aspartic acid, $C_4 H_7 N O_4$. Dextro-rotatory in acid solutions; lævo-rotatory in alkaline solutions.

Dilute nitric acid $p = 4.711$. $t = 20$. $[a]_D = + 25.16$ Ritth. (*a*)
Hydrochloric acid of 9·5° Baumé $p = 5.094$. $t = 22$. $[a]_j = + 27.68$
Soda solution containing 4·84 per cent. $Na_2 O$. $p = 9.99$. $[a]_j = - 2.22$
Ammonia with 10 per cent. $N H_3$. $p = 4.02$. $[a]_j = - 11.67$

Pasteur
(*b*)

(*a*) Ritthausen's preparation from legumin (*Journ. für prakt. Chem.* [1] **107**, 227).

(*b*) Pasteur's preparation from asparagin (A. C. P. [3] **31**, 78).

Asparagin, $C_4 H_8 N_2 O_3$. Dextro-rotatory in acid solutions; lævo-rotatory in alkaline.

Nitric acid of specific gravity 1·1102 at 22°. $p = 11.08$. $t = 22$. $[a]_j = + 35.09$.
Hydrochloric acid „ 1·0706 at 23°. $p = 11.125$. $t = ?$ $[a]_j = + 34.40$.

Soda solution with 4·84 per cent. $Na_2 O$. $p = 8·89.$ $t = 8·5.$ $[a]_j = - 7·50.$
,,   ,, 4·84   ,,   ,, $p = 17·9.$ $t = 22.$ $[a]_j = - 7·84.$
,,   ,, 12·69 ,,   ,, $p = 15·21.$ $t = 22.$ $[a]_j = - 7·31.$
Ammonia (per cent. ?) $p = 12·72$.$t = 18.$ $[a]_j = - 11·18.$

Pasteur (A. C. P. [3] **31**, 75).

Water $p = 1·66.$ $[a]_D = - 6·23.$
Water and 10 grammes $N H_3$ in100 cub. cent. solution $p = 1·66.$ $[a]_D = - 10·68.$
                                                                      $[a]_j = - 11·38.$
,,   10 grammes H Cl   ,,   ,,   ,, $p = 1·66.$ $[a]_D = + 37·45.$
,,   10 grammes acetic acid in 100 cub. cent. solution $p = 1·66.$ Inactive.

Champion and Pellet (C. R. **82**, 819).

Quinic acid, $C_7 H_{12} O_6$. Lævo-rotatory.

Water $c = 2$ to 10. $t = 15.$ $[a]_D = - 43·9.$
Alcohol of 80 per cent. by vol. $c = 5.$ $[a]_D = - 39·2.$
1 mol. quinic acid + 1 mol. $Na_2 O$ + Water. $c = 2.$ $[a]_D = - 47·0.$

Hesse (L. A. **176**, 124).

§ 116. Terpenes, $C_{10} H_{16}$.

I. Dextro-rotatory Oil of Turpentine, Australene.

[The latter name is applied to the preparation obtained by distillation after neutralization by soda solution.]

(a) From *Pinus australis* and *P. Taeda*. American or so-called English oil of turpentine.

$a_j = 13·5$ (crude) ; 14·6 (rectified). Luboldt.
$[a]_j = 18·6.$ Guibourt and Bouchardat.
$[a]_D = 14·15$ at 20°. $d = 0·9108.$ Landolt (§ 31, p. 70).

Wiedemann (*Pogg. Ann.* **82**, 222) obtained the following angles of rotation for different rays :—

Lines	C	D	E	b	F	G
a	10·9	14·0	18·7	19·6	23·2	32·7

Australene. Distilled at 100° with reduced pressure. Portion I. $[a]_j = 24·3.$ Portion II. $[a]_j = 21·4.$ (Boiling-point, 161°.) Berthelot (A. C. P. [3] **40**, 5).

(b) From *Pinus sylvestris* and *P. Abies*. Russian or so-called German oil of turpentine.

$[a]_j = 14·6$ to 16·3. Luboldt.

Australene from *Pinus sylvestris*, $[a]_D = 32·4.$ $[a]_j = 40·3$ with $t = 24·5°,$ ($a_D = 27·7,$ $d = 0·8547.$ Boiling-point, 155·5 to 156·5). Flawitzky (D. C. G. 1878, 1846).

Australene from Swedish wood-tar of *Pinus sylvestris*, $[a]_D = 36·3$ ($d = 0·8631$ at 16°. Boiling-point, 156·5 to 157·5.) Atterberg (D. C. G. 1877, 1203).

II. Lævo-rotatory turpentine oil. Terebenthene. (Neutralized and distilled.)

(*a*) From *Pinus Pinaster* (*P. maritima*). French oil of turpentine.

Commercial oils, $a_j = -18 \cdot 2$, Luboldt; $a_j = -31 \cdot 1$, Gladstone; $a = -35 \cdot 4$ to $36 \cdot 5$, Berthelot; $35 \cdot 6$, Gladstone; $36 \cdot 5$, Gernez; $40 \cdot 0$, Biot; $42 \cdot 2$, Mayer; $43 \cdot 0$, Deville; $43 \cdot 4$, Soubeiran and Capitaine; $45 \cdot 5$, Buignet.

For different rays, Wiedemann (A. P. **82**, 222) found:—

Lines	B	C	D	E	b	F	G
$a = -$	21·5	23·4	29·3	36·8	38·3	43·6	55·9

Terebenthene. $[a]_j = 42 \cdot 3$. (Distilled with reduced pressure, at temperatures between 80° and 100°. Boiling-point, 161°.) Berthelot (A. C. P. [3] **40**, 5). $[a]_D = -37 \cdot 01$ at 20° ($d = 0 \cdot 8629$ at 20°. Boiling-point, 161°). Landolt (§ 30, p. 66). $[a]_D = -40 \cdot 30$ ($d = 0 \cdot 8685$ at 10°. Boiling-point, 156°). Riban (A. C. P. [5] **6**, 15).

Iso-terebenthene. Obtained by exposing terebenthene for two hours to a temperature of 300°. $[a]_D = -9 \cdot 45$ ($d = 0 \cdot 8431$ at 20°. Boiling-point, 175°). Riban (A. P. C. [5] **6**, 218).

(*b*) From *P. Larix*. Venetian oil of turpentine.

$[a]_j = -6 \cdot 0$, Luboldt. $[a]_j = -5 \cdot 24$, Guibour[+] and Bouchardat.

(*c*) *Templin oil. "Krummholz" oil.*

1. From the cones of *Pinus picea*. Crude: $[a]_j = -85 \cdot 2$. $[a]_j = -98 \cdot 8$. Rectified: $[a]_j = -92 \cdot 5$. $[a]_j = -107 \cdot 6$. Flückiger, Berthelot (J. B. 1855, 643).
2. From young shoots of *Pinus pumilio*.[1] $[a]_j = -8 \cdot 2$. Jolly, Buchner (L. A. **116**, 328).

By exposure to the air, turpentine oils experience a decrease of rotatory power by oxidation (§ 30), whence probably these discrepancies in the data. When distilled, the unaltered and more highly rotatory portions are carried over first.

For the alterations of the rotatory power of oil of turpentine at temperatures above boiling-point, see § 16, p. 35.

For information on the rotatory properties of the numerous derivatives of oil of turpentine, as they have been investigated by Deville (A. C. P. [2] **75**, 37) and Berthelot (A. C. P. [3] **38**, 38; **39**, 10; **40**, 5), reference must be made to the original memoirs. In the case of most of the solid substances examined the concentrations of

---

[1] *Pinus pumilio* is the dwarf pine (Krummholz), a species with recumbent stem found in the Alps and Pyrenees.—D.C.R.

the solutions employed have not been recorded. The complete data are given only for :—

Lævo-rotatory camphene. Terecamphene, $C_{10} H_{16}$, obtained by heating terebenthene-chlorhydrate with an alcoholic solution of potash or with stearate of soda. Riban (*Bull. soc. chim.* **24**, 10) found for alcoholic solutions with $q = 62$ to 90 per cent. alcohol, at a temperature of $13°$ to $14°$: $[a]_D = -(53·80 - 0·03081 \; q)$.

§ 117. Ethereal Oils.

Commercial ethereal oils, being invariably mixtures of substances often of opposite optical powers, exhibit so much diversity in respect of rotation values, that these are of no use as a test of their purity, and special data are accordingly worthless.

§ 118. Resins.

Euphorbone, $C_{15} H_{24} O$ (from euphorbia resin). Dextro-rotatory.

Chloroform. $c = 4.$ $t = 15.$ $[a]_D = 18·8.$ Hesse (L. A. **192**, 195).
Ether. $(d = 0·72).$ $c = 4.$ $t = 15.$ $[a]_D = 11·7.$

Podocarpic acid, $C_{17} H_{22} O_3.$ Dextro-rotatory.

Alcohol. $c = 4$ to 9. $[a]_D = 136.$
Ether. $c = 4$ to 7. $[a]_D = 130.$

Sodium salt, $C_{17} H_{21} Na O_3 + 8 H_2 O.$

Water. $c = 4·6.$ $[a]_D = 82.$
" $c = 6·4.$ $[a]_D = 79.$ Oudemans
" $c = 13·8.$ $[a]_D = 73.$ (L. A. 166, 65)
Alcohol of 93 per cent. $c = 9$ $[a]_D = 86.$

§ 119. Camphors.

Laurel-camphor, $C_{10} H_{16} O.$ Dextro-rotatory.

For specific rotation in various solvents (§ 36, p. 84). The following others have been obtained :—

Alcohol of 80 per cent. $c = 2$ 6 10
$[a]_D = 40.9$ 39·25 38·65 Hesse
Chloroform. $c = 5.$ $[a]_D = 44·2$ (L. A. 176, 119).

Arndtsen (A. C. P. [3] **54**, 418) has determined the specific rotation in alcoholic solutions for different rays, and with different weight-percentages of alcohol (q). $t = 22 \cdot 9°$. The formulæ hold for $q = 50$ to 95.

$$[a]_c = 38 \cdot 549 - 0 \cdot 0852 \, q$$
$$[a]_D = 51 \cdot 945 - 0 \cdot 0964 \, q$$
$$[a]_E = 74 \cdot 331 - 0 \cdot 1343 \, q$$
$$[a]_b = 79 \cdot 348 - 0 \cdot 1451 \, q$$
$$[a]_F = 99 \cdot 601 - 0 \cdot 1912 \, q$$
$$[a]_e = 149 \cdot 696 - 0 \cdot 2346 \, q$$

Patchouli camphor, $C_{15} H_{26} O$.   Lævo-rotatory.

Fused.   (t above 59°). $[a]_D = -118.$ }   Montgolfier
In Alcohol. $[a]_D = -(124 \cdot 5 + 0 \cdot 21 \, c)$. }   (*Bull. soc. chim.* **28**, 414).

Camphoric acid, from laurel-camphor, $C_{10} H_{16} O_4$.   Dextro-rotatory.

Water	$c = 0 \cdot 64.$	$t = 20.$	$[a]_D = 46 \cdot 2.$
Alcohol of 98 per {	$c = 2 \cdot 562.$	$t = 20.$	$[a]_D = 47 \cdot 5.$
cent. by weight. {	$c = 19 \cdot 294.$	$t = 20.$	$[a]_D = 47 \cdot 4.$
Acetic acid {	$c = 3 \cdot 026.$	$t = 20.$	$[a]_D = 46 \cdot 3.$
50 per cent. {	$c = 6 \cdot 052.$	$t = 20.$	$[a]_D = 46 \cdot 2.$
by weight. {	$c = 12 \cdot 100.$	$t = 21 \cdot 5.$	$[a]_D = 46 \cdot 0.$ Landolt.

Salts of dextro-camphoric acid.   Solutions in water.

$K_2 . C_{10} H_{14} O_4$ $c = 4$ to 16. $t = 20.$ $[a]_D = 14 \cdot 39 + 0 \cdot 06 \, c.$
$Na_2 . C_{10} H_{14} O_4$ $c = 2$,, 9. $t = 20.$ $[a]_D = 16 \cdot 62 + 0 \cdot 06 \, c.$
$(NH_4)_2 . C_{10} H_{14} O_4$ $c = 4$,, 17. $t = 20.$ $[a]_D = 16 \cdot 98 + 0 \cdot 13 \, c.$ Landolt.

Methyl-camphoric acid, $CH_3 . C_{10} H_{15} O_4$.   Dextro-rotatory.

Alcohol. $p = 143.04.$ $t = 19 \cdot 3.$ $[a]_j = 51 \cdot 4.$ Loir (A. C. P. [3] **38**, 485).

For the rotatory power of other camphors and their derivatives no sufficient data exist.

§ 120. Alkaloids.

Conine, $C_8 H_{15} N$.   Dextro-rotatory.

$[a]_D = 17 \cdot 9.$ ($d = 0 \cdot 873$ at 15°. a_D for 1 decimetre $= 15 \cdot 6$). Schiff (L. A. **166**, 94).
$[a]_D = 10 \cdot 6.$ ($d = 0 \cdot 846$,, 12·5°). Alcohol reduces specific rotation ; ether, benzene, and oil have no effect. Petit (D. C. G. 1877, 896).

Nicotine, $C_{10} H_{14} N_2$.   Lævo-rotatory.   (See § 32, p. 72.)

CINCHONA ALKALOIDS.

Besides those already given, §§ 103, 104, the following data have been further recorded, mostly by Hesse.[1]

---

[1] Numerous earlier observations on the rotatory powers of alkaloids were made by Bouchardat (*Ann. chim phys.* [3] **9**, 213), Bouchardat and Boudet (*Journ. de Pharm. et de Chim.* [3] **23**, 288), Buignet (*Journ. de Pharm. et de Chim.* [3] **40**, 268), De Vrij and Alluard (*Compt. rend.* **59**, 201). These data are for Biot's red ray ; but they are not now of much use, as the nature of solvent employed is scarcely ever recorded with sufficient exactitude.

[The alcoholic chloroform mixture, much used by Hesse as solvent, consists of one volume alcohol of 97 per cent. (by volume) with two volumes chloroform.]

Quinine hydrate, $C_{20} H_{24} N_2 O_2 + 3 H_2 O$.　Lævo-rotatory.

Ether $(d = 0.7296)$　　　$c = 1.5$ to 6.　$t = 15°$.　$[a]_D = - (158.7 - 1.911\ c)$.
Alcohol of 97 per cent. (by vol.)　$c = 1$　,, 10.　$t = 15°$.　$[a]_D = - (145.2 - 0.657\ c)$.
　,,　　80 per cent.　　,,　$c = 1$　,, 6.　$t = 15°$.　$[a]_D = - (165.81 - 8.203\ c$
　　　　　　　　　　　　　　　　　　　　　　$+ 1.0654\ c^2 - 0.04644\ c^3)$.
Alcoholic chloroform.　$c = 2$.　$t = 15°$.　$[a]_D = - 141.0$.
　　　,,　　　　　$c = 5$.　$t = 15°$.　$[a]_D = - 140.5$.　Hesse (L. A. **176**, 206).

Anhydrous quinine, $C_{20} H_{24} N_2 O_2$.　Lævo-rotatory.

In absolute and aqueous alcohol.　Oudemans, p. 203.
Alcohol of 97 per cent. (by vol.)　$t = 15°$.　$c = 1$.　$[a]_D = - 170.5$.　$c = 2$.　$[a]_D = - 169.25$.
Chloroform.　$t = 15°$.　$c = 2$.　$[a]_D = - 116.0$.　$c = 5$.　$[a]_D = - 106.6$.　Hesse
　　　　　　　　　　　　　　　　　　　　　　　　(L. A. **176**, 208).

Absolute alcohol.　$c = 1.64$.　$t = 17°$.　$[a]_D = - 167.5$.
Benzene.　　　　$c = 0.61$.　$t = 17°$.　$[a]_D = - 136$.
Toluene.　　　　$c = 0.39$.　$t = 17°$.　$[a]_D = - 127$.　　Oudemans
Chloroform.　　$c = 1.465$.　$t = 17°$.　$[a]_D = - 117$.　　(L. A. **182**, 44).
　　,,　　　　　$c = 0.775$.　$t = 17°$. ·　$[a]_D = - 126$.

Quinine hydrochloride, $C_{20} H_{24} N_2 O_2$. $HCl + 2 H_2 O$.　Lævo-rotatory.

Water (p. 199).　Hydrochloric acid (p. 200).　Absolute alcohol (p. 204).
Alcohol of 97 per cent. (by vol.) $c = 1$ to 10.　$t = 15°$.　$[a]_D = - (147.30 - 1.958\ c$
　　　　　　　　　　　　　　　　　　　　　　$+ 0.1039\ c^2 - 0.00211\ c^3)$.

With aqueous alcohol, the specific rotation attains a maximum when the alcohol is in the proportion of 60 per cent. by volume. For $c = 2$, and $t = 15°$, the following values are given :—

Alcohol per cent. (by vol.)	97	90	85	80	70
$[a]_D = -$	143.86	160.75	168.25	174.75	182.27
Alcohol per cent. (by vol)	60	50	40	20	0 (Water)
$[a]_D = -$	187.75	187.50	182.82	166.59	138.75

Alcoholic chloroform.　$c = 2$.　$t = 15$.　$[a]_D = - 126.25$.
　　The anhydrous compound dissolved in chloroform shows with $c = 0.9$ to 9 and
　　$t = 15°$.　$[a]_D = - (81.81 - 23.756\ c + 3.9556\ c^2 - 0.2198\ c^3)$.

Solutions in dilute hydrochloric acid, with 1 molecule of the hydrochloride $(c = 2)$ in 100 cubic centimetre solution, we have as follows :—

Mols. of H Cl	0	1	2	4	10	16
$[a]_D = -$	138.75	223.2	225.7	223.6	213.9	209.5

Concentrated fuming acid.　$c = 2$.　$[a]_D =$　158.8.　Hesse (L. A. **176**, 210).

Diquinine sulphate (neutral), $2 (C_{20} H_{24} N_2 O_2) . H_2 S O_4 + 8 H_2 O$.
Lævo-rotatory.

Absolute alcohol (p. 200).   Aqueous alcohol (p. 200).
Alcoholic chloroform. $c = 1$ to 5. $t = 15°$. $[a]_D = (157\cdot5 - 0\cdot27 c)$.

Hesse (L. A. **176**, 213).

Quinine sulphate (*Mono-acid*), $C_{20} H_{24} N_2 O_2 . H_2 S O_4 + 7 H_2 O$.
Lævo-rotatory.

Water (p. 200).   Absolute alcohol (p. 204).
Alcohol of 97 per cent. (by vol.) $c = 2$. $t = 15°$. $[a]_D = - 134\cdot75$.
„   80   „   „ $c = 2$. $t = 15°$. $[a]_D = - 142\cdot75$. ⎫ Hesse (L. A.
„   60   „   „ $c = 2$. $t = 15°$. $[a] = - 155\cdot91$. ⎬ **176**, 215).
Alcoholic chloroform. $c = 2$. $t = 15°$. $[a]_D = - 138\cdot75$. ⎭

In alcoholic solutions with $c = 2$, the specific rotation decreases $0\cdot65°$ for each $1°$
rise of temperature.   Draper (*Silliman's American Journal*), [3] 11, 42).

Quinine disulphate (*di-acid*), $C_{20} H_{24} N_2 O_2 . 2 H_2 S O_4 + 4 H_2 O$.
Lævo-rotatory.

Water. $c = 2$ to 10. $t = 15°$. $[a]_D = - (170\cdot3 - 0\cdot94 c)$.
Alcohol of 80 per cent. (by vol.) $t = 15°$. $c = 1$. $[a]_D = - 154\cdot5$.
„   80   „   „ $t = 15°$. $c = 3$. $[a]_D = - 153\cdot3$.

Hesse (L. A. **176**, 217).

Quinine oxalate, $2 (C_{20} H_{24} N_2 O_2) . C_2 H_2 O_4 + 6 H_2 O$.   Lævo-
rotatory.

Absolute alcohol (p. 204).
Alcoholic chloroform. $c = 1$ to 3. $t = 15°$. $[a]_D = - (141\cdot58 - 0\cdot58 c)$.

Hesse (L. A. **176**, 218).

Cinchonidine, $C_{20} H_{24} N_2 O$.   Lævo-rotatory.

Alcohol 97 per cent. (by vol.) (p. 201).   Absolute and aqueous alcohols (p. 205).
Alcohol of 95 per cent. (by vol.) $c = 2$. $t = 15°$. $[a]_D = - (113\cdot53 - 0\cdot426 c)$.
„   80   „   „ $c = 2$. $t = 15$. $[a]_D = - 119\cdot5$.
Alcoholic chloroform $c = 2$. $t = 15$. $[a]_D = - 108\cdot9$.
Chloroform $c = 2$. $t = 15$. $[a]_D = - 83\cdot9$.

Hesse L. A. **176**, 219).

Absolute alcohol $c = 1\cdot54$. $t = 17$. $[a]_D = - 109\cdot6$. ⎫ Oudemans
Chloroform $c = 1\cdot545. t = 17$. $[a]_D = - 77\cdot3$. ⎬ (L. A. **182**,
„ $c = 3\cdot41. t = 17$. $[a]_D = - 74\cdot0$. ⎭   44).

Cinchonidine hydrochloride, $C_{20} H_{24} N_2 O . H Cl + H_2 O$.   Lævo-
rotatory.

Water (p. 201).   Hydrochloric acid (p. 201).   Absolute alcohol (p. 205).
Alcohol of 97 per cent. (by vol.) $c = $ 3. $t = 15°$. $[a]_D = - 108\cdot0$. ⎫   Hesse
„   80   „   „ $c = 2$. $t = 15°$. $[a]_D = - 135\cdot25$. ⎬ (L. A. **176**,
Anhydrous salt : Chloroform $c = 2\cdot85. t = 15°$. $[a]_D = - 24\cdot2$. ⎭   220).

Dicinchonidine sulphate, $2 \ (C_{20} H_{24} N_2 O) \ . \ H_2 S O_4 + 6 H_2 O.$ Lævo-rotatory.

Water. $c = 1.06.$ $t = 15°.$ $[a]_D = - 106.8.$
Salt with 3 mols. water: alcohol of 80 per cent. (by vol.) $c = 2.$ $t = 15°.$ $[a]_D = - 144.5.$ Hesse (L. A. **176**, 221).

Cinchonidine sulphate, $C_{20} H_{24} N_2 O \ . \ H_2 S O_4 + 5 H_2 O.$ Lævo-rotatory.

Water (p. 201).
Alcohol of 80 per cent. (by vol.) $c = 2.$ $t = 15°.$ $[a]_D = - 109.0.$ ⎫ Hesse (L.
Alcoholic chloroform. $c = 2.$ $t = 15°.$ $[a]_D = - 101.0.$ ⎬ A. **176**, 222).

Cinchonidine nitrate (p. 205).

Cinchonidine oxalate, $2 \ (C_{20} H_{24} N_2 O) . C_2 H_2 O_4 + 2 H_2 O.$ Lævo-rotatory.

Alcoholic chloroform. $= 1$ to 3. $t = 15°.$ $[a]_D = - 98.7.$ Hesse (L. A. **176**, 222).

Quinidine (conchinine) hydrate, $C_{20} H_{24} N_2 O + 2\frac{1}{2} H_2 O.$ Dextro-rotatory.

Alcohol of 97 per cent. (by vol.) (p. 201).
,, 80 ,, ,, $c = 2.$ $t = 15°.$ $[a]_D = 232.7.$
Alcoholic chloroform $c = 1.$ $t = 15°.$ $[a]_D = 244.5.$
,, ,, $c = 2.$ $t = 15°.$ $[a]_D = 241.75.$
Anhydrous alkaloid : Chloroform $c = 1.756.$ $t = 15°.$ $[a]_D = 230.35.$
Hesse (L. A. **176**, 223).
Absolute alcohol (anhydrous alkaloid): $c = 1.62.$ $t = 17°.$ $[a]_D = 255.4.$
Benzene ,, ,, $c = 1.62.$ $t = 17°.$ $[a]_D = 195.2.$
Toluene ,, ,, $c = 1.62.$ $t = 17°.$ $[a]_D = 206.6.$
Chloroform ,, ,, $c = 1.62.$ $t = 17°.$ $[a]_D = 228.8.$
Oudemans (L. A. **182**, 44).

Quinidine hydrochloride, $C_{20} H_{24} N_2 O_2 . H Cl + H_2 O.$ Dextro-rotatory.

Water, hydrochloric acid (p. 201). Absolute alcohol (p. 205).
Alcohol of 97 per cent. (by vol.) $c = 2$ to 5. $t = 15°.$ $[a]_D = 212.0 - 2.56 \ c.$
,, 80 ,, ,, $c = 2.$ $t = 15°.$ $[a]_D = 230.25.$
Acid salt: $C_{20} H_{24} N_2 O_2 . 2 H Cl + H_2 O.$ Water. $c = 2.$ $t = 15°.$ $[a]_D = 250.3.$
Hesse (L. A. **176**, 225).

Diquinidine sulphate, $2 C_{20} H_{24} N_2 O_2 . H_2 SO_4 + 2 H_2 O.$ Dextro-rotatory.

Water. Sulphuric acid. Hydrochloric acid (p. 202).
Alcohol of 80 per cent. (by vol.) $c = 2.$ $t = 15°.$ $[a]_D = 218.2.$ ⎫
,, 60 ,, ,, $c = 2.$ $t = 15°.$ $[a]_D = 227.0.$ ⎪
Alcoholic chloroform. $c = 2.$ $t = 15°.$ $[a]_D = 209.25.$ ⎬ Hesse
Anhydrous salt : Chloroform. $c = 3.$ $[a]_D = 184.2.$ ⎪ (L. A. **176**, 226).
,, ,, ,, $c = 5.$ $[a]_D = 180.1.$ ⎭

Quinidine sulphate, $C_{20} H_{24} N_2 O_2 . H_2 S O_4 + 4 H_2 O$. Dextro-rotatory.

Water. Sulphuric acid (p. 202).
Alcohol of 97 per cent. (by vol.) $c = 2$. $t = 15°$. $[a]_D = 183$. Hesse (L. A. 176, 227).

Quinidine nitrate (p. 205).

Quinidine oxalate, $2 (C_{20} H_{24} N_2 O_2) . C_2 H_2 O_4 + H_2 O$. Dextro-rotatory.

Alcoholic chloroform. $c = 1$ to 3. $t = 15°$. $[a]^D = 189·0 - 2·18 c$. Hesse (L. A. 176, 227).

Cinchonine, $C_{20} H_{24} N_2 O$. Dextro-rotatory.

Alcohol of 97 per cent. (by vol.) (p. 202).
Alcoholic chloroform. $c = 1$ to 5. $t = 15°$. $[a]_D = 238·8 - 1·46 c$. Hesse (L. A. 176, 228).

Absolute alcohol.	$c = 0·5$ to $0·75$.	$t = 17°$.	$[a]_D = 223·3.$
Chloroform.	$c = 0·455.$	$t = 17°$.	$[a]_D = 214·8.$
,,	$c = 0·535.$	$t = 17°$.	$[a]_D = 212·3.$
,,	$c = 0·560.$	$t = 17°$.	$[a]_D = 209·6.$

Oudemans (L. A. 182, 44).

Cinchonine hydrochloride, $C_{20} H_{24} N_2 O . H Cl + 2 H_2 O$. Dextro-rotatory.

Water. Hydrochloric acid (p. 202).
Alcohol of 97 per cent. (by vol.) $c = 1$ to 10. $t = 15°$. $[a]_D = 179·81 - 6·314 c + 0·8406 c^2 - 0·0371 c^3.$
,, 80 ,, ,, $c = 2$. $t = 15°$. $[a]_D = 188·9.$
,, 60 ,, ,, $c = 2$. $t = 15°$. $[a]_D = 195·5.$
Alcoholic Chloroform $c = 2$. $t = 15°$. $[a]_D = 152·0$. Hesse (L. A. 176, 231).

Dicinchonidine sulphate, $2 (C_{20} H_{24} N_2 O) . H_2 S O_4 + 2 H_2 O$. Dextro-rotatory.

Water (202). Sulphuric acid (p. 203).
Alcohol of 97 per cent. (by vol.) $c = 3$ to 10. $t = 15°$. $[a]_D = 193·29 - 0·374 c.$
,, 80 ,, ,, $c = 2$. $t = 15°$. $[a]_D = 202·95$ Hesse
,, 60 ,, ,, $c = 2$. $t = 15°$. $[a]_D = 204·14$ (L. A. 176,
Alcoholic chloroform $c = 2$. $t = 15°$: $[a]_D = 185·25$ 231).

Cinchonine oxalate, $2 (C_{20} H_{24} N_2 O) . C_2 H_2 O_4 + 2 H_2 O$. Dextro-rotatory.

Alcoholic chloroform $c = 1$ to 3. $t = 15°$. $[a]_D = 165·46 - 0·763 c.$
Hesse (L. A. 176, 232.)

The changes in the specific rotation of quinine, cinchonine, conchinine and cinchonidine in presence of different proportions of *hydrechloric, nitric, chloric, perchloric, formic, acetic, sulphuric, oxalic,* *and phosphoric acids* have been investigated by Oudemans (L. A. **182,** 51).

Cinchotenine, $C_{18} H_{20} N_2 O_3 + 3 H_2 O$. Dextro-rotatory.

Alcoholic chloroform. $c = 2$. $t = 15°$. $[a]_D = 115\cdot5$.
1 mol. substance + 2 mols. $H_2 S O_4$ + water. $c = 2$. $t = 15°$. $[a]_D = 175\cdot5$.
Hesse (L. A. **176,** 233).

Quinicine, $C_{20} H_{24} N_2 O_2$. Dextro-rotatory.

Chloroform. $c = 2$. $t = 15°$. $[a]_D = 44\cdot1$. Hesse (L. A. **178,** 260).

Quinicine oxalate, $2 (C_{20} H_{24} N_2 O_2) . C_2 H_2 O_4 + 9 H_2 O$. Dextro-rotatory.

Alcoholic chloroform. $c = 1$ to 3. $t = 15°$. $[a]_D = 20\cdot68 - 1\cdot14 c$.
Water. $c = 2$. $t = 15°$. $[a]_D = 9\cdot54$.
1 mol. salt + 2 mols. $H_2 S O_4$ + water. $c = 2$. $t = 15°$. $[a]_D = 15\cdot54$.
Hesse (L. A. **178,** 261).

Cinchonicine, $C_{20} H_{24} N_2 O$. Dextro-rotatory.

Alcohol of 95 per cent. (by vol.) $c = 1$. $= 15°$. $[a]_D = 48$.
Chloroform. $c = 2$. $= 15°$. $[a]_D = 46\cdot5$.
Hesse (L A. **178,** 262).

Cinchonicine oxalate, $2 (C_{20} H_{24} N_2 O) . C_2 H_2 O_4 + 3 H_2 O$. Dextro-rotatory.

Alcohol of 97 per cent. (by vol.) $c = 2$. $t = 15°$. $[a]_D = 23\cdot5$. ⎫
Alcoholic chloroform. $c = 1$ to 3. $t = 15°$. $[a]_D = 23\cdot1$. ⎬ Hesse
Water. $c = 2$. $t = 15°$. $[a]_D = 22\cdot6$. (L. A. **178,**
Water + 2 mols. $H_2 S O_4$. $c = 2$. $t = 15°$. $[a]_D = 25\cdot75$. ⎭ 263).

Quinamine, $C_{20} H_{26} N_2 O_2$. Dextro-rotatory.

Alcohol of 96 (?) per cent. (by vol.) $c = 0\cdot8378$. $[a]_D = 106\cdot8$. Hesse (L. A. **166,** 272).

Paytine, $C_{21} H_{24} N_2 O_2$. Lævo-rotatory.

Alcohol of 96 (?) per cent. (by vol.) $c = 0\cdot4542$. $[a]_D = 49\cdot5$.
Hesse (L. A. **166,** 272).

Quinidamine (conchinamine), $C_{19} H_{24} N_2 O_2$. Dextro-rotatory.

Alcohol of 97 per cent. (by vol.) $c = 1\cdot8$. $t = 15°$. $[a]_D = 200$.
Hesse (D. C. G. 1877, 2158).

Geissospermine, $C_{19} H_{24} N_2 O_2$. Lævo-rotatory.

Alcohol of 97 per cent. (by vol.) $c = 1\cdot5$. $t = 15°$. $[a]_D = - 93\cdot4$.
Hesse (D. C. G. 1877, 2164).

Homocinchonidine, $C_{19} H_{22} N_2 O$. Lævo-rotatory.

Alcohol of 97 per cent. (by vol.) $c = 2$. $t = 15°$. $[a]_D = - 109\cdot3$.
Hesse (D. C. G. 1877, 2156).

Cusconine, $C_{23} H_{26} N_2 O_4 + 2 H_2 O$.   Lævo-rotatory.

Ether ($d = 0.72$) $c = 1$. $t = 15°$. $[\alpha]_D = -27.1$
     ,, $c = 2$. $t = 15°$. $[\alpha]_D = -26.8$ Hesse
Alcohol of 97 per cent. (by vol.) $c = 2$. $t = 15°$. $[\alpha]_D = -54.3$ (L. A. 185,303.)
Water + 3 mols. H Cl $c = 0.5$. $t = 15°$. $[\alpha]_D = -71.8$

Aricine, $C_{33} H_{26} N_2 O_4$.   Lævo-rotatory.

Ether ($d = 0.72$) $c = 1$ to 2.5. $t = 15°$. $[\alpha]_D = -94.7$.
Alcohol of 97 per cent. (by vol.) $c = 1$. $t = 15°$. $[\alpha]_D = -54.1$.

                                      Hesse (L. A. 185, 313.)

Solution in hydrochloric acid gives no rotation.

OPIUM ALKALOIDS.

Morphine hydrate, $C_{17} H_{19} N O_3 + H_2 O$.   Lævo-rotatory.

Solutions in dilute soda. Hesse (L. A. 176, 190).
1 mol. alkaloid + 1 mol. $Na_2 O$. $c = 2$. $t = 22.5$. $[\alpha]_D = -67.5$.
1 ,,   ,,   + 5 ,,   ,, $c = 2$. $t = 22.5$. $[\alpha]_D = -70.2$.
1 ,,   ,,   + 2 ,,   ,, $c = 5$. $t = 22.5$. $[\alpha]_D = -71.0$.

Morphine hydrochloride, $C_{17} H_{19} N O_3 . H Cl + 3 H_2 O$.   Lævo-rotatory.

Water. $c = 1$ to 4. $t = 15°$. $[\alpha]_D = -(100.67 - 1.14 c)$.
Water + 10 mols. H Cl. $c = 2$. $t = 15°$. $[\alpha]_D = -94.3$.   Hesse (L. A. 176, 190).

Morphine sulphate, $2 C_{17} H_{19} N O_3 . H_2 S O_4 + 5 H_2 O$.   Lævo-rotatory.

Water. $c = 1$ to 4. $t = 15°$. $[\alpha]_D = -(100.47 - 0.96 c)$.   Hesse (L. A. 176, 190).

Morphine acetate, $C_{17} H_{19} N O_3 . C_2 H_4 O_2 + 3 H_2 O$.   Lævo-rotatory.

Absolute alcohol. $c = 1.2$. $[\alpha]_D = -100.4$.
Alcohol $d = 0.865$. $c = 0.97$. $[\alpha]_D = -98.9$.   Oudemans
Water. $c = 2.5$. $[\alpha] = -77$.   (L. A. 166, 77).
   ,, $c = 0.996$. $[\alpha]_D = -72$.

Codeine, $C_{18} H_{21} N O_3 + H_2 O$.   Lævo-rotatory.

Alcohol of 97 per cent. (by vol.) $c = 2$ to 8. $t = 15°$. $[\alpha]_D = -135.8$.
   ,,   80   ,,   ,, $c = 2$. $t = 15°$. $[\alpha]_D = -137.8$.
Chloroform $c = 2$. $t = 15°$. $[\alpha]_D = -111.5$.
                                      Hesse (L. A. 176, 191).

Codeine hydrochloride, $C_{18} H_{21} N O_3 . H Cl + 2 H_2 O$.   Lævo-rotatory.

Water. $c = 2$. $t = 22.5°$. $[\alpha]_D = -108.2$.
1 mol. salt + 10 mols. H Cl + water. $c = 2$. $t = 22.5°$. $[\alpha]_D = -105.2$.
Alcohol of 80 per cent. (by vol.) $c = 2$. $t = 22.5°$. $[\alpha] = -108$.
                                      Hesse (loc. cit.)

Codeine sulphate, $2 C_{18} H_{21} N O_3 . H_2 S O_4 + 5 H_2 O.$ Lœvo-rotatory.

Water. $c = 3.$ $t = 15°.$ $[\alpha]_D = -101·2$ ⎫
  ,, $c = 3.$ $t = 25°.$ $[\alpha]_D = -100·9$ ⎭   Hesse (L. A., 176, 191).

Narcotine, $C_{32} H_{25} N O_7.$ Lœvo-rotatory.

Alcohol of 97 per cent. (by vol.) $c = 0·74.$ $t = 22·5°.$ $[\alpha]_D = -185·0.$
Alcoholic chloroform $c = 2.$ $t = 22·5°.$ $[\alpha]_D = -191·5.$
Chloroform $c = 2$ to $5.$ $t = 22·5°.$ $[\alpha]_D = -207·35.$
  Hesse (L. A. 176, 192).

Narcotine dissolved in *hydrochloric acid.* Dextro-rotatory.

1 mol. alkaloid	+	2 mols. H Cl	+	water	$c = 2.$ $[\alpha]_D = +47·0.$
1 ,,	+	2 ,,	+	,,	$c = 5.$ $[\alpha]_D = +46·4.$
1 ,,	+	10 ,,	+	,,	$c = 2.$ $[\alpha]_D = +50·0.$
1 ,,	+	2 ,,	+	alkaloid of 80 per cent. $c = 2.$ $[\alpha]_D = +104·5.$

  (by vol.) Hesse (L. A. 176, 193).

Pseudomorphine hydrochloride, $C_{17} H_{19} N O_4 . H Cl + H_2 O.$ Lœvo-rotatory.

1 mol. salt $(c = 0·8$ to $1·6) + 1$ mol. H Cl + water. $t = 22·5 [\alpha]_D = - (114·76$
  $- 4·96 c).$

1 mol. salt $(c = 2) + 5\frac{1}{2}$ mols. $Na_2 O +$ water $= 1$ mol. ⎫
  alkaloid $+ 5$ mols. $Na_2 O + 1$ mol. Na Cl ⎬ $t = 22·5 [\alpha]_D = -198·9.$
  ⎭
  Hesse (L. A. 176, 195).

Thebaine, $C_{19} H_{21} N_2 O_3.$ Lœvo-rotatory.

Alcohol of 97 per cent. (by vol.) $c = 2.$ $t = 15°.$ $[\alpha]_D = -218·6$ ⎫
  ,,   97   ,,   ,, $c = 2.$ $t = 25°.$ $[\alpha]_D = -215·5$ ⎪ Hesse
  ,,   97   ,,   ,, $c = 1.$ $t = 22·5°.$ $[\alpha]_D = -216·4$ ⎬ (L. A. 176, 196).
Chloroform $c = 5.$ $t = 22·5°.$ $[\alpha]_D = -229·5$ ⎭

Thebaine hydrochloride, $C_{19} H_{21} NO_3 H Cl + H_2 O.$ Lœvo-rotatory.

Water. $c = 2$ to $4.$ $t = 15°.$ $[\alpha]_D = - (168·32 - 2·33 c).$
  ,, $c = 2.$ $t = 22·5°.$ $[\alpha]_D = -163·25.$
1 mol. salt $+ 10$ mols. H Cl + water. $c = 2.$ $t = 22·5.$ $[\alpha]_D = -158·6$
  Hesse (L. A. 176, 197).

Papaverine, $C_{21} H_{21} N O_4.$ Lœvo-rotatory.

Alcohol of 97 per cent. (by vol). $c = 2.$ $t = 15°.$ $[\alpha]_D = -4·0.$
Chloroform. $c = 5.$ $t = 15°.$ $[\alpha]_D = -5·7.$
  Hesse (L. A. 176, 198).

Hydrochloride is inactive.

Laudanine, $C_{20} H_{25} N O_4.$ Lœvo-rotatory.

Chloroform. $c = 2.$ $t = 22·5. [\alpha]_D = -13·5.$
1 mol. alkaloid $(c = 1) + 2$ mols. $Na_2 O +$ water. $t = 22·5. [\alpha]_D = -11·4.$
  Hesse (L. A. 176, 201).

Hydrochloride is inactive.

Laudanosine, $C_{21} H_{27} NO_4$. Dextro-rotatory.

Alcohol of 97 per cent. (by vol.) $c = 2.79$. $t = 15°$. $[\alpha]_D = 103.2$.
    „   97   „   „ $c = 2$. $t = 22.5$. $[\alpha]_D = 105.0$.
Chloroform. $c = 2$. $t = 22.5$. $[\alpha]_D = 56.0$.
1 mol. alkaloid ($c = 2$) + 2 mols. H Cl + water. $t = 22.5$. $[\alpha]_D = 108.4$.
                                    Hesse (L. A. **176**, 202).

Narceine, hydrocotarnine, cryptopine, meconine are inactive.

STRYCHNINE ALKALOIDS.

Strychnine, $C_{21} H_{22} N_2 O_2$. Lævo-rotatory.

Alcohol. $d = 0.865$. $c = 0.91$. $[\alpha]_D = -128$.
Chloroform. $c = 4$. $[\alpha]_D = -130$.   }
    „ $c = 2.25$. $[\alpha]_D = -137.7$.   } Oudemans
    „ $c = 1.5$. $[\alpha]_D = -140.7$.   } (L. A. **166**, 76).
Amyl alcohol. $c = 0.53$. $[\alpha]_D = -235$.   )

Brucine, $C_{23} H_{26} N_2 O_4$. Lævo-rotatory.

Alcohol. $c = 5.4$. $[\alpha]_D = -85$.   )
Chloroform. $c = 1.9$. $[\alpha]_D = -127$.   } Oudemans
    „ $c = 4.9$. $[\alpha]_D = -119$.   ) (L. A. **166**, 69).

§ 121. Unclassified Vegetable Substances.

Santonin, $C_{15} H_{18} O_3$. Lævo-rotatory.

Alcohol. $c = 2$. $t = 20$. $[\alpha]_j = -230$.   Buignet (*J. d. Pharm. et de chim.* [3]
40, 252).   This substance is characterized by very marked rotatory dispersion.
Alcohol of 97 per cent. (by vol.) $c = 2$. $t = 15°$. $[\alpha]_D = -174.0$.
    „   90   „   „ $c = 2$. $t = 15°$. $[\alpha]_D = -175.4$.   } Hesse (L. A.
    „   80   „   „ $c = 2$. $t = 15°$. $[\alpha]_D = -176.5$.   } **176**, 125).
Chloroform. $c = 2$ to 10. $t = 15°$. $[\alpha]_D = -171.5$.

Bichloro-santonin, $C_{15} H_{16} Cl_2 O_3$. Lævo-rotatory.

Alcohol of 97 per cent. (by vol.) $c = 1$. $t = 15°$. $[\alpha]_D = -23$.   Hesse (*loc. cit*).

Santonic acid, $C_{15} H_{20} O_4$. Lævo-rotatory.

Alcohol of 97 per cent. (by vol.) $c = 1$ to 3. $t = 22.5°$. $[\alpha]_D = -25.8$. } Hesse
    „   80   „   „ $c = 2$ „ 3. $t = 22.5°$. $[\alpha]_D = -26.5$. } (*loc. cit*).

Santonate of soda, $2 (C_{15} H_{19} Na O_4) + 7 H_2 O$. Lævo-rotatory.

Water. $c = 2$ to 6. $t = 22.5°$. $[\alpha]_D = -(18.70 + 0.33 c)$.
    The rotation decreases with increase of temperature.   At 25°, with $c = 3$ $[\alpha]_D$
$= -20.0$, and with $c = 10$ $[\alpha] = 21.7$. Hesse (*loc. cit*).

Picrotoxin, $C_{12} H_{14} O_5$. Lævo-rotatory.

Alcohol. $p = 3.125$. $[\alpha]_j = -28.1$.   Bouchardat and Boudet (*J. Pharm. chim.* [3]
                                    23, 288).

Echicerin, $C_{30}H_{48}O_2$. Dextro-rotatory.

Ether. $d = 0.73.$ $c = 2.$ $t = 15.$ $[a]_D = 63.75.$ } Jobst and Hesse (L. A.
Chloroform. $c = 2.$ $t = 15.$ $[a]_D = 65.75.$ } 178, 49).

Echitin, $C_{32}H_{52}O_2$. Dextro-rotatory.

Ether. $c = 2.$ $t = 15°.$ $[a]_D = 72.7.$ } Jobst and Hesse (loc. cit).
Chloroform. $c = 2.$ $t = 15°.$ $[a]_D = 75.3.$ }

Echitein, $C_{42}H_{70}O$. Dextro-rotatory.

Ether. $c = 2.$ $t = 15°.$ $[a]_D = 88.$ } Jobst and Hesse (loc. cit).
Chloroform. $c = 2.$ $t = 15°.$ $[a]_D = 85.5$ }

Echiretin, $C_{35}H_{56}O_2$. Dextro-rotatory.

Ether. $c = 2.$ $t = 15°.$ $[a]_D = 54.8.$ Jobst and Hesse (loc. cit).

§ 122. Bile Constituents.

Cholesterin, $C_{26}H_{44}O$, or $C_{25}H_{42}O$. Lævo-rotatory.

From gall stones. Anhydrous substance.
Ether. $d = 0.72.$ $c = 2.$ $t = 15.$ $[a]_D = -31.12.$
Chloroform. $c = 2$ to $8.$ $t = 15.$ $[a]_D = -(36.61 + 0.249 \, c).$
Hesse (L. A. 192, 178).

Lindenmeyer (*J. für prakt. Chem.* [1] 90, 323), examined by Broch's method solutions of cholesterin (with 1 mol. water?) in rock-oil, $c = 10$, and in ether, $c = 7.941$, and obtained the following specific rotations :—

Lines B C D E b F G
$[a] = -$ 20.63 25.54 31.59 39.91 41.92 48.65 62.37.

Phytosterin, $C_{26}H_{44}O$. Lævo-rotatory. From calabar beans or seed peas.

Chloroform. $c = 1.636.$ $t = 15°.$ $[a]_D = -34.2.$ Hesse (L. A. 192, 177).

The following determinations of specific rotation of bile acids were made by Hoppe-Seyler (*J. für prakt. Chem.* [1], 89, 257) by Broch's method, employing sunlight.

Glycocholic acid, $C_{26}H_{43}NO_6$. Dextro-rotatory.

From ox bile. Alcoholic solution. $c = 9.504.$ The concentration of the solution is without influence.

Lines C D E b F G
$[a] = +$ 21.6 29.0 37.9 40.0 48.7 56.8.

Glycocholate of soda, $C_{26} H_{42} Na NO_6$.   **Dextro-rotatory.**

Alcohol. $c = 20.143$. $[a]_D = 25.7$.   Concentration without influence.
Water. $c = 24.928$. $[a]_D = 20.8$.   ,,   ,,   ,,

Taurocholate of soda, $C_{26} H_{44} Na N S O_7$.   **Dextro-rotatory.**

Alcohol. $c = 9.898$. $[a]_D = 24.5$. $[a']_D = 39.0$.   Concentration without influence.
Water. $c = 8.856$. $[a]_D = 21.5$. $[a']_D = 34.0$.   ,,   ,,   ,,

Cholalic acid, anhydrous. $C_{24} H_{40} O_5$.   **Dextro-rotatory.**

From ox bile—

Alcohol. $c = 3.338$. $[a]_D = 50.2$.

From dogs' excrement—

Alcohol. $c = 2.942$. $[a]_D = 47.6$. $[a]_j = 50.4$.

Cholalic acid, crystallized with water, $C_{24} H_{40} O_5 + 2\frac{1}{2} H_2 O$. **Dextro-rotatory.**

From ox bile or dogs' excrement. Alcoholic solution. $c = 2.962$ ($= 2.659$ anhydrous substance).

Lines	B	C	D	E	b	F	G	H
Hydrated: $[a] =$	25.3	27.0	30.4	40.1	42.2	47.3	60.8	70.1
Anhydrous: $[a] =$	28.2	30.1	33.9	44.7	47.0	52.7	67.7	78.0

Other specimens of the crystallized compound gave the following rotations, calculated for the anhydrous acid :—

Alcohol. $c = 4.43$ anhydrous substance : $[a]_D = 34.8$.
   ,, $c = 6.0695$,,   ,, $[a']_D = 35.4$.
   ,, $\begin{cases} c = 2.7065 & ,, & ,, & [a]_j = 35.2. \\ c = 2.0298 & ,, & ,, & [a]_j = 34.5. \\ c = 1.8040 & ,, & ,, & [a]_j = 34.2. \end{cases}$

Cholalate of potash, $C_{24} H_{39} K O_5$.   **Dextro-rotatory.**

Solution in Alcohol, $c = 22$.

Lines	C	D	E	b	F
$[a] =$	23.7	30.8	38.5	40.9	47.5

Solution in water. $c = 29.775$. $[a]_D = 24.9$.
   ,,   ,,   ,, $c = 22.332$. $[a]_D = 24.1$.
   ,,   ,,   ,, $c = 16.749$. $[a]_D = 24.6$.
   ,,   ,,   ,, $c = 12.562$. $[a]_D = 25.9$.
   ,,   ,,   ,, $c = 7.000$. $[a]_D = 27.5$.
   ,,   ,,   ,, $c = 6.004$. $[a]_D = 28.2$.

Cholalate of soda, $C_{24} H_{39} Na O_5$.   **Dextro-rotatory.**

Alcohol. $c = 2.2296$. $[a]^D = 31.4$.
Solution in water. $c = 19.049$.

Lines	B	C	D	E	b	F
$[a] =$	19.7	21.0	26.0	33.1	34.9	42.0

Decrease of concentration raises the specific rotation.

Cholalate of methyl, $C_{24}H_{39}(CH_3)O_5$. Dextro-rotatory.

Alcohol. $c = 4.59$. $[a]_D = 31.9$.

Cholalate of ethyl, $C_{24}H_{39}(C_2H_5)O_5$. Dextro-rotatory.

Solution in alcohol. $c = 18.479$.

Lines	B	D	E	b
$[a] =$	25.4	32.4	40.5	42.3.

§ 123. Gelatinous Substances.

All these substances, and particularly chondrin, possess strong lœvo-rotatory powers. The following observations were all taken by de Bary (Hoppe-Seyler, *Med. chem. Untersuchungen*, 1, 71):—

Glutin. Aqueous solutions.

1. $c = 6.12$. $\begin{cases} t = 24° \text{ to } 25°. & [a]_D = -140.0. \\ t = 35° \text{ to } 40°. & [u]_D = -123.0. \end{cases}$

2. $c = 3.06$. $\begin{cases} t = 24° \text{ to } 25°. & [a]_D = -130.5. \\ t = 35°. & [a]_D = -125.0. \end{cases}$

The rotatory power of glutin solutions decreases with rise of temperature. Concentration, on the other hand, has no important influence.

The following experiments with aqueous solutions of concentration $c = 3.06$, show the effects of alkalies and acids:—

1. Solution mixed with an equal vol. ammonia $[a]_D = -130.5$.
2. „ „ a few drops solution of soda $[a]_n = -130.5$.
3. „ „ an equal vol. soda $[a]_D = -112.5$.
4. „ „ an equal vol. acetic acid $[a]_D = -114.0$.

Chondrin. Pure aqueous solutions of sufficient transparency cannot be prepared, but the cloudiness disappears on the addition of a few drops of soda solution. For such a solution with $c = 0.957$, it was found that $[a]_j = -213.5$.

After the addition of an equal volume of soda solution . $[a]_j = -552.0$.

The latter solution diluted by the addition of an equal volume of water . $[a]_j = -281.0$.

§ 124. Albumins.

All the albumins are lævo-rotatory. The following observations on their specific rotation are given by Hoppe-Seyler (*Zeitsch. für Chem. u. Pharm.* 1864, 737).

Serum-albumin.

Neutral aqueous solution	$[\alpha]_D = -56.$
Aqueous solution saturated with sodium chloride	$[\alpha]_B = -64.$
Aqueous solution saturated with addition of acetic acid	$[\alpha]_D = -71.$
Aqueous solution with addition of concentrated hydrochloric acid till the precipitate at first formed again disappears	$[\alpha]_D = -78.7.$

Potash and soda solutions, by forming alkali-albuminate, cause a considerable increase of rotatory power, even when present only in small quantity. Prolonged action of the alkalies, particularly at higher temperatures, again reduces the amount of rotation.

Egg-albumin.

Aqueous solution. Rotation independent of concentration	$[\alpha]_D = -35.5.$
Aqueous solution after addition of hydrochloric acid	$[\alpha]_D = -37.7.$

Casein.

Dissolved in magnesium sulphate solution	$[\alpha]_D = -80.$
Solution in dilute hydrochloric acid (4 cub. cent. of fuming acid per litre of water)	$[\alpha]_D = -87.$
Solution in smallest possible quantity of soda solution	$[\alpha]_D = -76.$

Albuminate (protein of Mulder), obtained by the action of concentrated potash upon albumins, always exhibits higher rotatory power than the latter. The following maxima have been observed :—

Albuminate from serum-albumin	$[\alpha]_D = -86.$
,, uncoagulated egg-albumin	$[\alpha]_D = -47.$
,, coagulated egg-albumin	$[\alpha]_D = -58.5.$
,, casein. Solution of casein in strong potash. Rotation varies with strength and amount of potash used	$[\alpha]_D = -91.$

Paralbumin, from ovarian cysts. Examination of the natural feebly alkaline solutions gave in several observations :—

$$[\alpha]_D = -59, -61, -64.$$

Syntonin, obtained from myosin of muscles by solution in very dilute hydrochloric acid, or by the action of concentrated hydrochloric acid upon albumins (coagulated egg-albumin or fibrin). Solution in very dilute hydrochloric acid. Rotation independent of concentration $[\alpha]_D = -72.$

In weak alkaline solutions the substance exhibits very nearly the same amount of rotation. By heating the hydrochloric acid solutions in a closed vessel to, 100°, the specific rotation rises to $[\alpha] = -84.8.$

REESE LIBRARY
OF THE
UNIVERSITY
OF
CALIFORNIA.

APPENDIX.

ON THE ESTIMATION OF MALTOSE AND DEXTRIN IN MALT WORTS
AND BEERS.

By J. STEINER, F.C.S.

WHEN an infusion of malt in cold water reacts on soluble starch under certain definite conditions, a chemical change takes place, which has attracted the attention of chemists for a considerable time. This chemical reaction being, moreover, of great practical importance in brewing, for example, numerous experiments have been performed to explain it. The results arrived at, although differing widely in many respects, lead nevertheless to the conclusion, that the starch is converted by the action of cold malt extract (diastase) into maltose and dextrins. These compounds are the only products under the most favourable conditions of temperature (55° to 63° C.), if the diastatic action continues no longer than two to three hours. But a more prolonged contact of the diastase leads to the partial conversion of maltose into dextrose, while the gradual saccharification of some of the dextrins into maltose, and the formation of dextrins of simpler molecular compositions seem to proceed during the whole time of the diastatic action.

If starch be boiled with dilute sulphuric or hydrochloric acid, dextrins, maltose, and finally dextrose are produced, but if the action be too protracted, or the acid too concentrated, the so-called neutral carbo-hydrates are formed simultaneously. These latter are not capable of fermentation, nor of reducing alkaline solution of metallic salts, and have no rotatory power. The progress of the conversion may be watched by an iodine solution, which gives the following colour reactions. First, a deep-blue (soluble starch); secondly, a violet (amylo-dextrin) ; thirdly, a red (erythro-dextrin) ; and finally, no change of colour (achro-dextrin, maltose and dextrose.)

All these substances possess a rotatory power, and in the following lines an account will be given of some experiments carried out for the purpose of ascertaining whether the optical properties of a malt wort, or beer, can be used for the determination of the relative proportions of maltose and dextrin contained in such solutions. But having to deal with a mixture, another characteristic property of these carbo-hydrates, *i.e.*, their deportment to alkaline solutions of cupric oxide (Fehling's test) and mercuric oxide (Knapp's and Sachsse's test) must also be considered here. The relative reducing power of different pure sugar solutions has recently been studied by Soxhlet, and the directions given by this chemist were strictly adhered to in the analysis of the samples given below.

The specific rotation of dextrin in all its modifications is considered by O'Sullivan to be $[a]_j = 220 - 222$ ($[a]_D = 195$), and the cupric oxide reducing power, *nil*. This view is, however, not generally accepted, for Musculus, as also Brown and Heron, contends that there exist a number of dextrins, each of which has its peculiar rotatory and reducing power. It is evident that if O'Sullivan's view be correct, the amount of reduction obtained by a mixture of maltose and dextrin corresponds merely to the proportion of maltose present. Then, on determining the angle of rotation of an unit volume of this mixture and deducing the angle corresponding to the estimated proportion of maltose, the remaining angle corresponds to the proportion of dextrin.

Experiments, carried out with this end in view, showed, however, that the relations existing between these substances are not so simple, as supposed above; and it appears that the dextrins in wort have properties differing from those in beer. These differences become especially marked, if not only Fehling's test, but also those of Knapp and Sachsse, be used for the determination of the reducing power.

The following are the data of the experiments referred to above :—

(A) A malt wort of an ordinary brew obtained from the taps of the mash tun of specific gravity 1·092, was diluted with water to specific gravity 1·0555 ; a wort of this concentration contains, according to W. Schultze's table, 14·7 grammes of solid matter in 100 cubic centimetres.

1. *Determination of Rotatory Power.*—50 cubic centimetres of this wort were clarified with 2 cubic centimetres basic acetate of lead, and made up to 100 cubic centimetres ; the observations were made in a 200 millimetre tube with a Soleil-Ventzke-Scheibler polariscope, for which white lamp-light is used. 100 cubic centi-

metres of this wort rotated 92·4 degrees (Ventzke scale, ray j), or 92·4 × ·346 = 31·97 angular measurement (ray D).[1]

2. *Determination of Reducing Power.*—Malt extract procured from the brewery has a cupric oxide reducing power equal to 60 to 70 per cent. expressed as maltose; but as, according to Soxhlet, the reducing power of a sugar solution alters with its concentration, a 1 per cent. solution as recommended by him was prepared as follows:—(100 cubic centimetres of wort contain 14·7 grammes extract, and therefore 14·7 × ·6 = 8·8 grammes maltose; consequently if 50 cubic centimetres are diluted to 450, a solution of required dilution is obtained.) The experiments were carried out with a filtered solution, (a) after clarification with basic acetate of lead and aluminium hydrate, and (β) after addition of aluminium hydrate only. The sugar solution was added all at once to 50 cubic centimetres Fehling's solution, the mixture boiled for three minutes in a porcelain dish, and then poured on a large filter. Rejecting the first few drops of the filtrate, about two-thirds were collected in a small beaker, cooled, and then tested for copper by the addition of a few drops of freshly prepared potassium ferro-cyanide solution, and the least excess of acetic acid. The titrations were repeated in this manner until in two successive tests one showed the slightest bronze reaction, while the other, using ·2 cubic centimetre more, indicated no trace of copper. The quantity of sugar solution required in both cases (a and β) was the same, 37·3 cubic centimetres, thus proving that the albuminous matter removable by the basic acetate has no effect on cupric oxide. Nevertheless, it is advisable to clarify such solution, as the final reaction is more marked after the removal of some of the albuminous matter which is liable to be precipitated by the ferro-cyanide. This same sugar solution was also estimated by Knapp's and Sachsse's solutions. The former was prepared according to the directions of Soxhlet, while in the latter mercuric chloride was substituted in equivalent quantities for mercuric iodide with an increase of 20 per cent. of potassium iodide used by Soxhlet: this modified solution has the same reducing properties, is more easily prepared, and keeps better. An alkaline solution of stannous chloride was used as the final re-agent. The working of Knapp's and Sachsse's tests differs in this, that on adding the sugar solution gradually, less is required for Sachsse's and more for Knapp's than if the total quantity is added all at once. The presence of albuminous matter in wort slightly influences both Knapp's and Sachsse's tests, and therefore only clarified worts have been compared.

[1] For the particulars of measurement the reader is referred to p. 169.—[V. H V.]

100 cc. of Sachsse's solution were reduced by 48·2 cc. of the diluted clarified wort above.
100 cc. of Knapp's ,, ,, ,, 34·2 cc. ,, ,, ,,

All three solutions (Fehling's, Knapp's, and Sachsse's) were standardized by a 1 per cent. solution of invert sugar (9·5 grammes cane-sugar inverted and made up to a litre). From the data obtained in this manner the values of dextrose and maltose were deduced in accordance with the proportions given by Soxhlet.[1] The amount of reduction expressed as maltose was then calculated for this wort to be :—

By Fehling's solution 37·3 : ·414 :: 450 : ($x = 4·968$ or 9·936 gr. in 100 cc. wort).
By Sachsse's ,, 48·2 : ·5097 :: 450 : ($x = 4·7586$ or 9·517 gr. ,, ,,).
By Knapp's ,, 34·2 : ·3134 :: 450 : ($x = 4·1237$ or 8·247 gr. ,, ,,).

It will be seen that the results obtained by Knapp's test are much lower than those by Fehling's and Sachsse's, and further, that the reducing power of this wort does not relate to maltose only.

3. In order to gain a clearer insight into the composition of the carbo-hydrates in this wort, two more experiments were carried out :—

(a) 50 cubic centimetres were diluted to about 80 cubic centimetres, and inverted with about 8 cubic centimetres of four times normal hydrochloric acid in a stoppered bottle of about 100 cubic centimetres capacity for five and a half hours in a water bath. The acid was then nearly neutralized by 7 cubic centimetres four times normal alkali, the solution clarified by a few drops basic acetate of lead, and the whole made up to 600 cubic centimetres, to make it approximately equivalent to 1 per cent. solution of dextrose. 50 cubic centimetres Fehling's test required 22·7 of this solution.

(β) 50 cubic centimetres of the clarified liquid used in the polariscope observation were treated in a similar manner and diluted to 300 cubic centimetres ; 50 cubic centimetres Fehling's test required 22·5 of this solution. In experiment (a) a dark flocculent precipitate appeared in the liquid after inversion, while in experiment (β) the liquid had only slightly deepened in colour. The calculation from these data is as follows :—

(a) 22·7 : ·253 :: 600 : ($x = 6·687$ gr. dextrose in 50 cub. cent. of original wort).
(β) 22·5 : ·253 : 300 : ($x = 3·373$ gr. ,, 25 cub. cent. ,,).

These results prove that the inversion was complete, and that there

[1] The proportions alluded to above are :—

	Fehling's.	Sachsse's.	Knapp's.
Invert sugar	100·00	100·00	100·00
Dextrose	96·15	124·24	101·00
Maltose	157·50	190·20	158·30

was no caramelization of the sugar, the dark precipitate in (a) being merely decomposed albuminous matter. 100 cubic centimetres of wort contain, therefore, after inversion, 13·4 grammes dextrose.

(B) A sample of the same brew taken from the cooler after having been boiled with hops was also submitted to analysis : specific gravity = 1·0588, 15·70 grammes solid matter in 100 cubic centimetres. (The concentration of this wort differs but little from that above.)

1. *Determination of Rotatory Power.*—50 cubic centimetres of wort were treated with about 2 cubic centimetres basic acetate of lead, and made up to 100 cubic centimetres. 100 cubic centimetres rotated 96·4 divisions equivalent to $a_D = 33·36$.

2. *Determination of Reducing Power.*—100 cubic centimetres of wort were clarified and diluted to 1000 cubic centimetres. Of this solution there were used for the reduction of

50 cub. cent.	Fehling's	38·5 cub. cent.		
100 ,,	Sachsse's	48·0 ,,		
100 ,,	Knapp's	32·4 ,,		

Expressed in terms of maltose, this equals for

Fehling's 38·5 : ·414 :: 1000 : ($x = 10·75$ gr. maltose in 100 cc. of wort).
Sachsse's 48 : ·5097 :: 1000 : ($x = 10·61$,, ,, ,,).
Knapp's 32·4 : ·3134 :: 1000 : ($x = 9·67$,, ,, ,,).

3. 30 cubic centimetres of the wort were inverted for five and a half hours with 7 cubic centimetres acid, then nearly neutralized, clarified, and finally made up to 600 cubic centimetres; 50 cubic centimetres of the Fehling's solution were reduced by 21·7 cubic centimetres, therefore—

21.7 : ·253 : : 600 : (x : 7·0 gr. dextrose in 50 cc. of wort)

or 100 cubic centimetres contained after inversion 14 grammes dextrose.

The results of the experiments described under A and B lead the writer to think that the dextrins in worts have the power of reducing Fehling's, Sachsse's, and Knapp's tests in varying degrees. It will be observed that the values obtained by Fehling's and Sachsse's differ but little, but much lower numbers are obtained by Knapp's. In the following table the results of similar analyses are given of different malt worts before and after inversion.

Sample	Unboiled Worts.				Boiled Worts.	
	I.	II.	III.	IV.	V.	VI.
Specific gravity	1·0745	1·0743	1 0916 diluted to 1·0655	1·014	1·0588	1·0472
Extract per 100 cc.	19·87 grms.	19·8 grms.	14·75 grms.	8·6 grms.	15·7 grms.	12·5 grms.
Before inversion.	100 cc. wort. / 100 gr. extract. As Maltose.	100 cc. wort. / 100 gr. extract. As Maltose.	100 cc. wort. / 100 gr. extract. As Maltose.	100 cc. wort. / 100 gr. extract. As Maltose.	100 cc. wort. / 100 gr. extract. As Maltose.	100 cc. wort. / 100 gr. extract. As Maltose.
	grms. / per cent.	grms. / per cent.	grms. / per cent.	grms. / per cent.	grms. / per cent.	grms. / per cent.
Fehling's	14.24 / 71·56	13·7 / 71·92	9·936 / 67·36	1·89 / 52·5	10·75 / 68·51	8·90 / 71·20
Sachsse's	13·24 / 66·53	12·9 / 66·86	9·517 / 64·50	1·758 / 48·83	10·61 / 67·58	8·50 / 68·00
Knapp's	11·65 / 53·54	? / ?	8·247 / 55·91	1·713 / 47·53	9·67 / 61·60	7·90 / 63·20
Angle of rotation a D	? / —	42 9° / —	31½7' / —	? / —	33·36° / —	27·3° / —
After inversion	? / —	{14·91 gr. dextrose}	13·4 gr. dextrose.	? / —	14 gr. dextrose.	11·4 gr. dextrose.

A comparison of these numbers shows, firstly, that the composition of a strong, or first wort (I. to III.), differs from that obtained during sparging (IV.) ; and secondly, that even if the cupric oxide reducing power of a boiled wort (VI.) closely agrees with that of an unboiled wort (I.), it must not, therefore, be concluded that their respective deportments towards Knapp's and Sachsse's will be alike.

These data are too few to permit of any general conclusion with regard to the nature of the different dextrins in worts ; but it appears that the dextrins in samples I. to III. possess a similar reducing power.

The properties of maltose are $[a]_D = 138·5$ (for a concentration of 10 grammes per 100 cubic centimetres), and a cupric oxide reducing power $R_F = 61$ per cent. dextrose (for a 1 per cent. solution, and if the test liquor is not diluted). Supposing, then, the dextrins in I. to III. have a specific rotation $[a]_D = 120$, and a reducing power $R_F = 37$ per cent. dextrose (or $R_S = 35$ per cent., and $R_K = 21$ per cent.), 100 grammes of extract would contain :—

In Sample III., 42 per cent. dextrin + 42 per cent. maltose.
In Sample II., 34·6 per cent. „ + 47·96 per cent. „

Whether these are the exact proportions of maltose and dextrin in such worts can only be proved by further investigations.

The following polariscope observations serve as an additional proof that the composition of a first or concentrated malt wort differs from a wort obtained after sparging :—Three samples of wort from the same brew of specific gravities 1·1050, 1·0483 (II.), and 1·0201 (V.), were collected at intervals from the taps of the mash tun. Part of the concentrated wort (specific gravity, 1·105) was diluted to 1·0489 (I.), and the rest to 1·0217 (IV.).

A sample of the same brew, after boiling with hops, of specific gravity 1·062, was diluted to specific gravity 1·0493. (III.) The amounts used for observation were carefully measured with a 50 cubic centimetre pipette, and after the addition of 2 cubic centimetres basic acetate of lead, made up to 100 and 110 cubic centimetres respectively.

Conditions of Sample.	Specific gravity.	Angle of rotation a_D corresponding to 100 cub. cent. of solution.
I. Concentrated wort diluted .	1·0489	27·68
II. Sample direct from mash tun	1·0483	28·23
III. Boiled wort	1·0493	28·03
IV. Concentrated wort	1·0217	12·06
V. Sample direct from mash tun	1·0201	13·05

The worts procured after sparging have therefore a greater rotatory power than the concentrated first wort.

(C) The beer resulting from the worts referred to under (A) and (B) was subsequently analyzed, on the eleventh day from the date of mashing (100 cubic centimetres contained 4 7 grammes extract). 250 cubic centimetres of this beer were treated with about 3 cubic centimetres basic acetate of lead, diluted to 500 cubic centimetres, and filtered. Beers act on Knapp's test much more violently before clarification than after, and it is therefore necessary to clarify for analysis. The final reaction of the mercury tests is in this case more precise than that of Fehling's, the working of which may, however, be rendered less difficult if to 50 cubic centimetres Fehling's, 200 cubic centimetres water are added, i.e., if the test liquor is diluted four times. The following results were obtained :—

(i.) Angle of rotation, $a_D = 10\cdot17$.

(ii.) Fehling's, Sachsse's, and Knapp's solutions were reduced thus :—

Fehling's = 39 cub. cent., therefore 39 : ·414 : : 500 : ($x = 5\cdot308$ gr.) or 2·123 gr. maltose in 100 cub. cent. beer.

Sachsse's = 42·5 cub. cent., therefore 42·5 : ·5097 : : 500 : ($x = 6\cdot00$ gr.) or 2·40 gr. maltose in 100 cub. cent. beer.

Knapp's = 38·5 cub. cent., therefore 38·5 : ·3134 : : 500 : ($x = 5\cdot50$ gr.) or 2·20 gr. maltose in 100 cub. cent. beer.

(iii.) 50 cubic centimetres of the diluted beer were inverted with 5 cubic centimetres acid for five and a half hours, and made up to 100 cubic centimetres. 50 cubic centimetres Fehling's required 24·9 cubic centimetres; therefore—

$$24·9 : ·253 : : 100 : (x = 1·016) \text{ or } 4·064 \text{ gr. dextrose in 100 cub. cent.}$$

In the analyses above, it appears that Fehling's gives the lowest and Sachsse's the highest result, and the same is noticeable in other samples of beer, the analyses of which are given in the table below.

Sample.	I.		II.		III.		IV.	
Extract per 100 cc.	4·8 grammes.		4·1 grammes.		3·2 grammes.		4·7 grammes.	
Before inversion	100 cc. beer.	100 gr. extract.	100 cc. beer.	100 gr. extract.	100 cc. beer.	100 gr. extract.	100 cc. beer.	100 gr. extract.
	As Maltose.		As Maltose.		As Maltose.		As Maltose.	
	grms.	per cent.	grms.	per cent.	grms.	per cent.	grms.	per cent.
Fehling	1·808	37·7	1·23	30·0	0·80	25·0	2·123	45·17
Sachsse . . . . . . ·.	1·998	41·6	1·50	36·6	1·185	37·0	2·400	51·06
Knapp . . ·	1·880	39·4	1·453	35·4	1·052	32·9	2·200	46·81
Angle of rotation a_D . .	11·52°	—	10·73°	—	7·6°	—	10·17°	—
After inversion	4·05 grms. dextrose.		3.85 grms. dextrose.		?	—	4·06 grms. dextrose.	

These results seem to indicate, not only that the dextrin in beer has a reducing power, but also that this dextrin differs in its properties from those in worts.

Supposing the dextrin in beer to have a specific rotation of $[a]_D = 131$, and a cupric oxide reducing power $R_F = 3$ per cent. dextrose, the composition would be of—

Sample IV., 1·72 grammes dextrin + 2·03 grammes maltose per 100 cubic centimetres.

Sample II., 2·43 ,, ,, + 1·11 ,, ,,
Sample I., 2·02 ,, ,, + 1·71 ,, ,,

These latter calculations are made merely for an illustration, but the actual proportions of maltose and dextrin in beers and worts remain still to be proved.

Before concluding these lines, the analyses of a few samples of starch-sugars are given to illustrate that the results of Fehling's, Knapp's, and Sachsse's solutions differ also in this instance, but in a different ratio from that of worts or beers. In the analysis below Sachsse's solution gives the lowest and Fehling's the highest result.

Sample.	I.	II.	III.	IV.	V.
	Sugar before inversion (as dextrose).				
	per cent.	per cent.	per cent.	per cent.	per cent.
Fehling's. . . .	75·93	78 63	76·00	71·52	62·00
Sachsse's. . . .	65·72	70·41	66·07	69·60	55·31
Knapp's	72·52	72·85	71·81	70·91	?
Angle of rotation a_D 10 gr. in 100 cc.	11·42°	9·48°	9·51°	10·93°	10·60
	Sugar after inversion (as dextrose).				
All three solutions.	89·70	86·05	82·83	81·50	83·26

Naturally the explanation of the results of starch-sugar analyses is still more difficult, on account of the presence of maltose and dextrose, besides dextrins, all three of which are formed by the mineral acid used in their production. A general comparison of the results given in this paper shows that :—

1st. In the analysis of worts Fehling's and Sachsse's tests lead to results which differ only little, and are much higher than those corresponding to Knapp's.

2nd. In testing beers Fehling's solution is reduced least and Sachsse's most, and it remains to be proved whether the higher results of the mercury tests are not partly due to the presence of the albuminous matter, which cannot be removed by lead acetate.

3rd. Solutions of starch sugars have the greatest action on Fehling's test, but least on Sachsse's, the numbers for the latter being in this instance even surpassed by those of Knapp's test.

The subject of these lines requires careful consideration, and I intend carrying on further investigations in order to elucidate the various points which here have only been touched upon.

INDEX.

A

$[a]$:—
   Formula for calculating, 50
Acid :—
   Arabic, 225
   Aspartic, 27, 232
   Camphoric, 29, 236
   Cholalic, 246
   Dextro-tartaric, 230
   Glutamic, 232
   Glutanic, 232
   Glycocholic, 245
   Lævo-tartaric, 231
   Malic, 27, 231
   Para-lactic, 228
   Podocarpic, 235
   Quinic, 233
   Santonic, 244
   Valerianic, 227
   Vegetable, 230
Acids :—
   Rotation influence of, 35
a_D :—
   Relation of, to a_j, 46
Albuminate, 248
Albumins, 247
   Egg, 248
   Serum, 248
Alcohol :—
   Amyl, 28, 226
Alkaloids, 236
   Cinchona, 198, 236
Analyzer, 4
Angle :—
   Critical, 8 [note]
Aricine, 242
Asparagin, 27, 232
Australene, 233

B

Beers :—
   Estimation of maltose and dextrin in, 249
Bile :—
   Constituents of, 245
Biot :—
   Experiments of, 64
Bi-rotation :—
   Phenomenon of, 61, 197
Broch :—
   Process of, 124
Brucine, 244

C

Camphor :—
   Laurel, 235
   Patchouli, 236
   True specific rotation of, 88
Carbo-hydrates, 224
Carbon :—
   Asymmetrical, 24, 26, 29, 31
Casein, 248
Cellulose, 224
Chemical constitution :—
   Dependence of optical activity on, 24
Cholesterin, 245
Cinchona bases :—
   Determination of, 198, 236
   Optical analysis of mixture of, 209
Cinchonicine, 241
   Oxalate, 241
Cinchonidine, 201, 238
   Disulphate, 201, 239
   Hydrochloride, 201, 238
   Nitrate, 239
   Oxalate, 239
   Sulphate, 201, 239
   Rotation constants of, 198, 203

Cinchonine, 202, 240
  Hydrochloride, 202, 240
  Disulphate, 202, 240
  Oxalate, 240
Cinchotenine, 241
Clerget :—
  Inversion method of, 187
Codeine, 242
  Hydrochloride, 242
  Sulphate, 243
Concentration :—
  Calculation of, 162
Conchinamine, 241
Conchinine, 205
Conine, 236
Cornu :—
  Instrument of, 117
Corrections :—
  Schmitz' table of, 163
Crystals :—
  Active structure of, 19
Cusconine, 242

D

Degrees :—
  Ventzke, 169
  Angular measurement, 169
Dextrin :—
  Estimation of, 249
  Specific rotation of, 250
Dextro-gyrate, 8
Dextro-rotatory, 8
Diabetometers, 193
Dispersion :—
  Formulæ of, 43
  Rotatory, 43

E

Echicerin, 245
Echiretin, 245
Echitein, 245
Echitin, 245
Errors :—
  Correction of, 169
Euphorbone, 235

F

Fehling :—
  Solution of, 251
Filtration :—
  Alteration during, 134

Flasks :—
  Graduated, 144
  Standardizing of, 145, 183
Fringes :—
  Parallel, 110

G

Galactose, 223
Geissosperminc, 241
Glucose :—
  Determination of, 190
Glucose anhydride :—
  True specific rotation of, 90
Glucose hydrate :—
  True specific rotation of, 99
Glutin, 247
Glycogen, 225
Gravity, specific :—
  Details of calculation of, 142
  Determination of, 138

H

Hesse :—
  Observations of, 55, 198
  Researches of, 209
Homocinchonidine, 241

I

Impurities :—
  Removal of optically active, 187
  Removal by filtration, 134
Instruments, *vide* Polariscopes.
Inulin, 225

K

Knapp :—
  Solution of, 251

L

Lævo-gyrate, 8
Lævo-rotatory, 8
Lævulose, 222
Lamp :—
  For instrument, 102
  For sodium flame, 99
Lang :—
  Process of, 125
Laudanine, 243
Laudanosine, 244
Laurent :—
  Polariscope of, 118
Le Bel :—
  Hypothesis of, 24

Light :—
  Ordinary, 1
  Polarized, 1
  Polarized by reflection, 1
  Polarized by refraction, 3
  Vertical band of, 100
Liquids :—
  Constitution of active, 20
  Specific gravity of, 138

M

Maltose :—
  Estimation of, 249
Mannitan, 224
Mannite, 223
  Group, 223
Melitose, 220
Melizitose, 220
Method :—
  Inversion, 187
Micose, 220
Milk sugar :—
  Determination of, 197
  Rotatory power of, 197
Mitscherlich :—
  Polariscope of, 98
Molecules :—
  Asymmetrical structure of, 21
  Optically different modification of, 21
Morphine, 242
  Acetate, 242
  Hydrochloride, 242
  Sulphate, 242

N

Narcotine, 243
Nicotine, 72, 236
Nitro-mannite, 224

O

Observation :—
  Method of double, 171
Oils :—
  Ethereal, 235
  Krummholz, 234
  Templin, 234
Oudemans :—
  Investigations of, 55, 203

P

Papaverine, 243
Para-albumin, 248
Paytine, 241
Phlorhizin, 225
Phytosterin, 245

Picrotoxin, 244
Plane :—
  Impassable, 9 [note]
Polarimeter :—
  Rotation, 98 [note]
Polarimètre, 98 [note]
Polariscope :—
  Cornu's, 117
  Jellett's, 117
  Laurent's, 118
    Actual observations by, 120
    Construction of, 119
    Mode of observation with, 121
  Mitscherlich's :—
    Construction of, 98
    Lamp for, 99
    Larger form of, 104
    Mode of observation with, 102
  Wild's :—
    Construction of, 108
    Mode of observation with, 110
    Observations by, 115
    With saccharimetric scale, 176
    With Savart's prism, 108
Polariscopes :—
  Comparison of, 122
Polaristrobometer, 98 [note]
Polarization :—
  By reflection, 1
  By refraction, 3
  Circular, 8
    Theory of, 19
  Note on, 8
  Plane of, 2, 5
Polarizer, 4
Populin, 225
Precipitates :—
  Error due to, 184
Prism :—
  Nicol's, 3
    Construction of, 8 [note]
  Savart's, 107
Pseudomorphine :—
  Hydrochloride, 243
Pycnometer :—
  Sprengel's, 139
    Method of filling, 140
    Specific gravity by, 141

Q

Quercite, 224
Quinamine, 241
Quinicine, 241
  Oxalate, 241

Quinidamine, 241
Quinidine :—
Rotation constant of, 198, 203
Hydrate, 201, 239
Hydrochloride, 201, 239
Nitrate, 240
Oxalate, 240
Sulphate, 202, 240
Quinine :—
Rotation constant of, 198, 203
Anhydrous, 203, 237
Estimation of, when mixed with cin-
chonidine, 212
Hydrate, 237
Hydrochloride, 199, 237
Oxalate, 238
Sulphate, 200, 238

R

Radicles active, 24
Ray :—
Determination of angles for different,
124
Extraordinary, 3
Linear polarized, 1
Transverse section of, 1
Reducing power (cupric oxide) :—
Of malt extract, 251
Of starch-sugars, 257
Reducing power (mercuric oxide) :—
Of malt extract, 251
Of starch-sugars, 257
Reducing powers (cupric and mercuric
oxides) :—
Comparisons of, 252, 253, 254, 255, 256,
257
Reflection : —
Polarization by, 1
Total, 4, 8 [note]
Refraction :—
Polarization by, 3
Refractive indices, 8
Representations, graphical, 68, 71, 76, 80,
86
Resins, 235
Resorcin, 30
Right-rotating, 8
Rotation :—
Magnetic, 18
Molecular, 93
Relative directions of, 113, 114
Specific, 49
Calculation of law of, 55
Data necessary for determining, 95

Rotation, specific :—
Dependent on amount and nature of
solvent, 53
Dependent on thickness of medium, 42
Dependent on wave length, 43
Determination of true, 64
Influence of observation-errors on, 148
Influence of temperature on, 51
Influence on, by acids, 35
Influence on, by alkalies, 35
Influence on, by salts, 35
Method of determining, 83
Practical application of, 154
Theories as to cause of variation of, 62
Values, worthlessness of, 92
Variability of, in solution, 53
Rotatory power :—
Nature of, 16
Of malt extract, 251
Of starch-sugars, 257
Qualitative examination of, 98

S

Saccharimeter :—
Soleil-Duboscq, 174
Scale of, 175
Soleil-Ventzke-Scheibler, 155
Construction of, 155
Correction of readings for, 163
External form of, 159
Mode of observation with, 160
With angular graduation, 178
Saccharimetry :—
Optical, 154
Salicin, 225
Santonin, 244
Scale :—
Soleil-Duboscq, 175
Ventzke, 162
Schmitz :—
Tables of, 161, 180
Solution :—
Fehling's, 251
Knapp's, 251
Preparation of, 133
Process of clearing, 186
Sachsse's, 251
Solutions :—
Decolorization of, 186
Sorbin, 223
Spar :—
Iceland, 3
Spiral, lead, 130

Sprengel:—
Pycnometer of, 139
Starch, 224
Starch-sugars:—
Estimation of, 257
Strychnine, 244
Substances:—
Active, 10
Artificial production of, 31
Classification of, 10
Derivatives of, 11
Rotation constants of, 214
Rotatory power of derivatives of, 35
Simultaneous influence of, 59
Gelatinous, 247
Inactive, 8
Inactivity of artificial, 32
Vegetable, 244
Sugar:—
Calculation of percentage of, 178
Derivatives of, 226
Detection of, in urine, 194
Detection of, in wine, 195
Estimation of, in urine, 194
Fruit, 222
Invert, 222
Of formula $C_6H_{12}O_6$, 221
Of formula $C_{12}H_{22}O_{11}$, 216

T

Tartaric acid, 21
Abnormal rotatory dispersion of, 47
Tartrate of ethyl:—
Specific rotation of, 77
Temperature:—
Arrangement for constant, 110
Influence of, 172

Terebenthene, 234
Terecamphene, 235
Terpenes, 233
Thickness:—
Standard of, for solids and liquids, 42
Tint:—
Transition, 103
Trehalose, 220
Tubes:—
Adjustment of, 128
Measurement of lengths of, 128, 130
Turpentine:—
Right-handed oil of, 70
Specific rotation of, 66

U

Urines:—
Diabetic, estimation of, 194

V

Vau't Hoff:—
Theory of, 24

W

Water bath, 129
Wave lengths, 125
Weighings:—
Reduction of, to vacuo, 135
Wine:—
Chaptalized, 195
Worts:—
Reducing power of, 253
Rotatory power of, 253

Pardon and Sons, Printers, Paternoster Row, London.

With Illustrations, 8vo., price 21s.

Studies on Fermentation:

THE DISEASES OF BEER, THEIR CAUSES AND THE MEANS OF PREVENTING THEM.

By L. PASTEUR, Member of the Institute of France, the Royal Society of London, &c. A Translation, made with the Author's sanction, of "Etudes sur la Bière," with Notes, Index, and original Illustrations, by FRANK FAULKNER, Author of the "Art of Brewing," &c., and D. CONSTABLE ROBB, B.A., late Scholar of Worcester College, Oxford.

--- --- --- --- --- --- ---

"Let brewers read and re-read 'Studies on Fermentation.' The book will repay much earnest study. Too warm praise cannot be accorded to these gentlemen for the able manner in which they have accomplished their undertaking, and thus helped to popularize Pasteur's book in this country. . . . . We cannot, however, too cordially recommend the English version to the notice of our readers. No brewer's library should be without it."—*Brewers' Journal.*

"This important work, which has been promised so long, has at last appeared, and will be hailed by scientific men and practical brewers as a valuable addition to the literature appertaining to one of our most important industries. . . . . The translation has been made in an admirable manner."—*Brewers' Guardian.*

"This book may be, in the first place, one of special interest to the practical brewer, but it has a nearly equal interest for every careful student of nature, and it is so clearly written, with all the technical expressions so well explained, that we doubt not that the ordinary reader who takes it up will not put it on the shelf again without a perusal. The chapter on the physiological theory of fermentation is one we would specially commend to the general reader, to whom it may open up a quite new field for thought."—*Nature.*

"We would advise all who wish to obtain clear ideas on fermentation and kindred subjects to read this work. It is well written and well translated, and the plates and diagrams have been prepared with the utmost care."—*Medical Times and Gazette.*

MACMILLAN & Co., LONDON, W.C.

TREATISE ON THE ORIGIN, NATURE, AND VARIETIES OF WINE. Being a complete Manual of Viticulture and Œnology. By J. L. W. THUDICUM, M.D., and AUGUST DUPRÉ, Ph.D. With numerous Illustrations. Medium 8vo. 25s.

A TREATISE ON CHEMISTRY. By H. E. ROSCOE, F.R.S., and C. SCHORLEMMER, F.R.S., Professors of Chemistry in the Victoria University, Owens College, Manchester. With Illustrations. 8vo. (Just published). Vol. III.—ORGANIC CHEMISTRY. Part I. 21s.
Vols. I. and II.—INORGANIC CHEMISTRY. Vol. I.—THE NON-METALLIC ELEMENTS. 21s. Vol. II.—METALS. Two Parts. 18s. each.

"Will deserve to rank for many years as among the best treatises on chemistry in the English language."—*British Medical Journal.*

"A series which we think may justly claim to be the best English handbook to the study of inorganic chemistry."—*Saturday Review.*

A MANUAL OF THE CHEMISTRY OF THE CARBON COMPOUNDS, OR ORGANIC CHEMISTRY. By C. SCHORLEMMER, F.R.S., Professor of Chemistry, the Victoria University, the Owens College, Manchester. With Illustrations. 8vo. 14s.

CONTRIBUTIONS TO SOLAR PHYSICS. By J. NORMAN LOCKYER, F.R.S. With numerous Illustrations. Royal 8vo, cloth, extra gilt. 31s. 6d.

THE FORCES OF NATURE. A Popular Introduction to the Study of Physical Phenomena. By AMÉDÉE GUILLEMIN. Translated from the French by Mrs. NORMAN LOCKYER; and edited, with Additions and Notes, by J. NORMAN LOCKYER, F.R.S. Illustrated by 455 Woodcuts. Royal 8vo, cloth extra. 21s.

THE APPLICATIONS OF PHYSICAL FORCES. By AMÉDÉE GUILLEMIN. Translated from the French by Mrs. NORMAN LOCKYER, and edited, with Additions and Notes, by J. NORMAN LOCKYER, F.R.S. With Coloured Plates and Illustrations. Royal 8vo. Cheaper issue, 21s.

THE SPECTROSCOPE AND ITS APPLICATIONS. By J. NORMAN LOCKYER, F.R.S. With Illustrations. Second Edition. Crown 8vo. 3s. 6d. (Nature Series.)

POLARISATION OF LIGHT. By W. SPOTTISWOODE, F.R.S. With Illustrations. Second Edition. Crown 8vo. 3s. 6d. (Nature Series.)

LIGHT: A Series of Simple Experiments in the Phenomena of Light. By A. M. MAYER and C. BARNARD. With Illustrations. Crown 8vo. 2s. 6d. (Nature Series.)

SOUND: A Series of Simple Experiments in the Phenomena of Sound. By A. M. MAYER, Professor of Physics in the Stevens Institute of Technology, &c. With Illustrations. Crown 8vo. 3s. 6d. (Nature Series.)

RESEARCHES ON THE SOLAR SPECTRUM, AND THE SPECTRA OF THE CHEMICAL ELEMENTS. By G. KIRCHHOFF. Professor of Physics in the University of Heidelberg. Second Part. Translated with the Author's sanction, from the Transactions of the Berlin Academy for 1862, by HENRY E. ROSCOE, B.A., PH.D., F.R.S., Professsor of Chemistry in the Owens College, Manchester. 4to. 7s. 6d.

MACMILLAN & Co., LONDON, W.C.

BEDFORD STREET, COVENT GARDEN, LONDON,
December, 1879.

*MACMILLAN & CO.'S CATALOGUE of WORKS
in MATHEMATICS and PHYSICAL SCIENCE;
including PURE and APPLIED MATHE-
MATICS; PHYSICS, ASTRONOMY, GEOLOGY,
CHEMISTRY, ZOOLOGY, BOTANY; and of
WORKS in MENTAL and MORAL PHILOSOPHY
and Allied Subjects.*

MATHEMATICS.

Airy.— Works by Sir G. B. AIRY, K.C.B., Astronomer Royal :—
ELEMENTARY TREATISE ON PARTIAL DIFFERENTIAL
EQUATIONS. Designed for the Use of Students in the Univer-
sities. With Diagrams. New Edition. Crown 8vo. 5*s.* 6*d.*

ON THE ALGEBRAICAL AND NUMERICAL THEORY OF
ERRORS OF OBSERVATIONS AND THE COMBINA-
TION OF OBSERVATIONS. Second Edition. Crown 8vo.
6*s.* 6*d.*

UNDULATORY THEORY OF OPTICS. Designed for the Use of
Students in the University. New Edition. Crown 8vo. 6*s.* 6*d.*

ON SOUND AND ATMOSPHERIC VIBRATIONS. With
the Mathematical Elements of Music. Designed for the Use of
Students of the University. Second Edition, revised and enlarged.
Crown 8vo. 9*s.*

A TREATISE ON MAGNETISM. Designed for the Use of
Students in the University. Crown 8vo. 9*s.* 6*d.*

Ball (R. S., A.M.).—EXPERIMENTAL MECHANICS. A
Course of Lectures delivered at the Royal College of Science for
Ireland. By ROBERT STAWELL BALL, A.M., Professor of Applied
Mathematics and Mechanics in the Royal College of Science for
Ireland (Science and Art Department). Royal 8vo. 16*s.*

    *" We have not met with any book of the sort in English. It eluci-
dates instructively the methods of a teacher of the very highest
rank. We most cordially recommend it to all our readers."—*
Mechanics' Magazine.

5,000.12.79.]

Bayma.—THE ELEMENTS OF MOLECULAR MECHA-NICS. By Joseph Bayma, S.J., Professor of Philosophy, Stonyhurst College. Demy 8vo. cloth. 10s. 6d.

Boole.—Works by G. Boole, D.C.L, F.R.S., Professor of Mathematics in the Queen's University, Ireland :—

A TREATISE ON DIFFERENTIAL EQUATIONS. Third Edition. Edited by I. Todhunter. Crown 8vo. cloth. 14s.

A TREATISE ON DIFFERENTIAL EQUATIONS. Supplementary Volume. Edited by I. Todhunter. Crown 8vo. cloth. 8s. 6d.

THE CALCULUS OF FINITE DIFFERENCES. Crown 8vo. cloth. 10s. 6d. New Edition revised.

Cheyne. — AN ELEMENTARY TREATISE ON THE PLANETARY THEORY. With a Collection of Problems. By C. H. H. Cheyne, M.A., F.R.A.S. Second Edition. Crown 8vo. cloth. 6s. 6d.

Clifford.—THE ELEMENTS OF DYNAMIC. An Introduction to the study of Motion and Rest in Solid and Fluid Bodies. By W. K. Clifford, F.R.S., Professor of Applied Mathematics and Mechanics at University College, London. Part I.—Kinematic. Crown 8vo. 7s. 6d.

Cumming.—AN INTRODUCTION TO THE THEORY OF ELECTRICITY. With numerous Examples. By Linnæus Cumming, M.A., Assistant Master at Rugby School. Crown 8vo. 8s. 6d.

Cuthbertson.--EUCLIDIAN GEOMETRY. By F. Cuthbertson, M.A., Head Mathematical Master of the City of London School. Extra fcap. 8vo. 4s. 6d.

Everett.—UNITS AND PHYSICAL CONSTANTS. By J. D. Everett, M.A., D.C.L., F.R.S., Professor of Natural Philosophy, Queen's College, Belfast. Extra fcap. 8vo. 4s. 6d.

Ferrers. —Works by the Rev. N.M. Ferrers, M.A., F.R.S., Fellow and Tutor of Gonville and Caius College, Cambridge :—

AN ELEMENTARY TREATISE ON TRILINEAR CO-ORDI-NATES, the Method of Reciprocal Polars, and the Theory of Projectors. Third Edition, revised. Crown 8vo. 6s. 6d.

SPHERICAL HARMONICS AND SUBJECTS CONNECTED WITH THEM. Crown 8vo. 7s. 6d.

Frost.—Works by PERCIVAL FROST, M.A., late Fellow of St. John's College, Mathematical Lecturer of King'sColl. Cambridge:—

THE FIRST THREE SECTIONS OF NEWTON'S PRIN-CIPIA. With Notes and Illustrations. Also a Collection of Problems, principally intended as Examples of Newton's Methods. Third Edition. 8vo. cloth. 12*s.*

AN ELEMENTARY TREATISE ON CURVE TRACING. 8vo. 12*s.*

SOLID GEOMETRY. Being a New Edition, revised and enlarged, of the Treatise by FROST and WOLSTENHOLME. Vol. I. 8vo. 16*s.*

Godfray.—Works by HUGH GODFRAY, M.A., Mathematical Lecturer at Pembroke College, Cambridge :—

A TREATISE ON ASTRONOMY, for the Use of Colleges and Schools. 8vo. cloth. 12*s.* 6*d.*

AN ELEMENTARY TREATISE ON THE LUNAR THEORY, with a Brief Sketch of the Problem up to the time of Newton. Second Edition, revised. Crown 8vo. cloth. 5*s.* 6*d.*

Green (George).—MATHEMATICAL PAPERS OF THE LATE GEORGE GREEN, Fellow of Gonville and Caius College, Cambridge. Edited by N. M. FERRERS, M.A., Fellow and Tutor of Gonville and Caius College. 8vo. 15*s.*

Hemming.—AN ELEMENTARY TREATISE ON THE DIFFERENTIAL AND INTEGRAL CALCULUS. For the Use of Colleges and Schools. By G. W. HEMMING, M.A., Fellow of St. John's College, Cambridge. Second Edition, with Corrections and Additions. 8vo. cloth. 9*s.*

Jackson.—GEOMETRICAL CONIC SECTIONS. An Ele-mentary Treatise in which the Conic Sections are defined as the Plane Sections of a Cone, and treated by the Method of Projections. By J. STUART JACKSON, M.A., late Fellow of Gonville and Caius College. Crown 8vo. 4*s.* 6*d.*

Kelland and Tait.—AN INTRODUCTION TO QUATER-NIONS. With numerous Examples. By P. KELLAND, M.A., F.R.S., and P. G. TAIT, M.A., Professors in the department of Mathematics in the University of Edinburgh. Crown 8vo. 7*s.* 6*d.*

Kempe.—HOW TO DRAW A STRAIGHT LINE. A lecture on Linkages. By A.B. KEMPE, B.A. Illustrated. Crown 8vo. 1*s.*6*d.*

A 2

Merriman.—ELEMENTS OF THE METHOD OF LEAST SQUARES. By MANSFIELD MERRIMAN, Professor of Civil and Mechanical Engineering, Lehigh University, Bethlehem, Penn., U.S.A. Crown 8vo. 7s. 6d.

Morgan.—A COLLECTION OF PROBLEMS AND EXAMPLES IN MATHEMATICS. With Answers. By H. A. MORGAN, M.A., Sadlerian and Mathematical Lecturer of Jesus College, Cambridge. Crown 8vo. cloth. 6s. 6d.

Newton's Principia.—4to. cloth. 31s. 6d.

It is a sufficient guarantee of the reliability of this complete edition of Newton's Principia that it has been printed for and under the care of Professor Sir William Thomson and Professor Blackburn, of Glasgow University.

Parkinson.—A TREATISE ON OPTICS. By S. PARKINSON, D.D., F.R.S., Fellow and Tutor of St. John's College, Cambridge. Third Edition, revised and enlarged. Crown 8vo. cloth. 10s. 6d.

Phear.—ELEMENTARY HYDROSTATICS. With Numerous Examples. By J. B. PHEAR, M.A., Fellow and late Assistant Tutor of Clare Coll. Cambridge. Fourth Edition. Cr. 8vo. cloth. 5s. 6d.

Pirrie.—LESSONS ON RIGID DYNAMICS. By the Rev. G. PIRRIE, M.A., Fellow and Tutor of Queen's College, Cambridge. Crown 8vo. 6s.

Puckle.—AN ELEMENTARY TREATISE ON CONIC SECTIONS AND ALGEBRAIC GEOMETRY. With numerous Examples and Hints for their Solution. By G. HALE PUCKLE, M.A. Fouth Edition, enlarged. Crown 8vo. 7s. 6d.

Rayleigh.—THE THEORY OF SOUND. By LORD RAYLEIGH, F.R.S., formerly Fellow of Trinity College, Cambridge. 8vo. Vol. I. 12s. 6d.; Vol. II. 12s. 6d. [Vol. III. *in preparation.*

Reuleaux.—THE KINEMATICS OF MACHINERY. Outlines of a Theory of Machines. By Professor F. REULEAUX. Translated and edited by A. B. W. KENNEDY, C.E., Professor of Civil and Mechanical Engineering, University College, London. With 450 Illustrations. Royal 8vo. 20s.

Routh.—Works by EDWARD JOHN ROUTH, M.A., F.R.S., late Fellow and Assistant Tutor of St. Peter's College, Cambridge; Examiner in the University of London :—

AN ELEMENTARY TREATISE ON THE DYNAMICS OF THE SYSTEM OF RIGID BODIES. With numerous Examples. Third Edition, enlarged. 8vo. 21s.

Routh—*continued.*

STABILITY OF A GIVEN STATE OF MOTION, PARTI-
CULARLY STEADY MOTION. The Adams' Prize Essay for
1877. 8vo. 8*s.* 6*d.*

Tait and Steele.—DYNAMICS OF A PARTICLE. With
numerous Examples. By Professor TAIT and Mr. STEELE. Fourth
Edition, revised. Crown 8vo. 12*s.*

Thomson.—PAPERS ON ELECTROSTATICS AND MAG-
NETISM. By Professor SIR WILLIAM THOMSON, F.R.S.
8vo. 18*s.*

Todhunter.—Works by I. TODHUNTER, M.A., F.R.S., of
St. John's College, Cambridge :—

> "*Mr. Todhunter is chiefly known to students of mathematics as the
> author of a series of admirable mathematical text-books, which
> possess the rare qualities of being clear in style and absolutely free
> from mistakes, typographical or other.*"—Saturday Review.

A TREATISE ON SPHERICAL TRIGONOMETRY. New
Edition, enlarged. Crown 8vo. cloth. 4*s.* 6*d.*

PLANE CO-ORDINATE GEOMETRY, as applied to the Straight
Line and the Conic Sections. With numerous Examples. New
Edition. Crown 8vo. 7*s.* 6*d.*

A TREATISE ON THE DIFFERENTIAL CALCULUS.
With numerous Examples. New Edition. Crown 8vo. 10*s.* 6*d.*

A TREATISE ON THE INTEGRAL CALCULUS AND ITS
APPLICATIONS. With numerous Examples. New Edition,
revised and enlarged. Crown 8vo. cloth. 10*s.* 6*d.*

EXAMPLES OF ANALYTICAL GEOMETRY OF THREE
DIMENSIONS. New Edition, revised. Crown 8vo. cloth. 4*s.*

A TREATISE ON ANALYTICAL STATICS. With numerous
Examples. New Edition, revised and enlarged. Crown 8vo.
cloth. 10*s.* 6*d.*

A HISTORY OF THE MATHEMATICAL THEORY OF
PROBABILITY, from the Time of Pascal to that of Laplace.
8vo. 18*s.*

RESEARCHES IN THE CALCULUS OF VARIATIONS,
Principally on the Theory of Discontinuous Solutions: An Essay
to which the Adams' Prize was awarded in the University of
Cambridge in 1871. 8vo. 6*s.*

Todhunter—*continued.*
A HISTORY OF THE MATHEMATICAL THEORIES OF ATTRACTION, and the Figure of the Earth, from the time of Newton to that of Laplace. Two vols. 8vo. 24*s.*

AN ELEMENTARY TREATISE ON LAPLACE'S, LAME'S, AND BESSEL'S FUNCTIONS. Crown 8vo. 10*s.* 6*d.*

Wilson (W. P.).—A TREATISE ON DYNAMICS. By W. P. WILSON, M.A., Fellow of St. John's College, Cambridge, and Professor of Mathematics in Queen's College, Belfast. 8vo. 9*s.* 6*d.*

Wolstenholme.—MATHEMATICAL PROBLEMS, on Subjects included in the First and Second Divisions of the Schedule of Subjects for the Cambridge Mathematical Tripos Examination. Devised and arranged by JOSEPH WOLSTENHOLME, late Fellow of Christ's College, sometime Fellow of St. John's College, and Professor of Mathematics in the Royal Indian Engineering College. New Edition, greatly enlarged. 8vo. 18*s.*

Young.—SIMPLE PRACTICAL METHODS OF CALCULATING STRAINS ON GIRDERS, ARCHES, AND TRUSSES. With a Supplementary Essay on Economy in suspension Bridges. By E. W. YOUNG, Associate of King's College, London, and Member of the Institution of Civil Engineers. 8vo. 7*s.* 6*d.*

PHYSICAL SCIENCE.

Airy (G. B.).—POPULAR ASTRONOMY. With Illustrations. By Sir G. B. AIRY, K.C.B., Astronomer Royal. New Edition. fcap. 8vo. 4s. 6d.

Balfour.—A TREATISE ON COMPARATIVE EMBRY-OLOGY. By F. M. BALFOUR, F.R.S. Illustrated. 8vo.
[*Shortly.*

Bastian.—Works by H. CHARLTON BASTIAN, M.D., F.R.S., Professor of Pathological Anatomy in University College, London, &c. :—

THE BEGINNINGS OF LIFE : Being some Account of the Nature, Modes of Origin, and Transformations of Lower Organisms. In Two Volumes. With upwards of 100 Illustrations. Crown 8vo. 28s.
"*It is a book that cannot be ignored, and must inevitably lead to renewed discussions and repeated observations, and through these to the establishment of truth.*"—A. R. Wallace *in* Nature.

EVOLUTION AND THE ORIGIN OF LIFE. Crown 8vo. 6s. 6d.
"*Abounds in information of interest to the student of biological science.*"- -Daily News.

Blake.—ASTRONOMICAL MYTHS. Based on Flammarion's "The Heavens." By John F. BLAKE. With numerous Illustrations. Crown 8vo. 9s.

Blanford (H. F.).—RUDIMENTS OF PHYSICAL GEO-GRAPHY FOR THE USE OF INDIAN SCHOOLS. By H. F. BLANFORD, F.G.S. With numerous Illustrations and Glossary of Technical Terms employed. New Edition. Globe 8vo. 2s. 6d.

Blanford (W. T.).—GEOLOGY AND ZOOLOGY OF ABYSSINIA. By W. T. BLANFORD. 8vo. 21s.

Bosanquet.—AN ELEMENTARY TREATISE ON MUSICAL INTERVALS AND TEMPERAMENT. With an Account of an Enharmonic Harmonium exhibited in the Loan Collection of Scientific Instruments, South Kensington, 1876 ; also of an Enharmonic Organ exhibited to the Musical Association of London, May, 1875. By R. H. Bosanquet, Fellow of St. John's College, Oxford. 8vo. 6s.

Clifford.—SEEING AND THINKING. By the late Professor W. K. CLIFFORD, F.R.S. With Diagrams. Crown 8vo. 3s. 6d.
[*Nature Series.*

Coal : ITS HISTORY AND ITS USES. By Professors GREEN, MIALL, THORPE, RÜCKER, and MARSHALL, of the Yorkshire College, Leeds. With Illustrations. 8vo. 12s. 6d.

" *It furnishes a very comprehensive treatise on the whole subject of Coal from the geological, chemical, mechanical, and industrial points of view, concluding with a chapter on the important topic known as the 'Coal Question.'*"—Daily News.

Cooke (Josiah P., Jun.).—FIRST PRINCIPLES OF CHEMICAL PHILOSOPHY. By JOSIAH P. COOKE, Jun., Ervine Professor of Chemistry and Mineralogy in Harvard College. Third Edition, revised and corrected. Crown 8vo. 12s.

Cooke (M. C.).—HANDBOOK OF BRITISH FUNGI, with full descriptions of all the Species, and Illustrations of the Genera. By M. C. COOKE, M.A. Two vols. crown 8vo. 24s.

" *Will maintain its place as the standard English book, on the subject of which it treats, for many years to come.*"—Standard.

Crossley.—HANDBOOK OF DOUBLE STARS, WITH A CATALOGUE OF 1,200 DOUBLE STARS AND EXTENSIVE LISTS OF MEASURES FOR THE USE OF AMATEURS. By E. CROSSLEY, F.R.A.S., J. GLEDHILL, F.R.A.S., and J. M. WILSON, F.R.A.S. With Illustrations. 8vo. 21s.

Dawkins.—Works by W. BOYD DAWKINS, F.R.S., &c., Professor of Geology at Owens College, Manchester.

CAVE-HUNTING : Researches on the Evidence of Caves respecting the Early Inhabitants of Europe. With Coloured Plate and Woodcuts. 8vo. 21s.

" *The mass of information he has brought together, with the judicious use he has made of his materials, will be found to invest his book with much of new and singular value.*"—Saturday Review.

EARLY MAN IN BRITAIN, AND HIS PLACE IN THE TERTIARY PERIOD. With Illustrations. 8vo. [*Shortly.*

Dawson (J. W.).—ACADIAN GEOLOGY. The Geologic Structure, Organic Remains, and Mineral Resources of Nova Scotia, New Brunswick, and Prince Edward Island. By JOHN WILLIAM DAWSON, M.A., LL.D., F.R.S., F.G.S., Principal and Vice-Chancellor of M'Gill College and University, Montreal, &c. With a Geological Map and numerous Illustrations. Third Edition, with Supplement. 8vo. 21s. Supplement, separately, 2s. 6d.

Fiske.—DARWINISM; AND OTHER ESSAYS. By JOHN FISKE, M.A., LL.D., formerly Lecturer on Philosophy in Harvard University. Crown 8vo. 7s. 6d.

Fleischer.—A SYSTEM OF VOLUMETRIC ANALYSIS. By Dr. E. FLEISCHER. Translated from the Second German Edition by M. M. Pattison Muir, with Notes and Additions. Illustrated. Crown 8vo. 7s. 6d.

Flückiger and Hanbury.—PHARMACOGRAPHIA. A History of the Principal Drugs of Vegetble Origin met with in Great Britain and India. By F. A. FLÜCKIGER, M.D., and D. HANBURY, F.R.S. Second Edition, revised. 8vo. 21s.

Forbes.—THE TRANSIT OF VENUS. By GEORGE FORBES, B.A., Professor of Natural Philosophy in the Andersonian University of Glasgow. With numerous Illustrations. Crown 8vo. 3s. 6d.

Foster and Balfour.—ELEMENTS OF EMBRYOLOGY By MICHAEL FOSTER, M.D., F.R.S., and F. M. BALFOUR, M.A., Fellow of Trinity College, Cambridge. With numerous Illustrations. Part I. Crown 8vo. 7s. 6d.

Galton.—Works by FRANCIS GALTON, F.R.S. :—
METEOROGRAPHICA, or Methods of Mapping the Weather. Illustrated by upwards of 600 Printed Lithographic Diagrams. 4to. 9s.
HEREDITARY GENIUS : An Inquiry into its Laws and Consequences. Demy 8vo. 12s.
The Times *calls it "a most able and most interesting book."*
ENGLISH MEN OF SCIENCE ; THEIR NATURE AND NURTURE. 8vo. 8s. 6d.
" The book is certainly one of very great interest."—Nature.

Gamgee.—A TEXT-BOOK, SYSTEMATIC and PRACTICAL, OF THE PHYSIOLOGICAL CHEMISTRY OF THE ANIMAL BODY. By ARTHUR GAMGEE, M.D., F.R.S., Professor of Physiology in Owens College, Manchester. With Illustrations. 8vo. [*In the press.*

Geikie.—Works by ARCHIBALD GEIKIE, LL.D., F.R.S., Murchison Professor of Geology and Mineralogy at Edinburgh :—
ELEMENTARY LESSONS IN PHYSICAL GEOGRAPHY. With numerous Illustrations. Fcap. 8vo. 4s. 6d. Questions, 1s. 6d.
OUTLINES OF FIELD GEOLOGY. With Illustrations. Crown 8vo. 3s. 6d.
PRIMER OF GEOLOGY. Illustrated. 18mo. 1s.
PRIMER OF PHYSICAL GEOGRAPHY. Illustrated. 18mo. 1s.

Gordon.—AN ELEMENTARY BOOK ON HEAT. By J. E. H. GORDON, B.A., Gonville and Caius College, Cambridge. Crown 8vo. 2s.

Gray.—STRUCTURAL BOTANY ON THE BASIS OF MORPHOLOGY. By Professor ASA GRAY. With Illustrations. 8vo.   [*In the press.*

Guillemin.—THE FORCES OF NATURE: A Popular Introduction to the Study of Physical Phenomena. By AMÉDÉE GUILLEMIN. Translated from the French by MRS. NORMAN LOCKYER; and Edited, with Additions and Notes, by J. NORMAN LOCKYER, F.R.S. Illustrated by Coloured Plates, and 455 Woodcuts. Third and cheaper Edition. Royal 8vo. 21*s*.

"*Translator and Editor have done justice to their task. The text has all the force and flow of original writing, combining faithfulness to the author's meaning with purity and independence in regard to idiom; while the historical precision and accuracy pervading the work throughout, speak of the watchful editorial supervision which has been given to every scientific detail. Nothing can well exceed the clearness and delicacy of the illustrative woodcuts. Altogether, the work may be said to have no parallel, either in point of fulness or attraction, as a popular manual of physical science.*"—Saturday Review.

THE APPLICATIONS OF PHYSICAL FORCES. By A. GUILLEMIN. Translated from the French by Mrs. LOCKYER, and Edited with Notes and Additions by J. N. LOCKYER, F.R.S. With Coloured Plates and numerous Illustrations. Cheaper Edition. Imperial 8vo. cloth, extra gilt. 36*s*.

Also in Eighteen Monthly Parts, price 1*s*. each. Part I. in November, 1878.

"*A book which we can heartily recommend, both on account of the width and soundness of its contents, and also because of the excellence of its print, its illustrations, and external appearance.*"—Westminster Review.

Hanbury.—SCIENCE PAPERS: chiefly Pharmacological and Botanical. By DANIEL HANBURY, F.R.S. Edited, with Memoir, by J. INCE, F.L.S., and Portrait engraved by C. H. JEENS. 8vo. 14*s*.

Henslow.—THE THEORY OF EVOLUTION OF LIVING THINGS, and Application of the Principles of Evolution to Religion considered as Illustrative of the Wisdom and Beneficence of the Almighty. By the Rev. GEORGE HENSLOW, M.A., F.L.S. Crown 8vo. 6*s*.

Hooker.—Works by Sir J. D. HOOKER, K.C.S.I., C.B., F.R.S., M.D., D.C.L. :—

THE STUDENT'S FLORA OF THE BRITISH ISLANDS. Second Edition, revised and improved. Globe 8vo. 10*s*. 6*d*.

"*Certainly the fullest and most accurate manual of the kind that has yet appeared. Dr. Hooker has shown his characteristic industry*

Hooker—*continued.*

and ability in the care and skill which he has thrown into the characters of the plants. These are to a great extent original, and are really admirable for their combination of clearness, brevity, and completeness."—Pall Mall Gazette.

PRIMER OF BOTANY. With Illustrations. 18mo. 1s. New Edition, revised and corrected.

Hooker and Ball.—JOURNAL OF A TOUR IN MAROCCO AND THE GREAT ATLAS. By Sir J. D. HOOKER, K.C.S.I., C.B., F.R.S., &c., and JOHN BALL, F.R.S. With Appendices, including a Sketch of the Geology of Marocco. By G. MAW, F.L.S., F.G.S. With Map and Illustrations. 8vo. 21s.
" This is, without doubt, one of the most interesting and valuable books of travel published for many years."—Spectator.

Huxley and Martin.—A COURSE OF PRACTICAL IN-STRUCTION IN ELEMENTARY BIOLOGY. By T. H. HUXLEY, LL.D., Sec. R.S., assisted by H. N. MARTIN, B.A., M.B., D.Sc., Fellow of Christ's College, Cambridge. Crown 8vo. 6s.
" This is the most thoroughly valuable book to teachers and students of biology which has ever appeared in the English tongue."—London Quarterly Review.

Huxley (Professor).—LAY SERMONS, ADDRESSES, AND REVIEWS. By T. H. HUXLEY, LL.D., F.R.S. New and Cheaper Edition. Crown 8vo. 7s. 6d.
Fourteen Discourses on the following subjects:—(1) *On the Advisable-ness of Improving Natural Knowledge:—*(2) *Emancipation—Black and White :—*(3) *A Liberal Education, and where to find it :—*(4) *Scientific Education :—*(5) *On the Educational Value of the Natural History Sciences:—*(6) *On the Study of Zoology:—*(7) *On the Physical Basis of Life:—*(8) *The Scientific Aspects of Positivism:—*(9) *On a Piece of Chalk:—*(10) *Geological Contem-poraneity and Persistent Types of Life:—*(11) *Geological Reform:—*(12) *The Origin of Species:—*(13) *Criticisms on the " Origin of Species:"—*(14) *On Descartes' "Discourse touching the Method of using One's Reason rightly and of seeking Scientific Truth."*

ESSAYS SELECTED FROM "LAY SERMONS, AD-DRESSES, AND REVIEWS." Second Edition. Crown 8vo. 1s.

CRITIQUES AND ADDRESSES. 8vo. 10s. 6d.
*Contents:—*1. *Administrative Nihilism.* 2. *The School Boards : what they can do, and what they may do.* 3. *On Medical Edu-cation.* 4. *Yeast.* 5. *On the Formation of Coal.* 6. *On Coral and Coral Reefs.* 7. *On the Methods and Results of Ethnology.* 8. *On some Fixed Points in British Ethnology.* 9. *Palæontology*

Huxley (Professor)—*continued.*
  and the Doctrine of Evolution. 10. *Biogenesis and Abiogenesis.*
  11. *Mr. Darwin's Critics.* 12. *The Genealogy of Animals.*
  13. *Bishop Berkeley on the Metaphysics of Sensation.*

LESSONS IN ELEMENTARY PHYSIOLOGY. With numerous
  Illustrations. New Edition. Fcap. 8vo. 4s. 6d.
  " *Pure gold throughout.*"—Guardian. " *Unquestionably the clearest
and most complete elementary treatise on this subject that we possess in
any language.*"—Westminster Review.

AMERICAN ADDRESSES: with a Lecture on the Study of
  Biology. 8vo. 6s. 6d.

PHYSIOGRAPHY : An Introduction to the Study of Nature. With
  Coloured Plates and numerous Woodcuts. New Edition. Crown
  8vo. 7s. 6d.
  " *It would be hardly possible to place a more useful or suggestive
  book in the hands of learners and teachers, or one that is better
  calculated to make physiography a favourite subject in the science
  schools.*"—Academy.

Jellet (John H., B.D.).—A TREATISE ON THE
  THEORY OF FRICTION. By John H. Jellet, B.D.,
  Senior Fellow of Trinity College, Dublin ; President of the Royal
  Irish Academy. 8vo. 8s. 6d.

Jones.—Works by Francis Jones, F.R.S.E., F.C.S., Chemical
  Master in the Grammar School, Manchester.

THE OWENS COLLEGE JUNIOR COURSE OF PRAC-
  TICAL CHEMISTRY. With Preface by Professor Roscoe.
  New Edition. 18mo. With Illustrations. 2s. 6d.

QUESTIONS ON CHEMISTRY. A Series of Problems and
  Exercises in Inorganic and Organic Chemistry. 18mo. 3s.

Kingsley.—GLAUCUS : OR, THE WONDERS OF THE
  SHORE. By Charles Kingsley, Canon of Westminster.
  New Edition, with numerous Coloured Plates. Crown 8vo. 6s.

Landauer.—BLOWPIPE ANALYSIS. By J. Landauer.
  Authorised English Edition, by James Taylor and W. E. Kay, of
  the Owens College, Manchester. With Illustrations. Extra fcap.
  8vo. 4s 6d.

Langdon.—THE APPLICATION OF ELECTRICITY TO
  RAILWAY WORKING. By W. E. Langdon, Member of the
  Society of Telegraph Engineers. With numerous Illustrations.
  Extra fcap. 8vo. 4s. 6d.
  " *There is no officer in the telegraph service who will not profit by
  the study of this book.*"—Mining Journal.

Lockyer (J. N.).—Works by J. NORMAN LOCKYER, F.R.S.—
ELEMENTARY LESSONS IN ASTRONOMY. With numerous Illustrations. New Edition. Fcap. 8vo. 5s. 6d.
"*The book is full, clear, sound, and worthy of attention, not only as a popular exposition, but as a scientific 'Index.'*"—Athenæum.
THE SPECTROSCOPE AND ITS APPLICATIONS. By J. NORMAN LOCKYER, F.R.S. With Coloured Plate and numerous Illustrations. Second Edition. Crown 8vo. 3s. 6d.
CONTRIBUTIONS TO SOLAR PHYSICS. By J. NORMAN LOCKYER, F.R.S. I. A Popular Account of Inquiries into the Physical Constitution of the Sun, with especial reference to Recent Spectroscopic Researches. II. Communications to the Royal Society of London and the French Academy of Sciences, with Notes. Illustrated by 7 Coloured Lithographic Plates and 175 Woodcuts. Royal 8vo. cloth, extra gilt, price 31s. 6d.
"*The book may be taken as an authentic exposition of the present state of science in connection with the important subject of spectroscopic analysis. . . . Even the unscientific public may derive much information from it.*"—Daily News.
PRIMER OF ASTRONOMY. With Illustrations. 18mo. 1s.

Lockyer and Seabroke.—STAR-GAZING: PAST AND PRESENT. An Introduction to Instrumental Astronomy. By J. N. LOCKYER, F.R.S. Expanded from Shorthand Notes of a Course of Royal Institution Lectures with the assistance of G. M. SEABROKE, F.R.A.S. With numerous Illustrations. Royal 8vo. 21s.
"*A book of great interest and utility to the astronomical student.*"
—Athenæum.

Lubbock.—Works by SIR JOHN LUBBOCK, M.P., F.R.S., D.C.L.:
THE ORIGIN AND METAMORPHOSES OF INSECTS. With Numerous Illustrations. Second Edition. Crown 8vo. 3s. 6d.
"*As a summary of the phenomena of insect metamorphoses his little book is of great value, and will be read with interest and profit by all students of natural history. The whole chapter on the origin of insects is most interesting and valuable. The illustrations are numerous and good.*"—Westminster Review.
ON BRITISH WILD FLOWERS CONSIDERED IN RELATION TO INSECTS. With Numerous Illustrations. Second Edition. Crown 8vo. 4s. 6d.
SCIENTIFIC LECTURES. With Illustrations. 8vo. 8s. 6d.
CONTENTS:—*Flowers and Insects—Plants and Insects—The Habits of Ants—Introduction to the Study of Prehistoric Archæology, &c.*

Macmillan (Rev. Hugh).—For other Works by the same Author, see THEOLOGICAL CATALOGUE.
HOLIDAYS ON HIGH LANDS; or, Rambles and Incidents in search of Alpine Plants. Globe 8vo. cloth. 6s.

Macmillan (Rev. Hugh)—*continued.*
FIRST FORMS OF VEGETATION. Second Edition, corrected
and enlarged, with Coloured Frontispiece and numerous Illustrations. Globe 8vo. 6s.
*The first edition of this book was published under the name of
"Footnotes from the Page of Nature; or, First Forms of Vegetation. Probably the best popular guide to the study of mosses,
lichens, and fungi ever written. Its practical value as a help to
the student and collector cannot be exaggerated."*—Manchester
Examiner.

Mansfield (C. B.).—Works by the late C. B. MANSFIELD :—
A THEORY OF SALTS. A Treatise on the Constitution of
Bipolar (two-membered) Chemical Compounds. Crown 8vo. 14s.
AËRIAL NAVIGATION. The Problem, with Hints for its
Solution. Edited by R. B. MANSFIELD. With a Preface by J.
M. LUDLOW. With Illustrations. Crown 8vo. 10s. 6d.

Mayer.—SOUND : a Series of Simple, Entertaining, and In-
expensive Experiments in the Phenomena of Sound, for the Use of
Students of every age. By A. M. MAYER, Professor of Physics
in the Stevens Institute of Technology, &c. With numerous Illustrations. Crown 8vo. 3s. 6d.

Mayer and Barnard.—LIGHT. A Series of Simple, Entertaining, and Useful Experiments in the Phenomena of Light, for
the use of Students of every age. By A. M. MAYER and C.
BARNARD. With Illustrations. Crown 8vo. 2s. 6d.

Miall.—STUDIES IN COMPARATIVE ANATOMY. No. 1,
The Skull of the Crocodile. A Manual for Students. By L. C.
MIALL, Professor of Biology in Yorkshire College. 8vo. 2s. 6d.
No. 2, The Anatomy of the Indian Elephant. By L. C. MIALL
and F. GREENWOOD. With Plates. 5s.

Miller.—THE ROMANCE OF ASTRONOMY. By R. KALLEY
MILLER, M.A., Fellow and Assistant Tutor of St. Peter's College, Cambridge. Second Edition, revised and enlarged. Crown
8vo. 4s. 6d.

Mivart (St. George).—Works by ST. GEORGE MIVART, F.R.S.
&c., Lecturer in Comparative Anatomy at St. Mary's Hospital:—
ON THE GENESIS OF SPECIES. Second Edition, to which
notes have been added in reference and reply to Darwin's "Descent
of Man." With numerous Illustrations. Crown 8vo. 9s.
*" In no work in the English language has this great controversy
been treated at once with the same broad and vigorous grasp of
facts, and the same liberal and candid temper."*—Saturday Review.

Mivart (St. George)—*continued.*

THE COMMON FROG. With Numerous Illustrations. Crown 8vo. 3*s.* 6*d.* (Nature Series.)

"*It is an able monogram of the Frog, and something more. It throws valuable crosslights over wide portions of animated nature. Would that such works were more plentiful.*"—Quarterly Journal of Science.

Moseley.—NOTES BY A NATURALIST ON THE "CHAL-LENGER," being an account of various observations made during the voyage of H.M.S. "Challenger" round the world in the years 1872—76. By H. N. MOSELEY, M.A.. F.R.S., Member of the Scientific Staff of the "Challenger." With Map, Coloured Plates, and Woodcuts. 8vo. 21*s.*

"*This is certainly the most interesting and suggestive book, descriptive of a naturalist's travels, which has been published since Mr. Darwin's 'Journal of Researches' appeared, now more than forty years ago. That it is worthy to be placed alongside that delightful record of the impressions, speculations, and reflections of a master mind, is, we do not doubt, the highest praise which Mr. Moseley would desire for his book, and we do not hesitate to say that such praise is its desert.*"—Nature.

Muir.—PRACTICAL CHEMISTRY FOR MEDICAL STU-DENTS. Specially arranged for the first M. B. Course. By M. M. PATTISON MUIR, F.R.S.E. Fcap. 8vo. 1*s.* 6*d.*

Murphy.—HABIT AND INTELLIGENCE: a Series of Essays on the Laws of Life and Mind. By JOSEPH JOHN MURPHY. Second Edition, thoroughly revised and mostly re-written. With Illustrations. 8vo. 16*s.*

Nature.—A WEEKLY ILLUSTRATED JOURNAL OF SCIENCE. Published every Thursday. Price 6*d.* Monthly Parts, 2*s.* and 2*s.* 6*d.* ; Half-yearly Volumes, 15*s.* Cases for binding Vols. 1*s.* 6*d.*

"*This able and well-edited Journal, which posts up the science of the day promptly, and promises to be of signal service to students and savants. . . . . Scarcely any expressions that we can employ would exaggerate our sense of the moral and theological value of the work.*"—British Quarterly Review.

Newcomb.—POPULAR ASTRONOMY. By SIMON NEW-COMB, LL.D., Professor U.S. Naval Observatory. With 112 Engravings and Five Maps of the Stars. 8vo. 18*s.*

"*As affording a thoroughly reliable foundation for more advanced reading, Professor Newcomb's 'Popular Astronomy' is deserving of strong recommendation.*"—Nature.

Oliver.—Works by DANIEL OLIVER, F.R.S., F.L.S., Professor of Botany in University College, London, and Keeper of the Herbarium and Library of the Royal Gardens, Kew :—

Oliver—*continued.*
LESSONS IN ELEMENTARY BOTANY. With nearly Two
Hundred Illustrations. New Edition. Fcap. 8vo. 4s. 6d.
FIRST BOOK OF INDIAN BOTANY. With numerous
Illustrations. Extra fcap. 8vo. 6s. 6d.
"*It contains a well-digested summary of all essential knowledge
pertaining to Indian Botany, wrought out in accordance with the
best principles of scientific arrangement.*"—Allen's Indian Mail.

Pasteur.—STUDIES ON FERMENTATION. The Diseases
of Beer; their Causes and Means of Preventing them. By L.
PASTEUR. A Translation of "Études sur la Bière," With Notes,
Illustrations, &c. By F. FAULKNER and D. C. ROBB, B.A.
8vo. 21s.

Pennington.—NOTES ON THE BARROWS AND BONE
CAVES OF DERBYSHIRE. With an account of a Descent
into Elden Hole. By ROOKE PENNINGTON, B.A., LL.B.,
F.G.S. 8vo. 6s.

Penrose (F. C.)—ON A METHOD OF PREDICTING BY
GRAPHICAL CONSTRUCTION, OCCULTATIONS OF
STARS BY THE MOON, AND SOLAR ECLIPSES FOR
ANY GIVEN PLACE. Together with more rigorous methods
for the Accurate Calculation of Longitude. By F. C. PENROSE,
F.R.A.S. With Charts, Tables, &c. 4to. 12s.

Perry.—AN ELEMENTARY TREATISE ON STEAM. By
JOHN PERRY, B.E., Professor of Engineering, Imperial College of
Engineering, Yedo. With numerous Woodcuts, Numerical Examples, and Exercises. 18mo. 4s. 6d.
"*Mr. Perry has in this compact little volume brought together an
immense amount of information, new told, regarding steam and
its application, not the least of its merits being that it is suited to
the capacities alike of the tyro in engineering science or the better
grade of artisan.*"—Iron.

Pickering.—ELEMENTS OF PHYSICAL MANIPULATION.
By E. C. PICKERING, Thayer Professor of Physics in the Massa-
chusetts Institute of Technology. Part I., medium 8vo. 10s. 6d.
Part II., 10s. 6d.
"*When finished 'Physical Manipulation' will no doubt be considered the best and most complete text-book on the subject of
which it treats.*"—Nature.

Prestwich.—THE PAST AND FUTURE OF GEOLOGY.
An Inaugural Lecture, by J. PRESTWICH, M.A., F.R.S., &c.,
Professor of Geology, Oxford. 8vo. 2s.

Radcliffe.—PROTEUS : OR UNITY IN NATURE. By. C.
B. RADCLIFFE, M.D., Author of "Vital Motion as a mode of
Physical Motion. Second Edition. 8vo. 7s. 6d.

Rendu.—THE THEORY OF THE GLACIERS OF SAVOY. By M. LE CHANOINE RENDU. Translated by A. WELLS, Q.C., late President of the Alpine Club. To which are added, the Original Memoir and Supplementary Articles by Professors TAIT and RUSKIN. Edited with Introductory remarks by GEORGE FORBES, B.A., Professor of Natural Philosophy in the Andersonian University, Glasgow. 8vo. 7s. 6d.

Roscoe.—Works by HENRY E. ROSCOE, F.R.S., Professor of Chemistry in Owens College, Manchester :—

LESSONS IN ELEMENTARY CHEMISTRY, INORGANIC AND ORGANIC. With numerous Illustrations and Chromolitho of the Solar Spectrum, and of the Alkalis and Alkaline Earths. New Edition. Fcap. 8vo. 4s. 6d.

CHEMICAL PROBLEMS, adapted to the above by Professor THORPE. Fifth Edition, with Key. 2s.
"*We unhesitatingly pronounce it the best of all our elementary treatises on Chemistry.*"—Medical Times.

PRIMER OF CHEMISTRY. Illustrated. 18mo. 1s.

Roscoe and Schorlemmer.—A TREATISE ON INORGANIC CHEMISTRY. With numerous Illustrations. By PROFESSORS ROSCOE and SCHORLEMMER.
Vol. I., The Non-metallic Elements. 8vo. 21s.
Vol. II., Part I. Metals. 8vo. 18s.
Vol. II., Part II. Metals, 8vo. 18s.
"*Regarded as a treatise on the Non-metallic Elements, there can be no doubt that this volume is incomparably the most satisfactory one of which we are in possession.*"—Spectator.
"*It would be difficult to praise the work too highly. All the merits which we noticed in the first volume are conspicuous in the second. The arrangement is clear and scientific; the facts gained by modern research are fairly represented and judiciously selected; and the style throughout is singularly lucid.*"—Lancet.

Rumford (Count).—THE LIFE AND COMPLETE WORKS OF BENJAMIN THOMPSON, COUNT RUMFORD. With Notices of his Daughter. By GEORGE ELLIS. With Portrait. Five Vols. 8vo. 4l. 14s. 6d.

Schorlemmer.—A MANUAL OF THE CHEMISTRY OF THE CARBON COMPOUNDS OR ORGANIC CHEMISTRY. By C. SCHORLEMMER, F.R.S., Lecturer in Organic Chemistry in Owens College, Manchester. 8vo. 14s.
"*It appears to us to be as complete a manual of the metamorphoses of carbon as could be at present produced, and it must prove eminently useful to the chemical student.*"—Athenæum.

B

Shann.—AN ELEMENTARY TREATISE ON HEAT, IN RELATION TO STEAM AND THE STEAM ENGINE. By G. SHANN, M.A. With Illustrations. Crown 8vo. 4s. 6d.

Smith.—HISTORIA FILICUM : An Exposition of the Nature, Number, and Organography of Ferns, and Review of the Principles upon which Genera are founded, and the Systems of Classification of the principal Authors, with a new General Arrangement, &c. By J. SMITH, A.L.S., ex-Curator of the Royal Botanic Garden, Kew. With Thirty Lithographic Plates by W. H. FITCH, F.L.S. Crown 8vo. 12s. 6d.

" *No one anxious to work up a thorough knowledge of ferns can afford to do without it.*"—Gardener's Chronicle.

South Kensington Science Lectures.

Vol. I.—Containing Lectures by Captain ABNEY, F.RS., Professor STOKES, Professor KENNEDY, F. J. BRAMWELL, F.R.S., Professor G. FORBES, H. C. SORBY, F.R.S., J. T. BOTTOMLEY, F.R.S.E., S. H. VINES, B.Sc., and Professor CAREY FOSTER. Crown 8vo. 6s.   [Vol. II. *nearly ready.*

Vol. II.—Containing Lectures by W. SPOTTISWOODE, P.R.S., Prof. FORBES, H. W. CHISHOLM, Prof. T. F. PIGOT, W. FROUDE, F.R.S., Dr. SIEMENS, Prof. BARRETT, Dr. BURDEN-SANDERSON, Dr. LAUDER BRUNTON, F.R.S., Prof. McLEOD, Prof. ROSCOE, F.R.S., &c. Crown 8vo. 6s.

Spottiswoode.—POLARIZATION OF LIGHT. By W. SPOTTISWOODE, President of the Royal Society. With numerous Illustrations. Second Edition. Cr. 8vo. 3s. 6d. (Nature Series.)

" *The illustrations are exceedingly well adapted to assist in making the text comprehensible.*"—Athenæum. " *A clear, trustworthy manual.*"—Standard.

Stewart (B.).—Works by BALFOUR STEWART, F.R.S.,[Professor of Natural Philosophy in Owens College, Manchester:—

LESSONS IN ELEMENTARY PHYSICS. With numerous Illustrations and Chromolithos of the Spectra of the Sun, Stars, and Nebulæ. New Edition. Fcap. 8vo. 4s. 6d.

The Educational Times *calls this the beau-idéal of a scientific text-book, clear, accurate, and thorough.*"

PRIMER OF PHYSICS. With Illustrations. New Edition, with Questions. 18mo. 1s.

Stewart and Tait.—THE UNSEEN UNIVERSE: or, Physical Speculations on a Future State. By BALFOUR STEWART, F.R.S., and P. G. TAIT, M.A. Sixth Edition. Crown 8vo. 6s.

" *The book is one which well deserves the attention of thoughtful and religious readers. . . . It is a perfectly sober inquiry, on scientific grounds, into the possibilities of a future existence.*"—Guardian.

Stone.—ELEMENTARY LESSONS ON SOUND. By Dr. W. H. STONE, Lecturer on Physics at St. Thomas' Hospital. With Illustrations. Fcap. 8vo. 3*s.* 6*d.*

Tait.—LECTURES ON SOME RECENT ADVANCES IN PHYSICAL SCIENCE. By P. G. TAIT, M.A., Professor of Philosophy in the University of Edinburgh. Second edition, revised and enlarged, with the Lecture on Force delivered before the British Association. Crown 8vo. 9*s.*

Tanner.—FIRST PRINCIPLES OF AGRICULTURE. By HENRY TANNER, F.C.S., Professor of Agricultural Science, University College, Aberystwith, Examiner in the Principles of Agriculture under the Government Department of Science. 18mo. 1*s.*

Taylor.—SOUND AND MUSIC : A Non-Mathematical Treatise on the Physical Constitution of Musical Sounds and Harmony, including the Chief Acoustical Discoveries of Professor Helmholtz. By SEDLEY TAYLOR, M.A., late Fellow of Trinity College, Cambridge. Large crown 8vo. 8*s.* 6*d.*

"*In no previous scientific treatise do we remember so exhaustive and so richly illustrated a description of forms of vibration and of wave-moti n in fluids.*"— Musical Standard.

Thomson.—Works by SIR WYVILLE THOMSON, K.C.B., F.R.S.
THE DEPTHS OF THE SEA : An Account of the General Results of the Dredging Cruises of H.M.SS. "Porcupine" and "Lightning" during the Summers of 1868-69 and 70, under the scientific direction of Dr. Carpenter, F.R.S., J. Gwyn Jeffreys, F.R.S., and Sir Wyville Thomson, F.R.S. With nearly 100 Illustrations and 8 coloured Maps and Plans. Second Edition. Royal 8vo. cloth, gilt. 31*s.* 6*d.*

The Athenæum *says :* "*The book is full of interesting matter, and is written by a master of the art of popular exposition. It is excellently illustrated, both coloured maps and woodcuts possessing high merit. Those who have already become interested in dredging operations will of course make a point of reading this work ; those who wish to be pleasantly introduced to the subject, and rightly to appreciate the news which arrives from time to time from the 'Challenger,' should not fail to seek instruction from it.*"

THE VOYAGE OF THE "CHALLENGER."—THE ATLANTIC. A Preliminary account of the Exploring Voyages of H.M.S. "Challenger," during the year 1873 and the early part of 1876. With numerous Illustrations, Coloured Maps & Charts, & Portrait of the Author, engraved by C. H. JEENS. 2 Vols. Medium 8vo. 42*s.*

The Times *says :*—"*It is right that the public should have some authoritative account of the general results of the expedition, and*

Thomson—*continued.*

> *that as many of the ascertained data as may be accepted with confidence should speedily find their place in the general body of scientific knowledge. No one can be more competent than the accomplished scientific chief of the expedition to satisfy the public in this respect. . . . The paper, printing, and especially the numerous illustrations, are of the highest quality. . . . We have rarely, if ever, seen more beautiful specimens of wood engraving than abound in this work. . . . Sir Wyville Thomson's style is particularly attractive; he is easy and graceful, but vigorous and exceedingly happy in the choice of language, and throughout the work there are touches which show that science has not banished sentiment from his bosom."*

Thudichum and Dupré.—A TREATISE ON THE ORIGIN, NATURE, AND VARIETIES OF WINE. Being a Complete Manual of Viticulture and Œnology. By J. L. W. THUDICHUM, M.D., and AUGUST DUPRÉ, Ph.D., Lecturer on Chemistry at Westminster Hospital. Medium 8vo. cloth gilt. 25s.

> *"A treatise almost unique for its usefulness either to the wine-grower, the vendor, or the consumer of wine. The analyses of wine are the most complete we have yet seen, exhibiting at a glance the constituent principles of nearly all the wines known in this country."* —Wine Trade Review.

Wallace (A. R.).—Works by ALFRED RUSSEL WALLACE. CONTRIBUTIONS TO THE THEORY OF NATURAL SELECTION. A Series of Essays. New Edition, with Corrections and Additions. Crown 8vo. 8s. 6d.

> *The Saturday Review says: "He has combined an abundance of fresh and original facts with a liveliness and sagacity of reasoning which are not often displayed so effectively on so small a scale."*

THE GEOGRAPHICAL DISTRIBUTION OF ANIMALS, with a study of the Relations of Living and Extinct Faunas as Elucidating the Past Changes of the Earth's Surface. 2 vols. 8vo. with Maps, and numerous Illustrations by Zwecker, 42s.

> *The Times says: "Altogether it is a wonderful and fascinating story, whatever objections may be taken to theories founded upon it. Mr. Wallace has not attempted to add to its interest by any adornments of style; he has given a simple and clear statement of intrinsically interesting facts, and what he considers to be legitimate inductions from them. Naturalists ought to be grateful to him for having undertaken so toilsome a task. The work, indeed, is a credit to all concerned—the author, the publishers, the artist—unfortunately now no more—of the attractive illustrations—last but by no means least, Mr. Stanford's map-designer."*

Wallace (A. R.)—*continued.*

TROPICAL NATURE; with other Essays. 8vo. 12s.

"*Nowhere amid the many descriptions of the tropics that have been given is to be found a summary of the past history and actual phenomena of the tropics which gives that which is distinctive of the phases of nature in them more clearly, shortly, and impressively.*"—Saturday Review.

Warington.—THE WEEK OF CREATION; OR, THE COSMOGONY OF GENESIS CONSIDERED IN ITS RELATION TO MODERN SCIENCE. By GEORGE WARINGTON, Author of "The Historic Character of the Pentateuch Vindicated." Crown 8vo. 4s. 6d.

Wilson.—RELIGIO CHEMICI. By the late GEORGE WILSON, M.D., F.R.S.E., Regius Professor of Technology in the University of Edinburgh. With a Vignette beautifully engraved after a design by Sir NOEL PATON. Crown 8vo. 8s. 6d.

Wilson (Daniel).—CALIBAN: a Critique on Shakespeare's "Tempest" and "Midsummer Night's Dream." By DANIEL WILSON, LL.D., Professor of History and English Literature in University College, Toronto. 8vo. 10s. 6d.

"*The whole volume is most rich in the eloquence of thought and imagination as well as of words. It is a choice contribution at once to science, theology, religion, and literature.*"—British Quarterly Review.

Wright.—METALS AND THEIR CHIEF INDUSTRIAL APPLICATIONS. By C. ALDER WRIGHT, D.Sc., &c., Lecturer on Chemistry in St. Mary's Hospital School. Extra fcap. 8vo. 3s. 6d.

Wurtz.—A HISTORY OF CHEMICAL THEORY, from the Age of Lavoisier down to the present time. By AD. WURTZ. Translated by HENRY WATTS, F.R.S. Crown 8vo. 6s.

"*The discourse, as a résumé of chemical theory and research, unites singular luminousness and grasp. A few judicious notes are added by the translator.*"—Pall Mall Gazette. "*The treatment of the subject is admirable, and the translator has evidently done his duty most efficiently.*"—Westminster Review.

SCIENCE PRIMERS FOR ELEMENTARY SCHOOLS.

Under the joint Editorship of Professors HUXLEY, ROSCOE, and BALFOUR STEWART.

Introductory. By Professor HUXLEY, F.R.S.   [*Nearly ready.*

Chemistry.—By H. E. ROSCOE, F.R.S., Professor of Chemistry in Owens College, Manchester. With numerous Illustrations. 18mo. 1s. New Edition. With Questions.

Physics.— By BALFOUR STEWART, F.R.S., Professor of Natural Philosophy in Owens College, Manchester. With numerous Illustrations. 18mo. 1s. New Edition. With Questions.

Physical Geography. — By ARCHIBALD GEIKIE, F.R.S., Murchison Professor of Geology and Mineralogy at Edinburgh. With numerous Illustrations. New Edition with Questions. 18mo. 1s.

Geology.—By Professor GEIKIE, F.R.S. With numerous Illustrations. New Edition. 18mo. cloth. 1s.

Physiology.—By MICHAEL FOSTER, M.D., F.R.S. Wit numerous Illustrations. New Edition. 18mo. 1s.

Astronomy.—By J. NORMAN LOCKYER, F.R.S. With numerous Illustrations. New Edition. 18mo. 1s.

Botany.—By Sir J. D. HOOKER, K.C.S.I., C.B., F.R.S. With numerous Illustrations. New Edition. 18mo. 1s.

Logic.—By Professor STANLEY JEVONS, F.R.S. New Edition. 18mo. 1s.

Political Economy.—By Professor STANLEY JEVONS, F.R.S. 18mo. 1s.

Others in preparation.

ELEMENTARY SCIENCE CLASS-BOOKS.

Astronomy.—By the ASTRONOMER ROYAL. POPULAR ASTRONOMY. With Illustrations. By Sir G. B. AIRY, K.C.B., Astronomer Royal. New Edition. 18mo. 4s. 6d.

Astronomy.—ELEMENTARY LESSONS IN ASTRONOMY. With Coloured Diagram of the Spectra of the Sun, Stars, and Nebulæ, and numerous Illustrations. By J. NORMAN LOCKYER, F.R.S. New Edition. Fcap. 8vo. 5s. 6d.

Elementary Science Class-books—*continued.*

QUESTIONS ON LOCKYER'S ELEMENTARY LESSONS IN ASTRONOMY. For the Use of Schools. By JOHN FORBES ROBERTSON. 18mo, cloth limp. 1s. 6d.

Physiology.—LESSONS IN ELEMENTARY PHYSIOLOGY. With numerous Illustrations. By T. H. HUXLEY, F.R.S., Professor of Natural History in the Royal School of Mines. New Edition. Fcap. 8vo. 4s. 6d.

QUESTIONS ON HUXLEY'S PHYSIOLOGY FOR SCHOOLS. By T. ALCOCK, M.D. 18mo. 1s. 6d.

Botany.—LESSONS IN ELEMENTARY BOTANY. By D. OLIVER, F.R.S., F.L.S., Professor of Botany in University College, London. With nearly Two Hundred Illustrations. New Edition. Fcap. 8vo. 4s. 6d.

Chemistry.—LESSONS IN ELEMENTARY CHEMISTRY, INORGANIC AND ORGANIC. By HENRY E. ROSCOE, F.R.S., Professor of Chemistry in Owens College, Manchester. With numerous Illustrations and Chromo-Litho of the Solar Spectrum, and of the Alkalies and Alkaline Earths. New Edition. Fcap. 8vo. 4s. 6d.

A SERIES OF CHEMICAL PROBLEMS, prepared with Special Reference to the above, by T. E. THORPE, Ph.D., Professor of Chemistry in the Yorkshire College of Science, Leeds. Adapted for the preparation of Students for the Government, Science, and Society of Arts Examinations. With a Preface by Professor ROSCOE. New Edition, with Key. 18mo. 2s.

Practical Chemistry.—THE OWENS COLLEGE JUNIOR COURSE OF PRACTICAL CHEMISTRY. By FRANCIS JONES, F.R.S.E., F.C.S., Chemical Master in the Grammar School, Manchester. With Preface by Professor ROSCOE, and Illustrations. New Edition. 18mo. 2s. 6d.

Chemistry.—QUESTIONS ON. A Series of Problems and Exercises in Inorganic and Organic Chemistry. By F. JONES, F.R.S.E., F.C.S. 18mo. 3s.

Political Economy.—POLITICAL ECONOMY FOR BEGINNERS. By MILLICENT G. FAWCETT. New Edition. 18mo. 2s. 6d.

Logic.—ELEMENTARY LESSONS IN LOGIC ; Deductive and Inductive, with copious Questions and Examples, and a Vocabulary of Logical Terms. By W. STANLEY JEVONS, M.A., Professor of Political Economy in University College, London. New Edition. Fcap. 8vo. 3s. 6d.

Elementary Science Class-books—*continued.*

Physics.—LESSONS IN ELEMENTARY PHYSICS. By BALFOUR STEWART, F.R.S., Professor of Natural Philosophy in Owens College, Manchester. With numerous Illustrations and Chromo-Litho of the Spectra of the Sun, Stars, and Nebulæ. New Edition. Fcap. 8vo. 4s. 6d.

Anatomy.—LESSONS IN ELEMENTARY ANATOMY. By ST. GEORGE MIVART, F.R.S., Lecturer in Comparative Anatomy at St. Mary's Hospital. With upwards of 400 Illustrations. Fcap. 8vo. 6s. 6d.

Mechanics.—AN ELEMENTARY TREATISE. By A. B. W. KENNEDY, C.E., Professor of Applied Mechanics in University College, London. With Illustrations. [*In preparation.*

Steam.—AN ELEMENTARY TREATISE. By JOHN PERRY, Professor of Engineering, Imperial College of Engineering, Yedo. With numerous Woodcuts and Numerical Examples and Exercises. 18mo. 4s. 6d.

Physical Geography.—ELEMENTARY LESSONS IN PHYSICAL GEOGRAPHY. By A. GEIKIE, F.R.S., Murchison Professor of Geology, &c., Edinburgh. With numerous Illustrations. Fcap. 8vo. 4s. 6d.
QUESTIONS ON THE SAME. 1s. 6d.

Geography.—CLASS-BOOK OF GEOGRAPHY. By C. B. CLARKE, M.A., F.R.G.S. Fcap. 8vo. 2s. 6d.

Natural Philosophy.—NATURAL PHILOSOPHY FOR BEGINNERS. By I. TODHUNTER, M.A., F.R.S. Part I. The Properties of Solid and Fluid Bodies. 18mo. 3s. 6d. Part II. Sound, Light, and Heat. 18mo. 3s. 6d.

Sound.—AN ELEMENTARY TREATISE. By Dr. W. H. STONE. With Illustrations. 18mo. 3s. 6d.

Others in Preparation.

MANUALS FOR STUDENTS.

Crown 8vo.

Dyer and Vines.—THE STRUCTURE OF PLANTS. By Professor THISELTON DYER, F.R.S., assisted by SYDNEY VINES, B.Sc., Fellow and Lecturer of Christ's College, Cambridge. With numerous Illustrations. [*In preparation.*

Manuals for Students—*continued.*

Fawcett.—A MANUAL OF POLITICAL ECONOMY. By Professor FAWCETT, M.P. New Edition, revised and enlarged. Crown 8vo. 12s. 6d.

Fleischer.—A SYSTEM OF VOLUMETRIC ANALYSIS. Translated, with Notes and Additions, from the second German Edition, by M. M. PATTISON MUIR, F.R.S.E. With Illustrations. Crown 8vo. 7s. 6d.

Flower (W. H.).—AN INTRODUCTION TO THE OSTE-OLOGY OF THE MAMMALIA. Being the Substance of the Course of Lectures delivered at the Royal College of Surgeons of England in 1870. By Professor W. H. FLOWER, F.R.S., F.R.C.S. With numerous Illustrations. New Edition, enlarged. Crown 8vo. 10s. 6d.

Foster and Balfour.—THE ELEMENTS OF EMBRY-OLOGY. By MICHAEL FOSTER, M.D., F.R.S., and F. M. BALFOUR, M.A. Part I. crown 8vo. 7s. 6d.

Foster and Langley.—A COURSE OF ELEMENTARY PRACTICAL PHYSIOLOGY. By MICHAEL FOSTER, M.D., F.R.S., and J. N. LANGLEY, B.A. New Edition. Crown 8vo. 6s.

Hooker (Dr.)—THE STUDENT'S FLORA OF THE BRITISH ISLANDS. By Sir J. D. HOOKER, K.C.S.I., C.B., F.R.S., M.D., D.C.L. New Edition, revised. Globe 8vo. 10s. 6d.

Huxley.—PHYSIOGRAPHY. An Introduction to the Study of Nature. By Professor HUXLEY, F.R.S. With numerous Illustrations, and Coloured Plates. New Edition. Crown 8vo. 7s. 6d.

Huxley and Martin.—A COURSE OF PRACTICAL IN-STRUCTION IN ELEMENTARY BIOLOGY. By Professor HUXLEY, F.R.S., assisted by H. N. MARTIN, M.B., D.Sc. New Edition, revised. Crown 8vo. 6s.

Huxley and Parker.—ELEMENTARY BIOLOGY. PART II. By Professor HUXLEY, F.R.S., assisted by — PARKER. With Illustrations. [*In preparation.*

Jevons.—THE PRINCIPLES OF SCIENCE. A Treatise on Logic and Scientific Method. By Professor W. STANLEY JEVONS, LL.D., F.R.S., New and Revised Edition. Crown 8vo. 12s. 6d.

Manuals for Students—*continued.*

Oliver (Professor).—FIRST BOOK OF INDIAN BOTANY. By Professor DANIEL OLIVER, F.R.S., F.L.S., Keeper of the Herbarium and Library of the Royal Gardens, Kew. With numerous Illustrations. Extra fcap. 8vo. 6s. 6d.

Parker and Bettany.—THE MORPHOLOGY OF THE SKULL. By Professor PARKER and G. T. BETTANY. Illustrated. Crown 8vo. 10s. 6d.

Tait.—AN ELEMENTARY TREATISE ON HEAT. By Professor TAIT, F.R.S.E. Illustrated.   [*In the Press.*

Thomson.— ZOOLOGY. By Sir C. WYVILLE THOMSON, F.R.S. Illustrated.   [*In preparation.*

Tylor and Lankester.—ANTHROPOLOGY. By E. B. TYLOR, M.A., F.R.S., and Professor E. RAY LANKESTER, M.A., F.R.S. Illustrated.   [*In preparation.*

Other volumes of these Manuals will follow.

WORKS ON MENTAL AND MORAL
PHILOSOPHY, AND ALLIED SUBJECTS.

Aristotle.—AN INTRODUCTION TO ARISTOTLE'S RHETORIC. With Analysis, Notes, and Appendices. By E. M. COPE, Trinity College, Cambridge. 8vo. 14s.

ARISTOTLE ON FALLACIES; OR, THE SOPHISTICI ELENCHI. With a Translation and Notes by EDWARD POSTE, M.A., Fellow of Oriel College, Oxford. 8vo. 8s. 6d.

Balfour.—A DEFENCE OF PHILOSOPHIC DOUBT: being an Essay on the Foundations of Belief. By A. J. BALFOUR, M.P. 8vo. 12s.

"Mr. Balfour's criticism is exceedingly brilliant and suggestive."— Pall Mall Gazette.
"An able and refreshing contribution to one of the burning questions of the age, and deserves to make its mark in the fierce battle now raging between science and theology."—Athenæum.

Birks.—Works by the Rev. T. R. BIRKS, Professor of Moral Philosophy, Cambridge :—
FIRST PRINCIPLES OF MORAL SCIENCE; or, a First Course of Lectures delivered in the University of Cambridge. Crown 8vo. 8s. 6d.

This work treats of three topics all preliminary to the direct exposition of Moral Philosophy. These are the Certainty and Dignity of Moral Science, its Spiritual Geography, or relation to other main subjects of human thought, and its Formative Principles, or some elementary truths on which its whole development must depend.

MODERN UTILITARIANISM; or, The Systems of Paley, Bentham, and Mill, Examined and Compared. Crown 8vo. 6s. 6d.

MODERN PHYSICAL FATALISM, AND THE DOCTRINE OF EVOLUTION; including an Examination of Herbert Spencer's First Principles. Crown 8vo. 6s.

SUPERNATURAL REVELATION; or, First Principles of Moral Theology. 8vo. 8s.

Boole. — AN INVESTIGATION OF THE LAWS OF THOUGHT, ON WHICH ARE FOUNDED THE MATHEMATICAL THEORIES OF LOGIC AND PROBABILITIES. By GEORGE BOOLE, LL.D., Professor of Mathematics in the Queen's University, Ireland, &c. 8vo. 14*s*.

Butler.—LECTURES ON THE HISTORY OF ANCIENT PHILOSOPHY. By W. ARCHER BUTLER, late Professor of Moral Philosophy in the University of Dublin. Edited from the Author's MSS., with Notes, by WILLIAM HEPWORTH THOMPSON, M.A., Master of Trinity College, and Regius Professor of Greek in the University of Cambridge. New and Cheaper Edition, revised by the Editor. 8vo. 12*s*.

Caird.—A CRITICAL ACCOUNT OF THE PHILOSOPHY OF KANT. With an Historical Introduction. By E. CAIRD, M.A., Professor of Moral Philosophy in the University of Glasgow. 8vo. 18*s*.

Calderwood.—Works by the Rev. HENRY CALDERWOOD, M.A., LL.D., Professor of Moral Philosophy in the University of Edinburgh :—

PHILOSOPHY OF THE INFINITE : A Treatise on Man's Knowledge of the Infinite Being, in answer to Sir W. Hamilton and Dr. Mansel. Cheaper Edition. 8vo. 7*s*. 6*d*.
> "*A book of great ability . . . . written in a clear stle, and may be easily understood by even those who are not versed in such discussions.*"—British Quarterly Review.

A HANDBOOK OF MORAL PHILOSOPHY. Sixth Edition. Crown 8vo. 6*s*.
> "*It is, we feel convinced, the best handbook on the subject, intellectually and morally, and does infinite credit to its author.*"—Standard.
> "*A compact and useful work, going over a great deal of ground in a manner adapted to suggest and facilitate further study. . . . His book will be an assistance to many students outside his own University of Edinburgh.* —Guardian.

THE RELATIONS OF MIND AND BRAIN. 8vo. 12*s*.
> "*It should be of real service as a clear exposition and a searching criticism of cerebral pyschology.*"—Westminster Review.
> "*Altogether his work is probably the best combination to be found at present in England of exposition and criticism on the subject of physiological psychology.*"—The Academy.

Clifford.—LECTURES AND ESSAYS. By the late Professor W. K. CLIFFORD, F.R.S. Edited by LESLIE STEPHEN and FREDERICK POLLOCK, with Introduction by F. POLLOCK. Two Portraits. 2 vols. 8vo. 25*s*.

Clifford—*continued.*

" The Times *of October 22nd says :—" Many a friend of the author on first taking up these volumes and remembering his versatile genius and his keen enjoyment of all realms of intellectual activity must have trembled, lest they should be found to consist of fragmentary pieces of work, too disconnected to do justice to his powers of consecutive reading, and too varied to have any effect as a whole. Fortunately these fears are groundless. . . . It is not only in subject that the various papers are closely related. There is also a singular consistency of view and of method throughout. . . . It is in the social and metaphysical subjects that the richness of his intellect shows itself, most forcibly in the rarity and originality of the ideas which he presents to us. To appreciate this variety it is necessary to read the book itself, for it treats in some form or other of all the subjects of deepest interest in this age of questioning."*

Fiske.—OUTLINES OF COSMIC PHILOSOPHY, BASED ON THE DOCTRINE OF EVOLUTION, WITH CRITICISMS ON THE POSITIVE PHILOSOPHY. By JOHN FISKE, M.A., LL.B., formerly Lecturer on Philosophy at Harvard University. 2 vols. 8vo. 25s.

" The work constitutes a very effective encyclopædia of the evolutionary philosophy, and is well worth the study of all who wish to see at once the entire scope and purport of the scientific dogmatism of the day."—Saturday Review.

Harper.—THE METAPHYSICS OF THE SCHOOL. By the Rev. THOMAS HARPER (S.J.). In 5 vols. 8vo.
[Vol I. in November.

Herbert.—THE REALISTIC ASSUMPTIONS OF MODERN SCIENCE EXAMINED. By T. M. HERBERT, M.A., late Professor of Philosophy, &c., in the Lancashire Independent College, Manchester. 8vo. 14s.

" Mr. Herbert's work appears to us one of real ability and importance. The author has shown himself well trained in philosophical literature, and possessed of high critical and speculative powers."—Mind.

Jardine.—THE ELEMENTS OF THE PSYCHOLOGY OF COGNITION. By ROBERT JARDINE, B.D., D.Sc., Principal of the General Assembly's College, Calcutta, and Fellow of the University of Calcutta. Crown 8vo. 6s. 6d.

Jevons.—Works by W. STANLEY JEVONS, LL.D., M.A., F.R.S., Professor of Political Economy, University College, London.

Jevons—*continued.*

THE PRINCIPLES OF SCIENCE. A Treatise on Logic and Scientific Method. New and Cheaper Edition, revised. Crown 8vo. 12s. 6d.

"*No one in future can be said to have any true knowledge of what has been done in the way of logical and scientific method in England without having carefully studied Professor Jevons' book.*"— Spectator.

THE SUBSTITUTION OF SIMILARS, the True Principle of Reasoning. Derived from a Modification of Aristotle's Dictum. Fcap. 8vo. 2s. 6d.

ELEMENTARY LESSONS IN LOGIC, DEDUCTIVE AND INDUCTIVE. With Questions, Examples, and Vocabulary of Logical Terms. New Edition. Fcap. 8vo. 3s. 6d.

PRIMER OF LOGIC. New Edition. 18mo. 1s.

Maccoll.—THE GREEK SCEPTICS, from Pyrrho to Sextus. An Essay which obtained the Hare Prize in the year 1868. By NORMAN MACCOLL, B.A., Scholar of Downing College, Cambridge. Crown 8vo. 3s. 6d.

M'Cosh.—Works by JAMES M'COSH, LL.D., President of Princeton College, New Jersey, U.S.

"*He certainly shows himself skilful in that application of logic to psychology, in that inductive science of the human mind which is the fine side of English philosophy. His philosophy as a whole is worthy of attention.*"—Revue de Deux Mondes.

THE METHOD OF THE DIVINE GOVERNMENT, Physical and Moral. Tenth Edition. 8vo. 10s. 6d.

"*This work is distinguished from other similar ones by its being based upon a thorough study of physical science, and an accurate knowledge of its present condition, and by its entering in a deeper and more unfettered manner than its predecessors upon the discussion of the appropriate psychological, ethical, and theological questions. The author keeps aloof at once from the à priori idealism and dreaminess of German speculation since Schelling, and from the onesidedness and narrowness of the empiricism and positivism which have so prevailed in England.*"—Dr. Ulrici, in "Zeitschrift für Philosophie."

THE INTUITIONS OF THE MIND. A New Edition. 8vo. cloth. 10s. 6d.

"*The undertaking to adjust the claims of the sensational and intuitional philosophies, and of the à posteriori and à priori methods, is accomplished in this work with a great amount of success.*"—Westminster Review. "*I value it for its large acquaintance with English Philosophy, which has not led him to neglect the great German works. I admire the moderation and clearness, as well as comprehensiveness, of the author's views.*"—Dr. Dörner, of Berlin.

M'Cosh—*continued.*

AN EXAMINATION OF MR. J. S. MILL'S PHILOSOPHY: Being a Defence of Fundamental Truth. Second edition, with additions. 10s. 6d.

"*Such a work greatly needed to be done, and the author was the man to do it. This volume is important, not merely in reference to the views of Mr. Mill, but of the whole school of writers, past and present, British and Continental, he so ably represents.*"—Princeton Review.

THE LAWS OF DISCURSIVE THOUGHT: Being a Text-book of Formal Logic. Crown 8vo. 5s.

"*The amount of summarized information which it contains is very great; and it is the only work on the very important subject with which it deals. Never was such a work so much needed as in the present day.*"—London Quarterly Review.

CHRISTIANITY AND POSITIVISM : A Series of Lectures to the Times on Natural Theology and Apologetics. Crown 8vo. 7s. 6d.

THE SCOTTISH PHILOSOPHY FROM HUTCHESON TO HAMILTON, Biographical, Critical, Expository. Royal 8vo. 16s.

Masson.—RECENT BRITISH PHILOSOPHY : A Review with Criticisms ; including some Comments on Mr. Mill's Answer to Sir William Hamilton. By DAVID MASSON, M.A., Professor of Rhetoric and English Literature in the University of Edinburgh. Third Edition, with an Additional Chapter. Crown 8vo. 6s

"*We can nowhere point to a work which gives so clear an exposition of the course of philosophical speculation in Britain during the past century, or which indicates so instructively the mutual influences of philosophic and scientific thought.*"—Fortnightly Review.

Maudsley.—Works by H. MAUDSLEY, M.D., Professor of Medical Jurisprudence in University College, London.

THE PHYSIOLOGY OF MIND ; being the First Part of a Third Edition, Revised, Enlarged, and in great part Re-written, of "The Physiology and Pathology of Mind." Crown 8vo. 10s. 6d.

THE PATHOLOGY OF MIND. Revised, Enlarged, and in great part Re-written. 8vo. 18s.

BODY AND MIND : an Inquiry into their Connexion and Mutual Influence, specially with reference to Mental Disorders. An Enlarged and Revised edition. To which are added, Psychological Essays. Crown 8vo. 6s. 6d.

Maurice.—Works by the Rev. FREDERICK DENISON MAURICE, M.A., Professor of Moral Philosophy in the University of Cambridge. (For other Works by the same Author, see THEOLOGICAL CATALOGUE.)

SOCIAL MORALITY. Twenty-one Lectures delivered in the University of Cambridge. New and Cheaper Edition. Crown 8vo. 10s. 6d.

"Whilst reading it we are charmed by the freedom from exclusiveness and prejudice, the large charity, the loftiness of thought, the eagerness to recognize and appreciate whatever there is of real worth extant in the world, which animates it from one end to the other. We gain new thoughts and new ways of viewing things, even more, perhaps, from being brought for a time under the influence of so noble and spiritual a mind."—Athenæum.

THE CONSCIENCE: Lectures on Casuistry, delivered in the University of Cambridge. New and Cheaper Edition. Crown 8vo. 5s.

The Saturday Review says: "We rise from them with detestation of all that is selfish and mean, and with a living impression that there is such a thing as goodness after all."

MORAL AND METAPHYSICAL PHILOSOPHY. Vol. I. Ancient Philosophy from the First to the Thirteenth Centuries; Vol. II. the Fourteenth Century and the French Revolution, with a glimpse into the Nineteenth Century. New Edition and Preface. 2 Vols. 8vo. 25s.

Morgan.—ANCIENT SOCIETY: or Researches in the Lines of Human Progress, from Savagery, through Barbarism to Civilisation. By LEWIS H. MORGAN, Member of the National Academy of Sciences. 8vo. 16s.

Murphy.—THE SCIENTIFIC BASES OF FAITH. By JOSEPH JOHN MURPHY, Author of "Habit and Intelligence." 8vo. 14s.

"The book is not without substantial value; the writer continues the work of the best apologists of the last century, it may be with less force and clearness, but still with commendable persuasiveness and tact; and with an intelligent feeling for the changed conditions of the problem."—Academy.

Paradoxical Philosophy.—A Sequel to "The Unseen Universe." Crown 8vo. 7s. 6d.

Picton.—THE MYSTERY OF MATTER AND OTHER ESSAYS. By J. ALLANSON PICTON, Author of "New Theories and the Old Faith." Cheaper issue with New Preface. Crown 8vo. 6s.

Picton—*continued.*

> CONTENTS :— *The Mystery of Matter—The Philosophy of Igno-rance—The Antithesis of Faith and Sight—The Essential Nature of Religion—Christian Pantheism.*

Sidgwick.—THE METHODS OF ETHICS. By HENRY SIDGWICK, M.A., Prælector in Moral and Political Philosophy in Trinity College, Cambridge. Second Edition, revised throughout with important additions. 8vo. 14*s.*

A SUPPLEMENT to the First Edition, containing all the important additions and alterations in the Second. 8vo. 2*s.*

> "*This excellent and very welcome volume. . . . . Leaving to meta-physicians any further discussion that may be needed respecting the already over-discussed problem of the origin of the moral faculty, he takes it for granted as readily as the geometrician takes space for granted, or the physicist the existence of matter. But he takes little else for granted, and defining ethics as 'the science of conduct,' he carefully examines, not the various ethical systems that have been propounded by Aristotle and Aristotle's followers downwards, but the principles upon which, so far as they confine themselves to the strict province of ethics, they are based.*"—Athenæum.

Thornton.—OLD-FASHIONED ETHICS, AND COMMON-SENSE METAPHYSICS, with some of their Applications. By WILLIAM THOMAS THORNTON, Author of "A Treatise on Labour." 8vo. 10*s. 6d.*

> *The present volume deals with problems which are agitating the minds of all thoughtful men. The following are the Contents :—I. Ante-Utilitarianism. II. History's Scientific Pretensions. III. David Hume as a Metaphysician. IV. Huxleyism. V. Recent Phase of Scientific Atheism. VI. Limits of Demonstrable Theism.*

Thring (E., M.A.).—THOUGHTS ON LIFE-SCIENCE. By EDWARD THRING, M.A. (Benjamin Place), Head Master of Uppingham School. New Edition, enlarged and revised. Crown 8vo. 7*s. 6d.*

Venn.—THE LOGIC OF CHANCE: An Essay on the Founda-tions and Province of the Theory of Probability, with especial reference to its logical bearings, and its application to Moral and Social Science. By JOHN VENN, M.A., Fellow and Lecturer of Gonville and Caius College, Cambridge. Second Edition, re-written and greatly enlarged. Crown 8vo. 10*s. 6d.*

> "*One of the most thoughtful and philosophical treatises on any sub-ject connected with logic and evidence which has been produced in this or any other country for many years.*"—Mill's Logic, vol. ii. p. 77. Seventh Edition.

NATURE SERIES.

THE SPECTROSCOPE AND ITS APPLICATIONS.
By J. N. LOCKYER, F.R.S. With Illustrations. *Second Edition.* Crown
8vo. 3s. 6d.

THE ORIGIN AND METAMORPHOSES OF IN-
SECTS. By Sir JOHN LUBBOCK, M.P., F.R.S. With Illustrations.
Crown 8vo. 3s. 6d. *Second Edition.*

THE TRANSIT OF VENUS. By G. FORBES, B.A.,
Professor of Natural Philosophy in the Andersonian University, Glasgow
With numerous Illustrations. Crown 8vo. 3s. 6d.

THE COMMON FROG. By ST. GEORGE MIVART,
F.R.S. Illustrated. Crown 8vo. 3s. 6d.

POLARISATION OF LIGHT. By W. SPOTTISWOODE,
LL.D., President of the Royal Society. Illustrated. *Second Edition.* Crown
8vo. 3s. 6d.

ON BRITISH WILD FLOWERS CONSIDERED IN
RELATION TO INSECTS. By Sir JOHN LUBBOCK, M.P., F.R.S.
Illustrated. *Second Edition.* Crown 8vo. 4s. 6d.

THE SCIENCE OF WEIGHING AND MEASURING.
By H. W. CHISHOLM, Warden of the Standards. Illustrated. Crown 8vo.
4s. 6d.

HOW TO DRAW A STRAIGHT LINE: A Lecture on
Linkages. By A. B. KEMPE, B.A. Illustrated. Crown 8vo. 1s. 6d.

LIGHT: A Series of Simple, Entertaining and Useful
Experiments in the Phenomena of Light for the Use of Students of every Age.
By ALFRED M. MAYER and CHARLES BARNARD. With Illustrations.
Crown 8vo. 2s. 6d.

SOUND: A Series of Simple, Entertaining and Inex-
pensive Experiments in the Phenomena of Sound, for the Use of Students of
every Age. By A. M. MAYER, Professor of Physics in the Stevens Institute
of Technology, &c. With numerous Illustrations. Crown 8vo. 3s. 6d.

SEEING AND THINKING. By Prof. W. K. CLIFFORD,
F.R.S. With Diagrams. Crown 8vo. 3s. 6d.

(*Others to follow.*)

MACMILLAN AND CO., LONDON.

Published every Thursday, price 6d.; Monthly Parts 2s. and 2s. 6d., Half-Yearly Volumes, 15s.

NATURE:

AN ILLUSTRATED JOURNAL OF SCIENCE.

NATURE expounds in a popular and yet authentic manner, the GRAND RESULTS OF SCIENTIFIC RESEARCH, discussing the most recent scientific discoveries, and pointing out the bearing of Science upon civilisation and progress, and its claims to a more general recognition, as well as to a higher place in the educational system of the country.

It contains original articles on all subjects within the domain of Science ; Reviews setting forth the nature and value of recent Scientific Works ; Correspondence Columns, forming a medium of Scientific discussion and of intercommunication among the most distinguished men of Science, Serial Columns, giving the gist of the most important papers appearing in Scientific Journals, both Home and Foreign ; Transactions of the principal Scientific Societies and Academies of the World, Notes, &c.

In Schools where Science is included in the regular course of studies, this paper will be most acceptable, as it tells what is doing in Science all over the world, is popular without lowering the standard of Science, and by it a vast amount of information is brought within a small compass, and students are directed to the best sources for what they need. The various questions connected with Science teaching in schools are also fully discussed, and the best methods of teaching are indicated.

www.ingramcontent.com/pod-product-compliance
Lightning Source LLC
Chambersburg PA
CBHW021504210326
41599CB00012B/1125

9 783337 321635